T0174129

PROCEEDINGS OF THE 6TH NIRMA UNIVERSITY INTERNATIONAL CONFERENCE ON ENGINEERING (NUICONE 2017), NOVEMBER 23–25, 2017, AHMEDABAD, INDIA

Technology Drivers: Engine for Growth

Editors

Alka Mahajan, B.A. Modi & Parul Patel

Institute of Technology, Nirma University, Ahmedabad, India

CRC Press is an imprint of the
Taylor & Francis Group, an **informa** business

A BALKEMA BOOK

CRC Press
Taylor & Francis Group
6000 Broken Sound Parkway NW, Suite 300
Boca Raton, FL 33487-2742

First issued in paperback 2020

© 2018 by Taylor & Francis Group, LLC
CRC Press is an imprint of Taylor & Francis Group, an Informa business

No claim to original U.S. Government works

Typeset by V Publishing Solutions Pvt Ltd., Chennai, India

ISBN-13: 978-1-138-56042-0 (hbk)
ISBN-13: 978-0-367-73309-4 (pbk)

**Visit the Taylor & Francis Web site at
http://www.taylorandfrancis.com**

**and the CRC Press Web site at
http://www.crcpress.com**

Technology Drivers: Engine for Growth – Mahajan, Modi & Patel (Eds)
© 2018 Taylor & Francis Group, London, ISBN 978-1-138-56042-0

Table of contents

Technology Drivers: Engine for Growth – Mahajan, Modi & Patel (Eds)
© 2018 Taylor & Francis Group, London, ISBN 978-1-138-56042-0

Preface

This volume of proceedings from the conference provides an opportunity for readers to engage with a selection of refereed papers that were presented during the 6th International conference NUiCONE'17. Researchers from industry and academia were invited to present their research work in the areas as listed below. The research papers presented in these tracks have been published in this proceeding with the support of CRC Press, Taylor & Francis. This proceeding will definitely provide a platform to proliferate new findings among researchers.

- ❖ Chemical Process Development and Design
- ❖ Technologies for Green Environment
- ❖ Advances in Transportation Engineering
- ❖ Emerging Trends in Water Resources and Environmental Engineering
- ❖ Construction Technology and Management
- ❖ Concrete and Structural Engineering
- ❖ Sustainable Manufacturing Processes
- ❖ Design and Analysis of Machine and Mechanism
- ❖ Energy Conservation and Management

Introduction

We are delighted to present the knowledge sharing journey of the 6th international conference on Engineering, organised by the Institute of Technology, Nirma University during 23–25 November 2017.

The main theme of the conference was Technology Drivers – The growth engine and was also focussed on green technology and innovations to sensitise the researchers, industry professionals and students towards the environment issues.

There were nineteen themes to cater the diversified needs of academics and industries:

❖ Autonomic Computing: Architectural Challenges and Solutions
❖ Networking Technologies: Performance and Security Challenges
❖ Next Generation Pattern Analysis
❖ Electronic Communications and Signal Processing
❖ VLSI Design
❖ Innovative Electronic Circuit Design
❖ Latest Trends in Electrical Power Systems
❖ Control of Power Electronics Converters and Drives
❖ Advanced Electrical Machines and Apparatus
❖ Control and Automation
❖ Chemical Process Development and Design
❖ Technologies for Green Environment
❖ Advances in Transportation Engineering
❖ Emerging Trends in Water Resources and Environmental Engineering
❖ Construction Technology and Management
❖ Concrete and Structural Engineering
❖ Sustainable Manufacturing Processes
❖ Design and Analysis of Machine and Mechanism
❖ Energy Conservation and Management.

Around 290 research papers were received from various disciplines and only 90 papers were accepted for oral presentation after stringent quality checks. To ensure the quality the papers were checked for plagiarism, double blind review and language correction. These papers are going to be published by the reputed publishers IEEE and Taylor & Francis.

The conference received active participation across the country and abroad from industry, academia and researchers from reputed organisations like IITs, NITs, IPR, ISRO, Northumbria University UK, University of New Brunswick Canada and University of Victoria Canada. Total participation including invited industry delegates, research organisations, post graduate students and faculty members was 589.

Nirma University International Conference on Engineering, NUiCONE is a flagship event of the Institute of Technology, Nirma University, Ahmedabad, Gujarat, India. This conference follows the successful organization of four national conferences and five international conferences in previous years. NUiCONE2017 was very successful with more than 290 research papers submitted, out of which approx. 90 papers were selected and presented after the rigorous reviews including plagiarism checks and blind-fold technical reviews by multiple reviewers. The expert speakers and conference delegates were from all across India and abroad covering researchers and eminent experts from academia, industry such as J. K. Lakshmi Cements, Mitsubishi, Grind Master, Jacobs, Pronesis Technologies, ARK

Info-Solution, Electrotherm, CapGemini etc., and government R&D organizations like IPR, ISRO, Western Railways. NUiCONE-2017 is being planned with major interesting changes in the mode of events/themes to make it like a multi-conference with many inter-disciplinary technical themes encompassing and enabling researchers from broad range of disciplines. Besides, NUiCONE 2017 will also have exciting new sets of events especially technical white paper presentation by Industry Professionals which are scheduled explicitly on the second day of the conference. These special sessions are arranged with talks on current topics of multidisciplinary interest. We are indeed grateful to have prominent speakers for the invited talks in each theme. Further, this year, the conference has been linked with many reputed associations of various engineering institutions like IEEE, Taylor & Francis, IE (India), ISA, IGBC, ASHRAE, CSCE and GUJCOST.

The conference was graced by the Chief Guest, **Padma Vibhushan Shri Dr. Anil Kakodkar**, President, National Academy of Sciences India; Chairman, Rajiv Gandhi Science & Technology Commission; Member & Former Chairman Atomic Energy Commission. **Dr. A. Jayaraman**, Vice President, International Committee on Space Research-COSPAR, Paris and Former, Director, National Atmospheric Research Laboratory, ISRO was a guest of honour and keynote speaker. **Padma Shri Dr. Karshanbhai Patel**, Chairman Nirma Group of Industries and President Nirma University and motivated the gathering through his august presence. **Mr. S. Sriram**, General Manager – Marketing at Mitsubishi Electric India Private Ltd. was the chief guest of the valedictory function and the function was presided over by Shri. K.K. Patel, Vice President, Nirma University.

Acknowledgement

Putting together NUiCONE'17 was a team effort. We first thank the authors for providing the papers presenting their research work. We are grateful to the program committee and the reviewers, who worked very hard in reviewing papers and providing feedback for authors. Due credit should be given to CRC Press, Taylor & Francis for publishing the proceeding of the research papers presented in the conference. Finally, we thank the hosting organization Nirma University, our sponsors, international advisory committee members, organising committee members as listed below.

ORGANIZING COMMITTEE

Chief Patrons:	Shri K.K. Patel, *Vice President, Nirma University*
	Dr. Anup Singh, *Director General, Nirma University*
Patron:	Dr. Alka Mahajan, *Director, IT, NU*
Conference Chair:	Dr. B.A. Modi, *IT, NU*
Conference Co-Chairs:	Dr. Parul Patel, *IT, NU*
	Dr. Y.P. Kosta, *Ex-Officio Expert, IEEE Gujarat Section*

MEMBERS

Dr. G.R. Nair (Exec. Registrar, NU)
Shri D.P. Chhaya
Dr. A.S. Patel
Dr. Sanjay Garg
Dr. P.V. Patel
Dr. P.N. Tekwani
Dr. R.N. Patel
Dr. V.J. Lakhera
Dr. J.P. Rupareliya
Dr. D.K. Kothari
Dr. Dipak Adhyaru
Dr. Madhuri Bhavsar
Dr. Sanjay Patel

INDUSTRY-INTERACTION COMMITTEE

Dr. N.P. Gajjar (Chair)
Mr. Jitendra Gadhavi
Dr. Urmil Dave
Dr. Parin Shah
Prof. Amit Patel
Prof. Jayesh Patel
Dr. J.B. Patel
Dr. Zunnun Narmawala
Prof. Ath Singhal

TECHNICAL COMMITTEE

Dr. Dipak Adhyaru (Chair)
Dr. Usha Mehta (Co-Chair)

Dr. A.M. Lakdawala	Dr. N.D. Ghetiya
Dr. J.P. Rupareliya	Dr. R.K. Mewada
Prof. D.D. Joshi	Prof. Hemang Dalwadi
Prof. Vijay Ukani	Prof. Gaurang Raval
Dr. Siddharth Chauhan	Dr. Akhilesh Nimje
Dr. Y.N. Trivedi	Prof. Bhavin Kakani
Prof. S.A. Mehta	Prof. J.B. Shah
Dr. Richa Mishra	Prof. Kunal Pathak

PUBLICATION COMMITTEE

Dr. Millind Joshipura (Chair)
Prof. Anand Bhatt
Prof. Sunil Raiyani
Prof. Leena Bora
Prof. Hormaz Amrolia
Prof. Vaishali Dhare
Prof. Harsh Kapadia
Prof. K.P. Agrawal

VENUE MANAGEMENT COMMITTEE

Dr. Tanish Zaveri (Chair)
Dr. Sanjay Jain
Prof. Jhanvi Suthar
Prof. Femina Patel
Prof. Sarika Kanojia
Prof. Ruchi Gajjar
Prof. N.S. Patel
Prof. Tejal Upadhyay
Dr. K. Ambika

REGISTRATION COMMITTEE

Dr. Manisha Upadhyay (Chair)
Dr. Aachari
Prof. Utsav Koshti
Prof. Nikita Chokshi
Prof. Shanker Godwal
Prof Vijay Savani
Prof. Vidita Tilva
Prof. Preeti Kathiria
Prof. Ujala Shamnani

ACCOMMODATION AND TRANSPORTATION COMMITTEE

Dr. Sharad Purohit (Chair)
Prof. N.K. Shah
Prof. Hemanth Kamplimath
Prof. R.N. Reddy
Prof. Mayur Gojiya
Prof. Rachna Sharma
Prof. Alpesh Patel
Prof. Rajan Datt
Mr. Alok Bhatnagar
Mr. Bhaichand Patel
Dr. Samir Mahajan
Smt. Ritaben Barot
Mr. Shailesh Patel

FINANCE COMMITTEE

Dr. K.M. Patel (Chair)
Shri B.J. Patel (Treasure)
Dr. B.K. Mawandiya
Prof. Sonal Thakkar
Dr. Nimish Shah
Prof. Chintan Mehta
Prof Hardik Joshi
Prof. H.K. Patel
Prof. Vishal Parikh
Mr. Jitendra Gadhavi
Prof. P.S. Vellala

CYBER COMMITTEE

Dr. Priyanka Sharma (Chair)
Prof. Ankit Thakkar

Prof. Devendra Vashi
Prof. Sapan Mankad

CATERING COMMITTEE

Prof. Vipul Chudasama (Chair)
Prof. Mayur Makhesana
Prof. Sandesh Sharma
Dr. Ankur Dwivedi
Prof. Dhiren Rathod
Prof. Twinkle Bhavsar
Prof. Vishal Vaidya
Prof. Rushabh Shah
Prof. Anuja Nair
Dr. Mahesh Yeolekar

ANCHORING COMMITTEE

Dr. Richa Mishra (Chair)
Dr. Sharda Velivati

MEDIA AND PUBLICITY COMMITTEE

Dr. S.S. Patel (Chair)
Prof. Tejas Joshi
Prof. Jaladhi Vyas
Prof. Darshita Shah
Dr. S.G. Pillai
Prof. Bhupendra Fatania
Mr. Sunil Nayak
Dr. Akhilesh Nimje

EXHIBITION COMMITTEE

Dr. Parin Shah (Chair)
Dr. M.B. Panchal
Prof. Hasan Rangwala
Prof. Priya Saxena
Dr. Gulshan Sharma
Prof. Rutul Patel
Prof. Ankit Sharma
Prof. Daiwat Vyas
Dr. Amit Mishra

INTERNATIONAL ADVISORY COMMITTEE

1. Dr. Rajiv Ranjan, *New Castle University, UK*
2. Dr. Jalel Ben Othman, *University of Paris, France*
3. Dr. Shailendra Mishra, *Majmaah University, Saudi Arabia*
4. Dr. M.M. Gore, *MNNIT, Allahabad, India*
5. Mrs. Sutapa Ranjan, *Institute of Plasma Research, Gandhinagar, India*
6. Prof. Santanu Chaudhury, *Director, CEERI Pilani and Professor, Department of Electrical Engineering. IIT Delhi*
7. Mr. Preetham S. Raj, *Senior Manager, RF/Wireless, Broadcom Communications Technologies Pvt. Ltd, Bangalore*
8. Dr. Lajos Hanzo, *Professor, School of Electronics and Computer Science, University of Southampton, UK*
9. Dr. Madhav Manjrekar, *Associate Professor, Electrical and Computer Engineering, The William States Lee College of Engineering, The University of North Carolina at Charlotte, NC, USA*
10. Prof. Dr. K. Gopakumar, *(FIEEE), Department of Electronic Systems Engineering, Indian Institute of Science (IISc), Bangalore, India*
11. Prof. Dr. Ramesh Bansal, *FIET (UK), FIE (India), FIEAust, SM IEEE (USA), CPEngineering (UK), Professor & Group Head (Power), Department of Electrical, Electronic and Computer Engineering, University of Pretoria, Pretoria, South Africa*
12. Dr. I.N. Kar, *Professor, IIT Delhi*
13. Dr. Sisil Kumarawadu, *Professor, University of Moratuwa, Sri Lanka*
14. Mr. Jayesh Gandhi, *Managing Director, Harikrupa Automation Pvt. Ltd., Ahmedabad, India*
15. Dr. D.B. Gurung, *Associate Professor, Department of Natural Sciences (Maths. Group), Kathmandu University, Dhulikhel, Kavre, Kathmandu, Nepal*
16. Dr. R.R. Bhargava, *Professor, Department of Mathematics, IIT-Roorkee*
17. Dr. K.R. Pardasani, *Professor, Maulana Azad National Institute of Technology, (MANIT) Bhopal (MP)*

18. Dr. M.N. Mehta, *Professor, Department of Applied Mathematics, SVNIT, Surat*
19. Dr. B. Chandra, *Professor, Department of Mathematics, IIT-Delhi, Delhi*
20. Dr. John Hoben, *Faculty of Education, Memorial University, St. John's, Canada*
21. Dr. Ketan Kotecha, *Provost, Parul University, Vadodara, Gujarat, India*
22. Prof. A.V. Thomas, *Director, Adani Institute of Infrastructure Engineering (AIIE), Ahmedabad, Gujarat, India*
23. Prof. Sanjay Choudhary, *Associate Dean, Ahmedabad University, Ahmedabad, Gujarat, India*
24. Prof. (Dr.) G.D. Yadav, *Director, Institute of Chemical Technology, Mumbai, India*
25. Prof. (Dr.) Suparna Mukharji, *Professor, Chemical Engineering Department, IIT-Bombay, Mumbai, India*
26. Prof. (Dr.) K.K. Pant, *Professor, Chemical Engineering Department, IIT-Delhi, Delhi, India*
27. Abinash Agrawal, Ph.D., *Department of Earth & Environmental Sciences, and Environmental Science Program, Wright State University, Dayton, USA*
28. Ram B. Gupta, Ph.D., *Associate Dean for Research, Professor, School of Engineering, Virginia Commonwealth University, USA*
29. Dr. Sudhirkumar Barai, *Dean (Undergraduate Studies) & Professor, Department of Civil Engineering, IIT Kharagpur*
30. Dr. Ajay Chourasia, *Principal Scientist, CSIR – Central Building Research Institute (CBRI), Roorkee*
31. Dr. Rishi Gupta, *Assistant Professor, Department of Mechanical Engineering, University of Victoria, Canada*
32. Dr. Kumar Neeraj Jha, *Associate Professor, Department of Civil Engineering, IIT Delhi*
33. Laurencas Raslavicius, *Vice Dean of Faculty of Mechanical Engineering Kaunas University of Technology, Lithuania*
34. Dr. P.B. Ravikumar, *Professor, Emeritus of University of Wisconsin-Platteville, USA*
35. Prof. Raghu Echempati, *Kettering University, Flint, MI USA*
36. Shri Sanjay Desai, *CEO, RBD Engineers, India*
37. Shri Rajesh Sampat, *Vice President, Inspiron Pvt. Ltd., Ahmedabad, India*

LIST OF REVIEWERS

Chemical Engineering Track

1. Dr. Jayesh Ruparelia, *Professor, Head of Chemical Department, Institute of Technology, Nirma University, Ahmedabad 382481*
2. Dr. Sanjay Patel, *Professor, Chemical Department, Institute of Technology, Nirma University, Ahmedabad 382481*
3. Dr. R.K. Mewada, *Professor, Chemical Department, Institute of Technology, Nirma University, Ahmedabad 382481*
4. Dr. M.H. Joshipura, *Professor, Chemical Department, Institute of Technology, Nirma University, Ahmedabad 382481*
5. Dr. Femina Patel, *Associate Professor, Chemical Department, Institute of Technology, Nirma University, Ahmedabad 382481*
6. Dr. P.D. Shah, *Associate Professor, Chemical Department, Institute of Technology, Nirma University, Ahmedabad 382481*
7. Prof. Nimish Shah, *Associate Professor, Chemical Department, Institute of Technology, Nirma University, Ahmedabad 382481*
8. Dr. A.P. Vyas, *Principal, Saffrony College of Engineering, Ahmedabad*
9. Dr. Sachin Parikh, *HoD, Chemical Engineering Department, L D College of Engineering Ahmedabad*

Civil Engineering Track

1. Dr. B.J. Shah, *Professor, Department of Applied Mechanics, Government Engineering College (GEC), Modasa, Shamlaji Road, Aravali District, Modasa - 383315, Gujarat*
2. Dr. D.P. Soni, *Professor & Head, Civil Engineering Department, Sardar Vallabhbhai Patel Institute of Technology (SVIT), B/h. S.T. Bus Depot, Vasad, Anand - 388306, Gujarat*
3. Dr. C.D. Modhera, *Professor, Department of Applied Mechanics, S. V. National Institute of Technology (SVNIT), Ichchhanath, Surat - 395007, Gujarat*
4. Dr. Y.D. Patil, *Assistant Professor, Department of Applied Mechanics, S. V. National Institute of Technology (SVNIT), Ichchhanath, Surat - 395007, Gujarat*
5. Dr. C.C. Patel, *Professor & Head, Civil Engineering Department, Marwadi University, Marwadi Education Foundation's Group of Institutions, Rajkot-Morbi Road, At & PO: Gauridad, Rajkot 360 003, Gujarat, India*
6. Dr. M.K. Srimali, *Professor, Department of Civil Engineering, Malaviya National Institute of Technology (MNIT), JawaharLal Nehru Marg, Indra Nagar, Basant Vihar, Malviya Nagar, Jaipur - 302017, Rajasthan*
7. Dr. S.G. Shah, *Professor & Director (Innovations), Babaria Institute of Technology, Vadodara - Mumbai NH 8, Varnama, Vadodara - 391240, Gujarat*
8. Dr. V.R. Panchal, *Professor & Head, Civil Engineering Department, Chandubhai S. Patel Institute of Technology, Charotar University of Science and Technology, Charusat Campus, Changa, Taluka: Petlad, Anand - 388421, Gujarat*
9. Dr. Niraj Shah, *Director, Civil Engineering Department, School of Engineering, P.P. Savani University, NH 8, GETCO, Near Biltech, Village: Dhamdod, Kosamba, Dist.: Surat - 394125*
10. Dr. J.A. Amin, *Professor, Civil Engineering Department, Sardar Vallabhbhai Patel Institute of Technology (SVIT), B/h. S.T. Bus Depot, Vasad, Anand - 388306, Gujarat*
11. Dr. P.V. Patel, *Professor & Head, Civil Engineering Department, Institute of Technology, Nirma University, S-G Highway, Post – Chandlodia, Ahmedabad - 382481, Gujarat*
12. Dr. U.V. Dave, *Professor, Civil Engineering Department, Institute of Technology, Nirma University, S-G Highway, Post – Chandlodia, Ahmedabad - 382481, Gujarat*
13. Dr. S.P. Purohit, *Professor, Civil Engineering Department, Institute of Technology, Nirma University, S-G Highway, Post – Chandlodia, Ahmedabad - 382481, Gujarat*
14. Dr. L.B. Zala, *Professor & Head, Civil Engineering Department, Birla Vishvakarma Mahavidyalaya (BVM Engineering College), Post Box No. 20, Vallabh Vidyanagar, Anand - 388120, Gujarat*
15. Dr. G.J. Joshi, *Associate Professor, Civil Engineering Department, S. V. National Institute of Technology (SVNIT), Ichchhanath, Surat - 395007, Gujarat*
16. Dr. S.S. Arkatkar, *Assistant Professor, Civil Engineering Department, S. V. National Institute of Technology (SVNIT), Ichchhanath, Surat - 395007, Gujarat*
17. Dr. Rakesh Kumar, *Associate Professor, Civil Engineering Department, S. V. National Institute of Technology (SVNIT), Ichchhanath, Surat - 395007, Gujarat*
18. Dr. Anjaneyappa, *Associate Professor, Civil Engineering Department, R.V. College of Engineering, R. V. Vidyanikethan, Post: Mysuru Road, Bengaluru - 560059, Karnataka*
19. Dr. Anjana Vyas, *Professor, Faculty of Civil Engineering, CEPT University, Kasturbhai Lalbhai Campus, University Road, Navrangpura, Ahmedabad - 380009, Gujarat*
20. Dr. J.N. Patel, *Professor, Civil Engineering Department, S. V. National Institute of Technology (SVNIT), Ichchhanath, Surat - 395007, Gujarat*
21. Dr. Debasis Sanjib Sarkar, *Associate Professor, School of Technology, Pandit Deendayal Petroleum University (PDPU), Raisan Village, Gandhinagar - 382007, Gujarat*
22. Dr. A.V. Thomas, *Director, Adani Institute of Infrastructure Engineering (AIIE), Shantigram Township, Nr. Vaishnodevi Circle, S-G Highway, Ahmedabad - 382421, Gujarat*
23. Dr. Shakil Malik, *Principal F.D. (Mubin), Institute of Engineering and Technology, Bahiyal, Ta. Dehgam, Dist. Gandhinagar - 382308, Gujarat*
24. Dr. P.H. Shah, *Dean, Faculty of Civil Engineering, CEPT University, Kasturbhai Lalbhai Campus, University Road, Navrangpura, Ahmedabad - 380009, Gujarat*

25. Dr. S.S. Trivedi, *Professor & Head, Civil Engineering Department, Gujarat Power Engineering and Research Institute, Near Toll Booth, Ahmedabad–Mehsana Expressway, Village - Mewad, Mehsana - 382710, Gujarat*
26. Dr. D.A. Patel, *Assistant Professor, Civil Engineering Department, S. V. National Institute of Technology (SVNIT), Ichchhanath, Surat - 395007, Gujarat*
27. Dr. P.R. Patel, *Professor, Civil Engineering Department, Institute of Technology, Nirma University, S-G Highway, Post – Chandlodia, Ahmedabad - 382481, Gujarat*

Mechanical Engineering Track

1. Dr. V.J. Lakhera, *Professor & Head, Mechanical Engineering Department, Institute of Technology, Nirma University, Ahmedabad 382481*
2. Dr. R.N. Patel, *Professor, Mechanical Engineering Department, Institute of Technology, Nirma University, Ahmedabad 382481*
3. Dr. K.M. Patel, *Professor, Mechanical Engineering Department, Institute of Technology, Nirma University, Ahmedabad 382481*
4. Dr. B.A. Modi, *Professor, Mechanical Engineering Department, Institute of Technology, Nirma University, Ahmedabad 382481*
5. Dr. B.K. Mawandiya, *Associate Professor, Mechanical Engineering Department, Institute of Technology, Nirma University, Ahmedabad 382481*
6. Dr. R.R. Trivedi, *Associate Professor, Mechanical Engineering Department, Institute of Technology, Nirma University, Ahmedabad 382481*
7. Dr. S.J. Joshi, *Associate Professor, Mechanical Engineering Department, Institute of Technology, Nirma University, Ahmedabad 382481*
8. Prof. N.K. Shah, *Associate Professor, Mechanical Engineering Department, Institute of Technology, Nirma University, Ahmedabad 382481*
9. Dr. A.M. Lakdawala, *Associate Professor, Mechanical Engineering Department, Institute of Technology, Nirma University, Ahmedabad 382481*
10. Dr. N.D. Ghetiya, *Associate Professor, Mechanical Engineering Department, Institute of Technology, Nirma University, Ahmedabad 382481*
11. Dr. S.V. Jain, *Associate Professor, Mechanical Engineering Department, Institute of Technology, Nirma University, Ahmedabad 382481*
12. Dr. M.B. Panchal, *Associate Professor, Mechanical Engineering Department, Institute of Technology, Nirma University, Ahmedabad 382481*
13. Dr. A.M. Achari, *Associate Professor, Mechanical Engineering Department, Institute of Technology, Nirma University, Ahmedabad 382481*
14. Dr. N.M. Bhatt, *Director, Gandhinagar Institute of Technology, Vil. Moti Bhoyan, Khatraj –, Tal. Kalol, Dist. Gandhinagar 382721*
15. Dr. Manoj Gour, *Associate Professor, MITS, Gwalior*
16. Dr. Manoj Chouksey, *Associate Professor, SGSITS, Indore*
17. Dr. Mitesh Shah, *Associate Professor, ADIT Anand*
18. Dr. Vinod N. Patel, *Associate Professor, G H of Engineering & Technology, Vallabh Vidhyanagar 388120*
19. Dr. Mayur Sutaria, *Associate Professor, CHARUSAT Changa*
20. Dr. D.P. Vakharia, *Professor, Mechanical Engineering Department, S. V. National Institute of Technology, Ichchhanath, Surat - 395 007*
21. Dr. A.A. Saikh, *Associate Professor, Mechanical Engineering Department, S. V. National Institute of Technology, Ichchhanath - 395 007*
22. Prof. S.B. Soni, *Retired Professor L.D. College of Engineering. Ahm'd & Visiting Faculty– Mech. Engineering Department IT, NU*
23. Dr. Vishvesh Badheka, *Assistant Professor, Pandit Deendayal Petroleum University, Raisan Village, District Gandhinagar 382007*
24. Dr. Mukul Shrivastava, *Associate Professor, SRM University, Chennai*
25. Vijay Duryodhan, *Assistant Professor, IIT Bhilai*
26. Dr. Hetal Shah, *Assistant Professor, Mechanical Engineering Department, Charutar Institute of Technology, Changa*

Chemical engineering track

Technology Drivers: Engine for Growth – Mahajan, Modi & Patel (Eds)
© 2018 Taylor & Francis Group, London, ISBN 978-1-138-56042-0

Studies on activity models to predict LLE for biodiesel system

M.H. Joshipura & K.J. Suthar
Department of Chemical Engineering, Nirma University, Gujarat, India

ABSTRACT: The accurate prediction of phase equilibria for the biodiesel system is essential for the design and optimization of biodiesel production, separation and purification steps. The UNIQUAC and UNIFAC activity coefficient models were studied to predict the liquid-liquid equilibrium for biodiesel-methanol-glycerol at two different temperatures. The Newton's forward iteration method was applied for the regression of interacting parameters which were fitted against the experimental data. The linear fit of the Hand and Bachman correction equation showed good consistency of the experimental data. The experimental and predicted data were compared and the root mean square deviations for both the models resulted within the deviation of 2% at 303.15 K and 323.15 K.

1 INTRODUCTION

Biodiesel is a promising alternative to the fuel and has gained significant attention during the last couple of decades due to large number well-known benefits. Biodiesel is usually produced by transesterification reaction which is one of the widely accepted techniques.[1] Transesterification is a reversible reaction and the problems of reversibility in most of the cases overcome using higher alcohol molar ratios which shift the equilibrium to the product side. The presence of alcohol increases the solubility of biodiesel and the main by-product glycerol.[2] Once the transesterification reaction is over successfully, multiphase liquid results which majorly includes fatty acid methyl ester along with by-product glycerol, excess alcohol, unreacted oil and other impurities depending on the process routes.

The phase equilibria data are essential not only in the mainstream but also in the downstream separation and purification units in biodiesel production processes involving oils, alcohols, esters, glycerol and water.[3] The phase equilibria data are the function of temperature, pressure and composition, Their better prediction provides better understanding of reaction process, improving the reaction rate, selectivity, simulation and optimization of the reactor and separation units.

The miscibility of involved compounds is reported to be an important factor in downstream processing during biodiesel production.[2] Fatty acids methyl esters and glycerol are partially miscible and form two liquid phases. Liquid-liquid equilibria are more depended on activity models and a small change in activity coefficient leads to a major change in phase equilibria. LLE does not depend on the vapour pressure of compounds involved as in the case of VLE. Even a small inaccuracy in predicting parameter using activity mode might lead to major errors while calculating liquid-liquid equilibria.[4]

The present work aims to investigate liquid-liquid equilibrium is predicted for biodiesel + methanol + glycerol at 303.15 K and 323.15 K temperature and the obtained LLE data were correlated using the UNIQUAC activity coefficient model[5] and original UNIFAC model.[6]

2 LITERATURE SURVEY

Liquid-liquid equilibrium is found in biodiesel production first while separating biodiesel-glycerol and later while purification of biodiesel if water washing is a part of production route. Also the unreacted oil and excess alcohol along with other liquid impurities have to be taken into consideration. Liquid-liquid equilibria are simple while they are experimentally predicted but its prediction using activity models particularly group contribution models may become complex even computationally. Experimental prediction can primarily be done by two methods; one is by finding solubility limit by gradual addition of known amounts of a substance to a solution of known composition until the turbidity changes. Other method used to find the equilibrium compositions is by first agitating a mixture of known overall composition thoroughly and then separating the phases. Once the phases are separated, sampling and analyzing them by appropriate means like chromatography or titration can predict phase equilibria.[7]

The accuracy in prediction using thermodynamic models has been a concern for their applicability over limited low range pressure and temperature particularly for pseudo compounds.[3] Few researchers have successfully attempted the prediction of phase equilibria dedicated to such biodiesel systems. Recently a comparative study[2] on experimental and theoretical prediction of phase equilibria is attempted for animal fat biodiesel-methanol-glycerol system. In other case[8], liquid-liquid equilibrium data was correlated using two models, NRTL and UNIFAC for different ternary system consisting of palm oil, its fraction, alcohol and water. The interacting parameters for NRTL were regressed by minimizing the objective function in terms of composition which is usually done in many cases. Palm oil was considered of having a single triacylglycerol with average molar mass. The results for the average deviation of various biodiesel systems reported within the range of 0.3 to 0.9% for NRTL and 2 to 3.25% for UNIFAC.

Similar trend of results were reported for NRTL and UNIFAC for ethyl biodiesel produced from Crambe fodder radish and Macauba pulp oil.[9] Though the results are promising, the inclusion of the major by product glycerol which is involved in the production process of biodiesel till last stage of purification is missing in both the studies. The studies involving glycerol and water both with alcohol and biodiesel suggested NRTL model for prediction of phase equilibria with root mean square deviation less than 1%.[10]

Four ternary systems including component like methyl esters, glycerol, methanol and water were predicted using NRTL, UNIQUAC, UNIFAC and its versions at three different temperatures. The reported results showed good agreement for UNIQUAC whereas UNIFAC and its versions were not reasonably able to predict phase equilibrium data.[11] Similar group of models were used to estimate 34 different biodiesel systems suggesting UNIFAC-DMD and UNIFAC ASOG models for better prediction of biodiesel system phase equilibria than NRTL and UNIQUAC.[12] Other than activity models, cubic equation of state, association equation of state and equation of state with mixing rule have been applied to predict phase equilibria of biodiesel system. It was reported[13,14] that SRK and PR equation of state with mixing rule were able to predict ternary phase equilibria consisting of ester, alcohol and glycerol. Cubic plus association equation of state is suggested for the prediction of biodiesel multicomponent systems involving water with pure fatty acid esters considered as biodiesel. The available literature on prediction of LLE is presented in table (Table 1) along with models used and the deviation in predicted values.

The literature suggest that the local composition models like NRTL, UNIQUAC and group contribution models are widely accepted in predicting thermodynamic properties for liquid systems due to the fact that they as capable of handling polar and a polar systems. It is also reported that the phase equilibria for compounds which widely differs in molecular size like is well predicted using UNIQUAC model.[7] The model like Van Laar and Margules are not recommended for multicomponent mixture due to the fact that they have two binary parameters. The Wilson activity model is not applicable for liquid phase equilibria as it is inherently unable to consider phase splitting.[4,7] The present study was focused on the studying phase equilibria using UNIQUAC local composition models and UNIFAC group contribution models for biodiesel systems.

The results of phase equilibria literature listed in Table 1 are contrasting with respect to the models used. Biodiesel is made from various sources and consideration of biodiesel as a compound while predicting phase equilibria might be one of the reason for such contrasting

4

Table 1. A review on equilibrium prediction for biodiesel system.

Biodiesel system	Predictive model	Deviation	Reference
Biodiesel (waste fish oil)-methanol-glycerol	NRTL (289.15–328.15 K)	2.05 (AD)	2
Biodiesel (waste fish oil)-methanol-glycerol	UNIQUAC (289.15–328.15 K)	2.85 (AD)	2
Biodiesel (waste fish oil)-methanol-glycerol	UNIFAC (289.15–328.15 K)	2.09 (AD)	2
Biodiesel-methanol-glycerol	UNIFAC-DMD (298.15 K)	3.29 (%rmsd)	2
Biodiesel-methanol-glycerol	UNIFAC-DMD (313.15 K)	2.02 (%rmsd)	2
Biodiesel-methanol-glycerol	UNIFAC-DMD (328.15 K)	0.81 (%rmsd)	2
Water-ethanol-FAEE (Crambe oil)	NRTL (298.15 K)	0.51 (%rmsd)	9
Water-ethanol-FAEE (Fodder radish oil)	NRTL (298.15 K)	0.66 (%rmsd)	9
Water-ethanol-FAEE (Macauba oil)	NRTL (298.15 K)	0.97 (%rmsd)	9
Water-ethanol-FAEE (Crambe oil)	UNIQUAC (298.15 K)	0.49 (%rmsd)	9
Water-ethanol-FAEE (Fodder radish oil)	UNIQUAC (298.15 K)	0.86 (%rmsd)	9
Water-ethanol-FAEE (Macauba oil)	UNIQUAC (298.15 K)	1.29 (%rmsd)	9
Biodiesel (palm)-ethanol-glycerol	NRTL (298.15 K)	0.172 (%rmsd)	10
Biodiesel (palm)-ethanol-glycerol	NRTL (323.15 K)	0.184 (%rmsd)	10
Biodiesel (palm)-ethanol-water	NRTL (298.15 K)	0.180 (%rmsd)	10
Biodiesel (palm)-ethanol-water	NRTL (323.15 K)	0.215 (%rmsd)	10
Biodiesel-methanol-glycerin	UNIQUAC (323.15 K)	1.85 (%rmsd)	15
Biodiesel-methanol-glycerin	NRTL (303.15 K)	1.18 (%rmsd)	15
Biodiesel-methanol-glycerin	UNIQUAC (323.15 K)	1.97 (%rmsd)	15
Ethyl linoleate-ethanol-water	CPA EoS (313.15 K)	2.14 (AD)	16
Ehhylpalmitate-ethanol-water	CPA EoS (308.15 K)	1.79 (AD)	16
Water-ethanol-soybean fatty acid oil	NRTL (298.15 K)	1.87 (%rmsd)	17
Water-ethanol-soybean fatty acid oil	UNIQUAC (323.15 K)	1.84 (%rmsd)	17
Water-ethanol-Ethyl ester (soybean)	NRTL (298.15 K)	1.71 (%rmsd)	17
Water-ethanol-Ethyl ester (soybean)	UNIQUAC (323.15 K)	1.49 (%rmsd)	17

Note: AD is Absolute deviation, rmsd: root meat square deviation, FAEE: fatty acid ethyl ester.

Table 2. Fatty acid composition of Jatropha oil.[19]

C16:0 (Palmitic acid)	18.22
C18:0 (Stearic acid)	5.40
C18:1 (Oleic acid)	28.46
C18:2 (Linoleic acid)	48.12

results. It is therefore necessary to investigate phase equilibria for compounds involved in process prior to developing and designing process. In most of the cases, pure methyl ester or fatty acids are considered for prediction of phase equilibria.

The studies on dedicated biodiesel as compound which consist of varying composition of fattyacid esters are scarce.[2,15] In the present study, Jatropha biodiesel is considered for the present study asoil form Jatropha is one of the few significant sources as a raw material for biodiesel production in Indiaand southern Asia.[3] It is reported that the Jatropha biodiesel shows lower solubility as compared to biodiesel produced from other sources like Castor oil.[18]

The accurately prediction of phase equilibria depends on highly on the composition selection and hence it has to be selected cautiously. In the present study, biodiesel was considered based on the composition of Jatropha oil as described in table (Table 2).

3 LIQUID-LIQUID EQUILIBRIUM PREDICTION

Some guidelines taken as heuristics while predicting phase equilibria are with usually with respect to activity models and the literature shows the same trend of using activity models. The composition and temperature are usually important variable considered while prediction of liquid phase equilibria whereas, pressure is not considered while prediction of liquid-liquid equilibria as it has negligible effects on condensed phases.

In the present study, attempt has been made to use three models for predicting liquid liquid equilibrium for biodiesel systems. The liquid-liquid equilibrium condition states that the overall compositions between two phases have the same partial fugacity and differ only in the proportions of the two phases having these compositions. While dealing with ternary liquid system, estimation of accurate parameters becomes important for the convergence. Certain helpful rules were reported[7] for finding accurate phase equilibria which have been followed in the present study. The first step is to identify miscible pair which in the case of biodiesel-methanol-glycerol is biodiesel-glycerol. Biodiesel and glycerol are called component–1 and component–3 respectively whereas methanol is name as component–2. The initial starting assumption for of composition all the models used is $x_1 = 0.9$, $x_2 = x_3 = 0.05$ and $x_1^* = 0.1$. The initial guessed values might change during iteration to improve the convergence. Walas[7] have described an algorithm to predict the phase equilibria. The expression involved in determining equilibria is dependent in distribution coefficient. It can be expressed for component i as described in equation 1A and 1B.

$$K_i = \frac{\gamma_i^I}{\gamma_i^{II}} = \frac{x_i^{II}}{x_i^I} \tag{1A}$$

$$x_i^{II} = x_i^I \, K_i \tag{1B}$$

The adjustable parameters are usually determined for the known experimental set of data by minimizing the suitable objective function which might be non-convex in terms of optimization variable.[20] In the present study, the parameters are obtained by regression using a five tie line data by evaluating the objective function described in equation 2 using solver in MS office excel.®

$$OF = \sum_{j=1}^{m} . \sum_{i=1}^{n} \left[\left[\left(\gamma_i^I x_i^I \right) - \left(\gamma_i^{II} x_i^{II} \right) \right] / \left[\left(\gamma_i^I x_i^I \right) + \left(\gamma_i^{II} x_i^{II} \right) \right] \right] \tag{2}$$

where m is the number of tie lines for data set and n is number of component, superscript I and II refer to the two liquid phases in equilibrium.

The regressed parameters are fitted in activity models and using an algebraic iterative method, equilibrium composition is predicted and compared with experimental values. The experimental LLE mass fraction data were converted to mole fraction before they were compared to the predicted values.[15] The converted LLE data at two different temperatures are tabulated in table (Table 3). The predicted values are usually compared to experimental values using root mean squared deviation as shown in the following equation 3.

$$RMSD(\%) = 100 \sqrt{\frac{\sum_{i=1}^{c} \sum_{j=1}^{n} \left[\left(x_{ij}^{I,exp} - x_{ij}^{I,calc} \right)^2 - \left(x_{ij}^{II,exp} - x_{ij}^{II,calc} \right)^2 \right]}{2\,c\,n}} \tag{3}$$

3.1 *Prediction of LLE using UNIQUAC model*

The UNIQUAC activity coefficient model is applicable for multicomponent liquid-liquid equilibria and it is a reported to show good agreement for mixtures with widely different molecular sizes. The major disadvantage of the model is its complexity for solving algebraic

Table 3. Experimental LLE data (mass fraction) for biodiesel-methanol-glycerol.

Biodiesel rich phase			Glycerol rich phase		
x_i	x_j	x_k	x_i	x_j	x_k
Temperature 303.15 K					
0.9871	0.0119	0.0010	0.0044	0.0335	0.9621
0.9843	0.0143	0.0014	0.0275	0.0778	0.8947
0.9796	0.0171	0.0033	0.0262	0.1190	0.8548
0.9747	0.0214	0.0039	0.0477	0.1756	0.7767
0.9697	0.0260	0.0043	0.0756	0.2204	0.7039
Temperature 323.15 K					
x_i	x_j	x_k	x_i	x_j	x_k
0.9822	0.0099	0.0078	0.0329	0.0624	0.9046
0.9770	0.0131	0.0099	0.0368	0.1265	0.8366
0.9717	0.0173	0.0110	0.0433	0.1992	0.7575
0.9698	0.0187	0.0115	0.1186	0.2584	0.6230
0.9665	0.0209	0.0126	0.2363	0.2981	0.4656

equation simultaneously. The UNIQUAC model is made up of two parts namely the configurational, often referred as combinatorial which accounts for sizes and shape of molecule and the other part is residual which accounts for the contribution due to energy interactions of molecules. The activity model is based in local composition concept and can be expressed for n components in the following equation 4.

$$ln\gamma_i = ln\left(\frac{\Phi_i}{x_i}\right) + 0.5Zq_i ln\left(\frac{\theta_i}{\Phi_i}\right) + L_i - \left(\frac{\Phi_i}{x_i}\right)\sum_{j=1}^{n} L_j x_j + q_i \left[1.0 - ln\sum_{j=1}^{n}\theta_j\tau_{ji} - \sum_{j=1}^{n}\left(\frac{\theta_j\tau_{ij}}{\sum_{k=1}^{n}\theta_k\tau_{ij}}\right)\right]$$

(4)

$$L_j = 0.5Z\left(r_j - q_j\right) - \left(r_j - 1\right)$$

$$\tau_{ij} = \exp(-a_{ij}/T)$$

where Z represents coordination number and can be defined as a function of temperature. It is taken as 10 usually for liquids at normal condition.[7] The volume fraction per mixture mole fraction and surface fraction per mixture mole fraction can be expressed as described in the following equation 5A and 5B.

$$\Phi_i = \frac{x_i r_i}{\sum_{i}^{n} x_i r_i}$$

(5A)

$$\theta_i = \frac{x_i q_i}{\sum_{i}^{n} x_i q_i}$$

(5B)

where, q_i and r_i denotes surface area and volume respectively for a molecule and can be expressed as shown in the equation 6A and 6B.

$$q_i = \sum_{k} v_k^i Q_k$$

(6A)

$$r_i = \sum_k v_k^i R_k \tag{6B}$$

The group surface area and volume parameters for biodiesel, methanol and glycerol are considered based on the composition shown in table. And the predicted values of q_i and r_i are shown in table (Table 4).

The accuracy in predicting liquid-liquid equilibria by only using binary parameters cannot be achieved. The three sets of binary parameters can be used as an approximate representation of the ternary system in case of unavailability of experimental ternary data. But the addition of ternary experimental data usually improves the results of phase equilibria.

The parameters are temperature dependent and used to represents the phase behavior over the entire set of tie lines taken from literature at 303.15 K temperature. The parameter equation is a linear variation with reciprocal temperature and represents the effect of temperature quite well.[7] The predicted parameters are listed in table (Table 5).

3.2 Prediction of LLE using UNIFAC model

The original UNIFAC model is used to predict phase equilibria as modified versions of UNIFAC is reported to give similar results for fatty acids and its ester.[8] The UNIFAC model can be expressed as shown in two parts namely combinatorial and residual as described in equation 7.

$$ln\gamma = ln\gamma^C + ln\gamma^R$$

$$ln\gamma^C = ln\left(\frac{\Phi_i}{x_i}\right) + 0.5Zq_i ln\left(\frac{\theta_i}{\Phi_i}\right) + L_i - \left(\frac{\Phi_i}{x_i}\right)\sum_{j=1}^{n} L_j x_j \tag{7}$$

The combinatorial part which accounts for sizes and shape of molecule can be predicted in the similar fashion to that of UNIQUAC. The residual part of UNIFAC should be studied cautiously due to its algebraic complexity. It can be expressed as

$$ln\gamma_i^R = \sum_k^C v_k^i \left(ln\Gamma_k^R - \Gamma_k^{Ro} \right)$$

where Γ_k^R is the residual activity coefficient for component k and Γ_k^{Ro} is a residual activity coefficient for an individual component which is given by the equation 8.

$$ln\Gamma_k = Q_k\left[1 - ln\sum_m \Theta_m \Psi_{mk} - \sum_m \frac{\Theta_m \Psi_{mk}}{\sum_n \Theta_n \Psi_{nm}} \right] \tag{8}$$

Table 4. UNIQUAC parameters r_i and q_i.

Component	r_i	q_i
Biodiesel	12.69	10.46
Methanol	1.90	2.048
Glycerol	4.79	4.908

Table 5. UNIQUAC interaction parameters for biodiesel-methanol-glycerol.

Pair i-j	a_{ij}	a_{ji}
Biodiesel-methanol	96.13	18888.64
Biodiesel-glycerol	1258.79	−165.20
Methanol-glycerol	−19932.6	735.02

8

$$\Theta_m = \frac{Q_m x_m}{\sum_n Q_n x_n}$$

$$x_m = \frac{\sum_j v_m^j x_j}{\sum_j \sum_n v_n^j x_j}$$

Θ_m and x_m are the area based surface mole fraction in solution and mole fraction in mixture respectively.

4 RESULTS AND DISCUSSION

The parameter regression took more than 150 iterations with Newton forward methods for both the models. It was observed that the better the model parameter less is the time taken for convergence. The calculated results of phase equilibria are shown in table (Table 6). The parameters fit the model well to predict the values of phase equilibria giving root mean squared deviation of 1.75% and 0.517% at 303.15 K and 323.15 K respectively for UNIQUAC model. The liquid-liquid equilibrium was plotted using Prosim ternary diagram and is shown in Figure 1 and Figure 2 for UNIQUAC model at 303.15 K and 323.15 respectively. The ternary plots for both the model shows that the two liquid phase region is large and the area of region decreases with increasing temperature.

The original UNIFAC model with adjusted parameter showed slightly better results than UNIQUAC. The initial values of considered subgroup $-CH_3-$, $-CH_2-$, $-CH-$, $-OH-$,

Table 6. Calculated RMSD in phase composition for biodiesel-methanol-glycerol system.

Model used	Temperature (K)	RMSD (%)
UNIQUAC	303.15	1.75
UNIQUAC	323.15	0.517
UNIFAC	303.15	1.052
UNIFAC	323.15	0.816

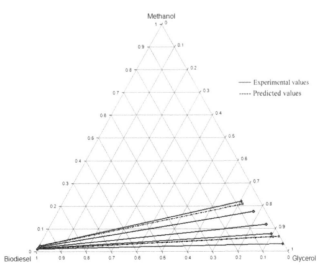

Figure 1. LLE results for biodiesel-methanol-glycerol using UNIQUAC at 303.15 K.

9

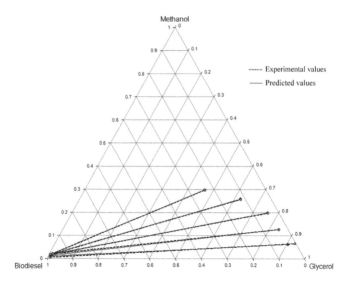

Figure 2.   LLE results for biodiesel-methanol-glycerol using UNIFAC at 323.15 K.

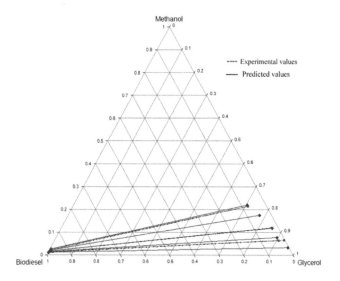

Figure 3.   LLE results for biodiesel-methanol-glycerol using UNIFAC at 303.15 K.

–CH = CH–, CH_2OOH are taken form literature prior to adjustment. The rmsd for biodiesel-methanol-glycerol was found to be 1.052% and 0.816% at 303.15 K and 323.15 K respectively. Figure 3 and Figure 4 shows the ternary diagram phase equilibrium data using UNIFAC model at 303.15 K and 323.15 respectively.

It is observed for the ternary plot that most of the tie-line end points of the biodiesel rich phase are located near the corner of biodiesel, whereas those are more widely distributed along the axis of glycerol rich phase which means that the solubility of methanol in biodiesel is more than glycerol. The predicted values are sensitive to a particular range of temperature for which the parameters are regressed. Extrapolating the phase equilibrium data for larger range might result in large deviation. Other model might be tested for the system

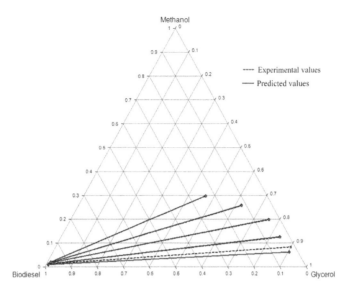

Figure 4. LLE results for biodiesel-methanol-glycerol using UNIFAC at 323.15 K.

5 SUMMARY

The results of vapour liquid equilibrium are often not sensitive to model parameters whereas the results of liquid-liquid equilibrium are extremely sensitive to the choice of parameters for the models. The adjusted parameters fit well and show good agreement for both the models used for determining phase equilibria. The original UNIFAC model showed slightly better results than UNIQUAC when specific subgroup parameters were adjusted for the system considered in the study. The root mean deviation for the predicted LLE resulted within 2% for the models at 303.15 K and 323.15 K. The prediction of liquid-liquid equilibria can be of useful for interpolating more tie lines data. It was also found that the mutual solubility between biodiesel and glycerol is quite minute. The information acquired in the study might be useful for the downstream processing of biodiesel production particularly in methanol recovery step where it gets distributed in the bi-phasic liquid and also in the separation of biodiesel-glycerol step which takes prolonged time despite large miscibility gap.

REFERENCES

[1] Lee, M.J., Lo, Y.C., & Lin, H.M. (2010) Liquid–liquid equilibria for mixtures containing water, methanol, fatty acid, methyl esters, and glycerol. Fluid Phase Equilibria, 299, 180–190.
[2] Maghami, M., Seyf, J.Y., Sadrameli, S.M. & Haghtalab, A. (2016) Liquid-liquid phase equilibria in ternary mixture of water fish oil biodiesel-methanol-glycerol: Experimental data and thermodynamics modelling. Fluid phase equilibria, 409, 124–130.
[3] Demirbas, A. (2009) Progress and recent trends in biodiesel fuels. Energy Conversion and Management, 50, 14–34.
[4] Poling, B.E., Prausnitz, J.M. & O'Connell, J.P. (2001) The properties of gases and liquids. Fifth edition, NewYork, McGrew Hills.
[5] Abrams, D.S. & Prausnitz, J.M. (1975) Statistical thermodynamics of liquid mixtures: a new expression for the excess Gibbs energy of partly or completely miscible systems. AIChE J., 21, 116–128.
[6] Fredenslund, A., Gmehling, J. & Rasmussen, P. (1977) Vapor–Liquid Equilibria Using UNIFAC: A Group-Contribution Method. Amsterdam, Elsevier.
[7] Walas, S.M. (1985) Phase equilibria in chemical engineering, Elsevier.
[8] X. Huang, X., Wang, J.B.J., Ouyang, J., Xiao, Y., Hao, Y., Bao, Y., Wang, Y. & Yin, Q. (2015) Liquid liquid equilibrium of binary and ternary systems composed by palm oil fractions with methanol/ethanol and water. Fluid phase equilibria, 404, 17–25.

[9] Basso, R.C., Miyake, F.H. & Meirelles, A.J.A. (2014) Liquid-liquid equilibrium data and thermodynamic modeling, at T/K = 298.2, in the washing step of ethyl biodiesel production from crambe, fodder radish and macauba pulp oils. Fuel 117, 590–597.

[10] Rocha, E.G.A, Romero, L.A.F., Duvoisin, S., & Anzar, M. (2014) Liquid-liquid equilibria for ternary systems containing ethylic palm oil biodiesel + ethanol + glycerol/water: Experimental data at 298.15 K and 323.15 K and thermodynamic modelling. Fuel, 128, 356–365.

[11] Lee, M.J., Lo, Y.C. & Lin, H.M. (2010) Liquid–liquid equilibria for mixtures containing water, methanol, fatty acid, methyl esters, and glycerol. Fluid Phase Equilibria, 299, 180–190.

[12] Carmo, F.R., Evangelista, N.S., Aguiar, R.S.S. & Fernandes, F.A.N. (2014) Evaluation of optimal activity coefficient models for modeling and simulation of liquid-liquid equilibrium of biodiesel + glycerol + alcohol systems. Fuel, 125, 57–65.

[13] Oliveira, M.B., Ribeiro, V., Queimada, A.J. & Coutinho, J.A.P. (2011) Modeling phase equilibria relevant to biodiesel production: A comparison of gE models, cubisEoS, EoS-gE and association EoS. Ind. Eng. Chem. Res. 50, 2348–2358.

[14] Oliveira, M.B., Barbedo, S., Soletti, J.I., Carvalho, S.H.V., Queimada, A.J. & Coutinho, J.A.P. (2011) Liquid-liquid equilibria for the Canola oil biodiesel + ethanol + glycerol system. Fuel, 90, 2738–2745.

[15] Goncalav, J.D., Aznar, M. & Santos, G.R. (2014) Liquid-liquid equilibrium data for systems containing Brazil nut biodiesel + methanol + glycerol at 303.15 K and 323.15 K Fuel, 133, 292–298.

[16] Follehatti-Romero, L.A., Oliveira, M.B., Batista, E.A.C., Coutinho, J.A.P. & Meirelles, A.J. (2012) Liquid-liquid equilibria for ethyl esters + ethanol + water systems: Experimental measurements and CPA EoSmodeling. Fuel 96, 327–334.

[17] Serres, J.D.S., Soares, D., Corazza, M.L., Krieger, N. & Mitchell, D.A. (2015) Liquid-liquid equilibria data and thermodynamic modeling for systems releted to the production of ethyl esters of fatty acids from soybean soap stock acid oil. Fuel 147, 147–154.

[18] France, B.B., Pinto, F.M., Pessoa, F.L.P. & Uller, A.M. (2009) Liquid-liquid equilibria for Castor oil biodiesel + glycerol + alcohol. J. Chem. Eng. Data, 54, 2359-2364.

[19] Sherbiny, S.A., Refaat, A.A. & Sheltawy, S.T, (2010) Production of biodiesel using the microwave technique. Journal of Advanced Research, 309–314.

[20] Marzougui, A., Petriciolet, A.B., Hasseine, A., Laiadi, D. & Labed, N. (2015) Modeling of liquid liquid equilibrium of systems relevant for biodiesel production using Backtracking search optimization. Fluid phase equilibria, 388, 84–92.

Biomethane and ammonia generation through treatment of rice mill wastewater

Keyur Raval
Department of Chemical Engineering, National Institute of Technology Karnataka, Surathkal, Mangalore, India

Ritu Raval
Department of Biotechnology, Manipal Institute of Technology, Manipal University, Karnataka, India

ABSTRACT: Small-scale rice industry produces about 7,000 l of effluent per day, having a Chemical Oxygen Demand (COD) ranging from 4,000 to 7,000 ppm. The effluent contains mainly long-chain carbohydrates, which is an ideal source for biogas production. In this research work, physical and chemical methods for treatment of rice mill waste were investigated and compared. Filtration, centrifugation and adsorption were used in the physical method, and chemical methods included lime and hydrogen peroxide treatments. Filtration and centrifugation did not reduce COD values because impurities were mainly in the form of dissolved solids. Adsorption using bottom ash obtained from the boiler located at a rice mill reduced the COD of effluent by c. 28%. Lime treatment ranging from 0.1 g to 2 g lime per 100 ml effluent reduced the COD by c. 23% to 43%, respectively. Hydrogen peroxide treatment gave the best results of all the treatments with c. 98% reduction in COD values. Sludge production was c. 60% lower in peroxide-treated effluent as compared to lime-treated effluent. The sludge obtained from lime and peroxide treatment methods was further added to a batch of anaerobic digester for biogas production. Lime-treated sludge reduced biogas production of a stable biogas-producing digester, due to an increase in pH from c. 7.2 to 11.5. Biogas production was enhanced markedly when hydrogen peroxide-treated sludge was added to an anaerobic digester, as compared to lime-treated sludge.

Keywords: rice mill waste; adsorption; lime; hydrogen peroxide; biogas

1 INTRODUCTION

The variety of wastes generated by industrial and domestic activities has increased tremendously worldwide. Estimated waste generation in the world was about 12 billion tons in 2002, which is expected to increase to 19 billion tons per year by 2025 (Yoshizawa et al., 2004). India alone generates about 350 million tons of solid waste, consisting of agricultural and organic waste (Sengupta, 2002). This waste is generated mainly by agricultural activities, municipal solid waste, waste from food processing industries, and industries handling agriculture products. A recent report of the Central Pollution Control Board of India estimated that about 38.254 million liters per day of sewage was generated by Class 1 cities and Class 2 towns: this sewage consists of municipal and industrial waste, out of which only 35% is treated, leaving a yawning gap of 65% untreated sewage (Kaur et al., 2012).

Rice is the prime cereal crop in India, and occupies an area of c. 42 million hectares with an annual production of 76 million tons. This amounts to nearly 42% of the country's food grain production. One of the dominant rice processing techniques is parboiling, which requires large quantities of groundwater. A typical small-scale parboiled-rice manufacturing industry requires groundwater in the range of 900 to 1,200 l per ton of rice paddy. During the process the water becomes unusable, with high Chemical Oxygen Demand (COD), Biological Oxygen Demand (BOD), and Total Suspended Solids (TSS) and Total Dissolved Solids (TDS) loads (Malik et al., 2011). This effluent is discharged directly to paddy fields or drained into rivers immediately after primary treatment. This rice mill effluent has a typical chemical oxygen demand ranging from c. 2000 mg/l to c. 7,000 mg/l. It mainly contains dissolved carbohydrates and minerals, and is acidic in nature with pH ranging from 4.5 to 5.5. The research in the field of rice mill wastewater treatment is nascent. Malik et al. (2011) used a biodegradation technique for rice mill effluent treatment, and reduced the COD level to 75% of the initial COD value. However, the technique needs about 15 days for this reduction. A small-scale industry has to have 15 days of effluent storage to use the biodegradation technique, which is not feasible. Another research group (Behera et al., 2010) utilized microbial fuel cells for treatment of rice mill waste, and reduced the COD levels by almost 97% using microbial fuel cells. However, microbial fuel cell technology demands high capital investment when scaled-up and again it is time-consuming, which requires larger storage capacity for effluent. There is a need for a fast and economical method of rice mill waste treatment. Moreover, rice mill effluent has a huge potential for energy generation owing to its high organic loading. Rao and Baral (2011) investigated the anaerobic co-digestion of various agricultural wastes. Their research revealed that the co-digestion of various food wastes enhances biogas production.

This present research focuses on the development of a fast, sustainable and feasible treatment method for rice mill effluent which can be easily scaled-up in existing small-scale industry. The second part of this research focuses on energy generation through anaerobic co-digestion using the sludge obtained from the treatment process.

2 MATERIALS AND METHODS

All the chemicals used in this research work were bought from Merck India unless otherwise specified. All experiments were performed in duplicate and error encountered was in the range of ±15%.

2.1 *Characterization of wastewater*

The rice mill wastewater was obtained from a nearby small-scale industry. The wastewater was characterized in terms of pH, turbidity and chemical oxygen demand within one hour of sample collection. The COD was measured as per BIS 3025 part 58: 2006.

2.2 *Physical methods*

2.2.1 *Adsorption*
Rice mill wastewater was treated with different quantities of bottom ash obtained from the boiler of a rice mill plant. Bottom ash amounts in the range of 0.1 to 1 g were added to100 ml of rice mill wastewater and the suspension was allowed to mix at room temperature for time intervals ranging from 1 to 3 hours. After the specific time interval, the solution was filtered with Whatman filter paper of size 40 microns and the filtrate was characterized in terms of its turbidity and COD.

2.3 *Chemical methods*

2.3.1 *Lime treatment*
Commercial-grade lime was used for the treatment of rice mill wastewater. Rice mill wastewater of 50 ml volume was pipetted out in five separate flasks. Lime was added to these five flasks in the quantities of 0.5, 1, 1.5 and 2 g and mixed thoroughly for one hour using a magnetic stirrer. The suspension was filtered using a vacuum filter and the filtrate was analyzed for pH and COD. The wet weight and dry weight of the filter cake were measured. The procedure was repeated for time intervals of 2,3,4,5 and 12 hours.

2.3.2 *Hydrogen peroxide treatment*

Different quantities of commercial-grade 30% hydrogen peroxide solution ranging from 5 to 50 ml was added to 50 ml rice mill wastewater. To this solution, about 0.1 g ferrous sulfate was added as a catalyst and stirred gently overnight. The solution was vacuum filtered, and the filtrate was analyzed in terms of its pH and COD. The wet weight and dry weight of the filter cake were measured.

2.4 *Anaerobic digestion of rice mill waste*

Experiments were carried out with a working volume of 200 ml in an airtight filter flask (BOROSIL) of volume 500 ml. The flask opening was closed by a rubber stopper. 100 ml of sludge obtained after rice mill wastewater treatment was mixed with 100 ml cow dung slurry and poured into airtight flasks. As a control experiment, another anaerobic digester was main tained in parallel with 200 ml of cow dung slurry in it. The flasks were static through- out, except for manual mixing three times a day. Biogas was collected through a gas outlet pipe located in the rubber stopper. The volume of biogas generated in the batch reactor was measured using a downward displacement technique.

2.5 *Gas chromatography*

Agas sample was analyzed by gas chromatography (CHEMITO) with the following operat- ing conditions: injection temperature –100°C; column – Porapak Q; flow rate –10 ml/min; oven temperature – 40 to 100°C; detector temperature – 250°C.

3 RESULTS AND DISCUSSION

Rice mill wastewater was collected every week for a period of 16 weeks. The samples had a pH value in the range of 4.2 to 5.2 and a COD in the range of 3,200 to 5,500 mg/l. The rice mill wastewater sample was free of suspended solids, and hence vacuum filtration did not reduce the COD of rice mill waste to a large extent. All rice mill industries have a boiler to generate steam or hot water, which generates bottom ash. A total of ten bottom ash samples were analyzed for Brunauer–Emmett–Teller (BET) specific surface area and results indicated the specific surface area as being in the range of 3.2 to 4.8 m^2/g. This bottom ash obtained from the boiler was used as an adsorbent. No chemical treatment was applied to enhance the adsorption capacity of the bottom ash.

Figure 1 shows the COD values obtained after adsorption for three hours. As bottom ash amount increased from 0.5 g per 100 ml to 2 g per 100 ml, the COD reduction was marginal. The maximum COD reduction of c. 28% after three hours was achieved using bottom ash as an adsorbent. No further COD reduction was achieved when adsorption was carried out for more than three hours (results not shown).

The BET specific surface area of bottom ash was much lower compared to the standard adsorbents available. It is known that bottom ash requires activation by chemical methods or steam to increase its adsorption capacity (Chiang et al., 2012; Wu et al., 2010). Because small-scale rice mill industries do not have the facility to activate bottom ash, the adsorbent used in this experiment was also not given any activation treatment and hence there was a marginal decrease in COD values, even after increasing adsorbent content and time. Because bottom ash was not available in large quantities, the maximum amount of bottom ash used was 2 g per 100 ml effluent.

Because physical treatment methods did not give the desired reduction in COD values of effluent, chemical methods were employed. Alum and lime are widely used as precipitants and flocculants in effluent treatment in many chemical process industries (Woodard, 2001). Hence, experiments were carried out with alum and lime. Surprisingly, no sludge formation was observed when alum was used in the same amounts as lime and the solution remained turbid. Therefore, lime was used for waste treatment. Figure 2 shows the results of lime

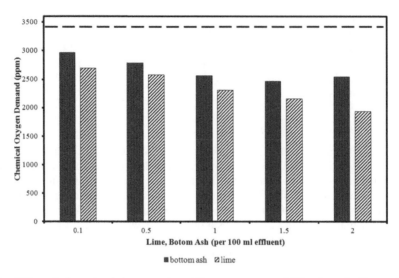

Figure 1. Effect of adsorption in bottom ash and lime treatment on COD values of effluent after three hours of treatment. Dotted line indicates the COD value of the original effluent.

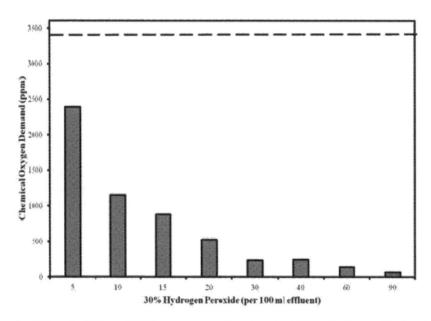

Figure 2. Effect of addition of 30% hydrogen peroxide solution on COD of effluent after three hours. 0.4 mM anhydrous ferrous sulfate was used as a catalyst. The dotted line indicates the COD value of the original effluent.

treatment on the COD value of the effluent after three hours of treatment. The decrease in COD was about 43% when 2 g lime was added to 100 ml effluent, whereas the addition of 0.1 g lime in effluent reduced COD by c. 23%. About 7,000 l of effluent per batch is generated in the rice mill from where the effluent sample was collected. Therefore, as per the experimental results obtained based on 2 g lime per 100 ml effluent, the industry would require about 140 kg of lime for one batch of effluent, which would not be feasible and would increase the hardness of treated effluent due to the presence of calcium ions. Moreover, the effluent pH increased to about 11.5 after the addition of lime. An increased amount of added lime would also increase the acid requirements to neutralize the effluent. Hence, the quantity of lime added was restricted to 2 g of lime per 100 ml effluent. Moreover, further reduction in COD was not observed when an extra 2 g of lime was added to effluent already treated with 2 g of lime. The sludge production was greater when lime was used because of its limited solubility in water.

According to the Central Pollution Control Board of India, the desirable COD limit for effluent to be discharged into rivers is c. 125 mg/l (Kaur et al., 2012). This limit was not achieved by any of the methods mentioned above. Therefore, a novel treatment method based on oxidation using hydrogen peroxide was investigated. Hydrogen peroxide is used in effluent treatment, for disinfection as well as for the reduction of organic loading of effluent (Bas et al., 2012; Ksibi, 2006; Mantzavinos, 2003; Pedahzur et al., 1997). However, it has not so far been used for organic loading removal in the treatment of food processing effluent. Figure 2 indicates the COD of the effluent after the addition of commercial-grade 30% hydrogen peroxide solution. There was a marked decrease in COD when hydrogen peroxide volume increased. A minimum of 31% and a maximum of 98% COD removal was achieved using 5 ml and 90 ml hydrogen peroxide, respectively.

The desirable COD limit of effluent was achieved when adding 60 ml hydrogen peroxide solution. Sludge production was c. 80% less than that obtained during lime treatment. The effluent pH remained in the range of 5.5 to 6.5 after the addition of hydrogen peroxide. The turbidity of the hydrogen peroxide-treated effluent (98% removal of COD) was c. 5, which is very close to the turbidity of tap water. Table 1 compares the effluent characteristics, having been treated with lime versus 30% commercial-grade hydrogen peroxide solution.

As Table 1 indicates, hydrogen peroxide proved to be a very important chemical for the treatment of rice mill wastewater. However, it should be noted that rice mill effluent contains a high amount of carbohydrates, which is a rich source of biomethane production through the anaerobic digestion process. In fact, Kothari et al. (2012) suggested the use of rice mill wastewater to produce hydrogen through a fermentative/anaerobic route. Hence, the sludge obtained after treatment with lime and hydrogen peroxide was investigated for biomethane production using an anaerobic digester.

Figure 3 shows the profile of biogas produced over time, using the sludge obtained by lime treatment as well as hydrogen peroxide treatment. Biogas production using hydrogen peroxide-treated sludge was almost five times more than that when treated by lime. Biogas production using lime-treated sludge was even less than that obtained by using only cow dung. The pH of lime-treated sludge was higher than 11 because of the strong basic nature of lime. It is known that an anaerobic digester functions with maximum efficiency in the pH range of 6 to 7. Biogas production reduces markedly at basic pH values (pH > 8) because of ammonia toxicity (Gerardi, 2003).

Table 1. Comparison of lime and hydrogen peroxide treatments.

	Lime treatment	Hydrogen peroxide treatment
Color	Dark yellow	Pale yellow to colorless
Odor	Foul	Odorless
pH	> 11	5–6
Sludge obtained per 100 ml effluent	c. 2–4 g depending upon addition of lime	c. 0.6–1.6 g
Maximum COD reduction*	c. 43%	c. 98%
Amount required to achieve maximum COD reduction	20 kg per 1,000 l effluent	600 l per 1,000 l effluent

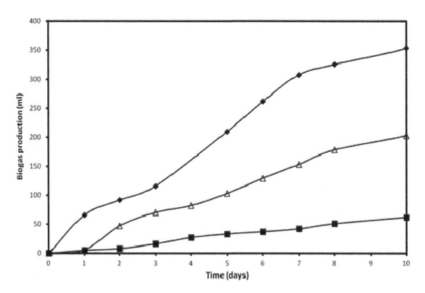

Figure 3. Biogas production in an anaerobic digester using (♦) cow dung + hydrogen peroxide-treated sludge, (△) cow dung, (■) cow dung + lime-treated sludge.

4 CONCLUSION

Rice mill wastewater was treated successfully with commercial-grade 30% hydrogen peroxide solution. Chemical oxygen demand of effluent was reduced to 98% after treatment with hydrogen peroxide solution and sludge production was minimal in all treatments, which reduces the load of solid waste management. Lime-treated effluent gave off a foul odor and a maximum COD reduction of c. 43% was achieved, which was not sufficient to allow discharge of effluent to a river or municipal drainage system. Biogas production increased twofold when sludge obtained after hydrogen peroxide treatment was fed into an anaerobic digester, whereas ammonia toxicity was observed during the anaerobic digestion of lime-treated sludge due to the alkaline nature of the sludge. Though a very clear and odorless effluent was obtained after hydrogen peroxide treatment, more research efforts are required to make the hydrogen peroxide treatment economical for small-scale industries.

REFERENCES

Bas, A.D., Yazici, E.Y. & Deveci, H. (2012). Recovery of silver from X-ray film processing effluents by hydrogen peroxide treatment. *Hydrometallurgy, 121*, 22–27.
Behera, M., Jana, P.S., More, T.T. & Ghangrekar, M.M. (2010). Rice mill wastewater treatment in microbial fuel cells fabricated using proton exchange membrane and earthen pot at different pH. *Bioelectrochemistry, 79*(2), 228–233.
Chiang, Y.W., Ghyselbrecht, K., Santos, R.M., Meesschaert, B. & Martens, J.A. (2012). Synthesis of zeolitic-type adsorbent material from municipal solid waste incinerator bottom ash and its application in heavy metal adsorption. *Catalysis Today, 190*(1), 23–30.
Gerardi, M.H. (2003). *The microbiology of anaerobic digesters.* Hoboken, NJ: John Wiley & Sons.
Kaur, R., Wani, S.P., Singh, A.K. & Lal, K. (2012). Wastewater production, treatment and use in India. In *National Report presented at the 2nd Regional Workshop on Safe Use of Wastewater in Agriculture, 16–18 May 2012, New Delhi.* Retrieved from http://www.ais.unwater.org/ais/pluginfile.php/356/mod_page/content/93/CountryReport_India.pdf.
Kothari, R., Singh, D.P., Tyagi, V.V. & Tyagi, S.K. (2012). Fermentative hydrogen production–An alternative clean energy source. *Renewable and Sustainable Energy Reviews, 16*(4), 2337–2346.
Ksibi, M. (2006). Chemical oxidation with hydrogen peroxide for domestic wastewater treatment. *Chemical Engineering Journal, 119*(2), 161–165.

Malik, K., Garg, F.C. & Nehra, K. (2011). Characterization and optimization of conditions for biodegradation of sella-rice mill effluent. *Journal of Environmental Biology*, *32*(6), 765.

Mantzavinos, D. (2003). Removal of benzoic acid derivatives from aqueous effluents by the catalytic decomposition of hydrogen peroxide. *Process Safety and Environmental Protection*, *81*(2), 99–106.

Pedahzur, R., Shuval, H.I. & Ulitzur, S. (1997). Silver and hydrogen peroxide as potential drinking water disinfectants: their bactericidal effects and possible modes of action. *Water Science and Technology*, *35*(11–12), 87–93.

Rao, P.V. & Baral, S.S. (2011). Experimental design of mixture for the anaerobic co-digestion of sewage sludge. *Chemical Engineering Journal*, *172*(2), 977–986.

Sengupta, J. (2002). Recycling of agro-industrial wastes for manufacturing of building materials and components in India. An overview. *Civil Engineering & Construction Review*, *15*(2), 23–33.

Woodard, F. (2001). *Industrial waste treatment handbook*. Boston, MA: Butterworth-Heinemann.

Wu, F.C., Wu, P.H., Tseng, R.L. & Juang, R.S. (2010). Preparation of activated carbons from unburnt coal in bottom ash with KOH activation for liquid-phase adsorption. *Journal of Environmental Management*, *91*(5), 1097–1102.

Yoshizawa, S., Tanaka, M. & Shekdar, A.V. (2004). Global trends in waste generation. In I. Gaballah, B. Mishar, R. Solozabal & M. Tanaka (Eds.), *Recycling, waste treatment and clean technology* (pp. 1541–1552). Madrid, Spain: TMS Mineral, Metals and Materials Publishers.

The Fresnel lens as a solar collector for various thermal energy applications

Jay Ranparia & Rajubhai Kanaiyalal Mewada
Department of Chemical Engineering, Institute of Technology, Nirma University, Ahmedabad, Gujarat, India

ABSTRACT: The Fresnel lens is a modified magnifying glass which can be made from various materials like glass, acrylic and so on. However, due to the lesser weight of acrylic materials, even very large Fresnel lenses can have considerably less weight. This is one of the biggest advantages in tracking of the system. In this paper, a Fresnel lens of about 0.5 m², retrieved from old projection TV was used as a solar collector and heat received from the sun was utilized for various applications like steam generation, high temperature reactions and waste water treatment. The results were highly encouraging.

Energy is a very crucial aspect for the growth of any country and at the same time environmental pollution is another big problem for the energy sector. Therefore, utilization of solar energy is an attractive path to solve both energy and environmental problems. This paper discusses the utilization of such solar energy by a using Fresnel lens recovered from a waste project TV. Some reported applications of Fresnel lenses are discussed in given references [1–4]. This work demonstrates a number of possibilities for the development of various solar thermal energy applications.

1 EXPERIMENTAL SETUP

In this experiment, a Fresnel lens was taken from an old television. It was a rectangular shape with a lens dimension of 884 mm × 663 mm × 3 mm, which provided an area of approximately 0.5 m².

As shown in Figure 1, a structure was made for the Fresnel lens with a dual axis tracking of the lens such that a receiver was also moving as per the lens so that the focus was always at a single point. In this setup, the Fresnel lens was mounted on a structure made of aluminum and the lens was surrounded by a photovoltaic panel of 12V having total a 100 W capacity, with a suitable battery and invertor to store energy. For this a different type of receiver was made based on application such as for chemical reaction a box of iron was made which was insulated from inside by the ceramic wool and the wool was coated with cement to minimize the heat loss and the box was covered with the glass to make it a closed system, for the steam generation and heating of oil a metal block was taken and a cylindrical groove was made in the central axis and nipple was attached to both ends of the metal block. In this, a hydraulic

Figure 1.   Fresnel lens set up.

jack was attached along with the receiver so that the focus point area was changed as per the requirements of the application such as for melting of metal a point focus was required, while for heating of metal block the area of focus is slightly more than a point.

Experiments: The following experiments were carried out for various applications:

1.1 *Low-pressure steam generation*

In this experiment, low-pressure steam was generated using a cylindrical copper tube. The tube was kept vertical with suitable insulation such that the focus of the Fresnel lens was at the bottom of the tube and the lower part of the copper tube was attached with a pipe through which a continuous supply of water was provided. The water supply tank was mounted at a higher level so that the water flows by the thermosyphone effect.

Amount of water (gm)	Delta T (°C)	cP (J/Kg/K)	Latent heat (J/(kg. K))	Sensible heat (J)	Latent heat (J)	Total energy (J)
48	75	4.86	2100	17496	100800	118296

Heat received (J)	Heat per time (J/s)	Heat (W/m^2)	Solar energy insolation (W/m^2)	% Efficiency
118296	94	188	520	36%

Applications of low-pressure steam: Such low-pressure steam can be used for various heating applications such as cooking or any process with a heating purpose up to 80 to 90°C.

1.2 *Batch wise high-pressure steam generation*

In this experiment, a block of cast iron of about 9.5 kg weight and a pipe in the center was used. A nipple, pressure gage and ball valves on both sides of the block were attached to make it a closed system. The total volume of water taken in the block was about 50 ml. Then, water was filled in the pipe and both the valves were closed and the block was kept at the focal point of the Fresnel lens, with 2 to 3 cm diameter of the focus adjusted on the surface of the block. The rise in temperature with time data is given in following table. After about two hours of exposure, the total pressure was about ten bar in the system.

Steam was produced through a Fresnel lens with a 0.5 m^2 surface.

Time (Min)	0	25	40	50	60	70	80	90	100	110
Pressure (bar)	0	0.5	0.8	1.2	1.4	2.1	2.9	4	5.5	7.3
Temperature (°C)	0	107	111	117	126	134	142	152	162	172

However, due to the limited capacity of the Fresnel lens, continuous high-pressure steam could not be produced. For that, we required a larger surface area of Fresnel lens to tap a greater amount of solar energy. For this, five pieces of 1.5 m^2 foldable Fresnel lens from the Heck Company, Denmark were purchased. However, due to time constraints, the necessary experiments could not be performed.

2 OXIDATION OF ZINC

In this experiment, zinc was oxidized at a temperature of 1100°C. In the reaction, zinc dust was taken in the ceramic crucible or zinc metal was taken in the receiver and water was sprayed continuously as H_2O was required for the reaction. At around 400°C, zinc melted and at 1100°C, zinc in the presence of water vapor reacts to form zinc oxide and hydrogen gas as a by-product. This reaction is known as a French or indirect method. In this reaction, we get hydrogen as the by-product which is a very useful product so there is no waste produce in the reaction and we can also convert zinc oxide to zinc by reduction (requiring a very high temperature about 2000°C). Therefore, by this cycle reaction, we can generate hydrogen gas. Zinc oxide is also largely used in many industries for different applications.

The French or indirect method is as follows:

$$Zn + 2\,H_2O \xrightarrow{\hspace{2cm}} Zn(OH)_2 + H_2 \tag{1}$$
Zinc   Zinc hydroxide

$$Zn(OH)_2 \xrightarrow{\hspace{2cm}} ZnO + H_2O \tag{2}$$
Zinc hydroxide   Zinc oxide

However, this applications shows the use of a Fresnel lens for some direct oxidation reactions where oxygen from the environment can be used. This reaction was carried out to demonstrate hydrogen production by oxidation of zinc using renewable energy. Hydrogen will be a green fuel for the future.

The experiment was carried out in the month of February between 10:00 *a.m.* to 2:00 *p.m.* At that time, solar insolation was 540 W/m^2.

Applications of ZnO: ZnO is used as a raw material for many applications including: adsorbents and absorbents; agricultural chemicals (non-pesticidal); corrosion inhibitors and anti-scaling agents; dyes and fillers.

3 OXIDATION OF COPPER

In this experiment a plate of copper was oxidized using atmospheric oxygen to produce copper oxide, which is black in color, at around 300°C to 800°C. The copper oxide which was formed during the reaction was tested by adding HCl, as copper oxide reacts with HCl to form copper chloride which is green in color.

$$2\,Cu + O_2 \xrightarrow{\hspace{2cm}} 2\,CuO \tag{3}$$
Copper   Copper(II)oxide

$$CuO + 2\,HCl \xrightarrow{\hspace{2cm}} CuCl_2 + H_2O \tag{4}$$
Copper(II)oxide   Copper(II)chloride

Applications: This application demonstrates the use of a Fresnel lens as a furnace for high temperature oxidation reactions. Copper oxide has many applications as listed below:

1. Agricultural chemicals (non-pesticidal)
2. Intermediates
3. Oxidizing/reducing agents
4. Process regulators
5. Processing aids not otherwise listed
6. Processing aids, specific to petroleum production

4 APPLICATION OF WASTE WATER TREATMENT

Experiments were carried out to remove the color of methylene blue dye using solar energy.

Experimental setup and materials: All experiments were carried out on a terrace between 12:00 to 4:00 when the intensity of solar energy is high. In the first experiment, 200 ml, 25 ppm concentrated methylene blue solution was prepared and 0.75 gm of titanium oxide catalyst was added. The solution was continuously stirred at 85 rpm with the help of a magnetic stirrer and kept under sunlight. The same experiment was also carried out using ZnO as a catalyst to find a better color removing catalyst. Based on the results, ZnO is better catalyst than titanium oxide.

Further experiments were performed using ZnO as a catalyst. Two simultaneous experiments were performed to degrade dye under solar energy. One with direct sunlight and the other using sunlight through a Fresnel lens. First, 25 ppm methylene blue solution of 500 ml was prepared and divided into two beakers of 200 ml. To both beakers, the same amount, 0.75 gm of ZnO catalyst, was added.

Both samples were continuously stirred at 850 rpm using a magnetic stirrer. Samples were collected after ten minutes and examined for color removal using UV spectroscopy.

The results obtained are shown below:

	% Color removed for initial 25 ppm solution	
Time	With Fresnel lens	Without Fresnel lens
0	0	0
10	97.55	94.82
20	99.41	99.06

The same experiment but with a high concentration of dye, 500 ppm, was also carried out. The results obtained are given below:

	% Color removed for initial 500 ppm solution	
Time (min)	With Fresnel lens	Without Fresnel lens
0	0	0
150	60	20

5 SUMMARY

1. Fresnel lenses were set up with a dual axis tracking system to collect solar thermal energy.
2. Atmospheric pressure steam generation on a continuous basis with 36% thermal efficiency was achieved. Compared to PV cell efficiency this is reasonably high.
3. High-pressure (10 bar) steam in a batch system (like a pressure cooker, closed system) was generated. However, to generate high-pressure steam on a continual basis a large surface area of Fresnel lens is required. New large and foldable five piece Fresnel lenses have been purchased which can be tested for such applications. However, due to time limitations, experiments with the newly purchased Fresnel lenses could not be performed.
4. A high temperature chemical reaction of zinc metal in the presence of water to form zinc oxide and hydrogen was tested. The reaction takes place at around 900°C. Here, zinc oxide as a product was verified based on its basic properties. However, due to setup limitations, hydrogen collection and testing could not be carried out. A suitable setup is required with better instrumentation to collect hydrogen and to control the reaction temperature.
5. Oxidation of copper to copper oxide at 300°C to 800°C in the presence of oxygen was carried out. This shows the use of such a collector as a furnace for oxidation reactions. However, here the source of copper should be from waste material to make this process economically more viable. Production of copper oxide and the recovery of copper from the mother board of electronic items can be done through this process.
6. Degradation of waste water was demonstrated using a Fresnel lens. A higher rate of degradation of pollutant in waste water was observed during the experiment with the Fresnel lens compared to without the Fresnel lens setup. For waste water having a 500 ppm methylene blue initial solution, about 60% reduction in color of waste water in the presence of a Fresnel lens compared around 20% reduction in color of waste water without the Fresnel lens setup was achieved.

REFERENCES

http://www.fresneltech.com, 21/02/2017, 15:00.

IEA SHC worldwide report. http://www.solarserver.com/solar-magazine/ solar-news/archive 2012/2012/ kw20/iea-solar-heating-and-cooling-programme-solar-thermal-markets-grew-14-in–2010.htm; 2012.

Miller, O.E., Mcleod, J.H. & Sherwood, W.T. (1951). Thin sheet plastic Fresnel lenses of high aperture. *Journal of the Optical Society of America*, *41*(11), 807–15.

Rabl, A. (1985). *Active solar collectors and their applications* (p. 503). New York: Oxford University Press, 1985.

Technology Drivers: Engine for Growth – Mahajan, Modi & Patel (Eds)
© *2018 Taylor & Francis Group, London, ISBN 978-1-138-56042-0*

Measurement and correlation of vapour-liquid equilibria for cyclopentylmethylether + acetic acid at atmospheric pressure

V.M. Parsana & S.P. Parikh
Department of Chemical Engineering, V.V.P. Engineering College, Gujarat Technological University, Gujarat, India
Department of Chemical Engineering, L.D. College of Engineering, Gujarat Technological University, Gujarat, India

ABSTRACT: Isobaric Vapour-Liquid Equilibrium (VLE) data for Cyclopentylmethylether (CPME) + Acetic acid were measured at atmospheric pressure using a dynamic recirculating apparatus. The measured experimental data of the binary system was confirmed to be thermodynamically consistent according to area test of Herington and point-to-point test of Van Ness. The experimentally measured binary VLE data were correlated by Wilson, Non-Random Two Liquids (NRTL) and Universal Quasi-Chemical (UNIQUAC) activity coefficient models. The Average Absolute Deviation (AAD) of temperature is 0.2008 K, 0.1983 K, and 0.2015 K and that of vapour phase composition is 0.0033, 0.0033, and 0.0034 for Wilson, NRTL, and UNIQUAC models respectively. The optimum binary interaction parameters (BIPs) for the system have been reported. The research findings would provide basic data for the design of separation process.

Keywords: Green solvents, Cyclopentyl methyl ether, Acetic acid, Vapour-liquid equilibrium

1 INTRODUCTION

Experimental vapour-liquid equilibrium (VLE) data are the primary requirement for equipment design for separation of mixtures, especially, the distillation operation, for the recovery of solvents for recycling. Cyclopentylmethylether (CPME), an ethereal solvent, being considered as one of the green solvents, exceeds other ethereal solvents regarding EHS (environmental, health and safety) aspects as per GlaxoSmithKline's solvent selection guide (Henderson, *et al.*, 2011). The properties such as the low formation of peroxides, high boiling point, lower evaporation energy and better stability in acidic and basic conditions make CPME a stable solvent for a wide variety of chemical production (Anastas, 2015). It also satisfies 8 out 12 principles of green chemistry developed by P.T. Anastas and J.C. Warner in 1991 (Sakamoto, 2013). Acetic acid is used to produce intermediates such as vinyl acetate and acetic anhydride which are used to make latex emulsion resins for paints, adhesives, paper coatings, textile finishing agents, cellulose acetate fibers, cellulosic plastic, etc. It is also an important organic raw material which is widely used in the organic synthesis, pharmaceuticals, dyes and intermediates, pesticides and other chemical industries (Kirk-Othmer, 1998).

The separation of acetic acid + water mixture is an important industrial problem because of economical and environmental aspects. H. Zhang et al. carried out liquid-liquid equilibrium (LLE) experiments at different temperatures to find out a good solvent for separation of acetic acid from water. After studying over 30 solvents, they ascertained that CPME would be a good substitute for conventional organic solvents (Zhang, *et al.*, 2012). If CPME is to be used as an organic solvent for the separation of acetic acid from water by LLE then it is to be recovered from the mixture for recycling. The application of CPME as the green solvent can be encouraged only if experimental VLE data for this binary system is available in the literature, but unfortunately, it is not. So this project work was aimed at providing experimental VLE data for the design of distillation operation for separation of the CPME and acetic acid mixture.

2 EXPERIMENTAL

2.1 Chemicals

The chemicals of the highest available purities were used in this study. All the reagents were of analytical purity grade. Cyclopentyl methyl ether and Acetic acid were purchased from LobaChemi Pvt. Ltd., laboratory reagents and fine chemicals, Mumbai. No further purification of any chemical was carried out. The purity of the chemicals reported by the supplier was verified by comparing the experimental refractive indexes with those reported in the literature (Table 1).

2.2 Apparatus and procedure

The experimental vapour-liquid equilibrium apparatus was purchased from Abhishek Scientific Company, Mumbai. The construction of this apparatus is based on the design of raal modification (1998) of the Yerazunis et al. (1964) dynamic VLE still. In this apparatus, liquid and condensate vapour both the phases are recirculated. The schematic diagram of actual VLE apparatus is given in Figure 1. The equilibrium chamber is insulated with a vacuum jacket. The Cottrell tube is also vacuum jacketed. The Cottrell tube discharges vapour-liquid mixture in the packed equilibrium chamber onto a temperature sensor (Pt-100). To eliminate concentration and temperature gradient in the condensate receiver, an effective magnetic stirrer has been provided.

About 700 cm³ of the mixture was taken into the boiling chamber initially and heating was started. When operating temperature was fully stabilized for at least 30 minutes without any fluctuation, the equilibrium state was considered to be attained. At the equilibrium point, about 1 ml of samples of liquid and condensate vapour were taken simultaneously with syringes through the septum. A built-in calibrated resistance sensor (Pt-100) connected to a digital temperature meter (SELEC PIC-101-N) with a precision of 0.1 K measured the equilibrium temperature. The accuracy of the Pt-100 sensor as stated by Selec Controls Pvt. Ltd. India is 0.25% of temperature span ± 0.1°C (after 20 min. warm-up).

2.3 Analysis and refractometer calibration

The composition of both liquid and condensate vapour phases were measured using a refractometer during the experimentation. The refractometer, used for analysis of samples, is a five-digit automatic digital refractometer RFM-950 supplied and calibrated by LABMAN. The range of refractive index for the instrument is 1.30000–1.70000 and the stated measuring accuracy is ± 0.00002. The measured refractive index (R.I.) values were converted to mole fractions of compounds of the respective phases. The calibration curve of refractive index vs. volume % of CPME is shown in Figure 2.

Table 1. Chemicals with their respective purities.

Chemical	Supplier	Stated minimum purity (mass%)	n_D (293.15 K) Exp.	Lit.
CPME	LOBA	99.9	1.41693	1.4189
Acetic acid	LOBA	99.7	1.36861	1.3717

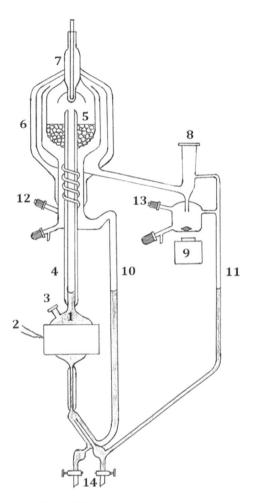

Figure 1.   Schematic diagram of actual VLE apparatus.
1) boiling chamber, 2) heater cartridge, 3) feed point, 4) cottrell pump, 5) equilibrium chamber, 6) vacuum jacket for the equilibrium chamber, 7) thermo well, 8) total condenser, 9) magnetic stirrer, 10) liquid return line, 11) vapour return line, 12) liquid sample point, 13) vapour sample point, and 14) drain valves.

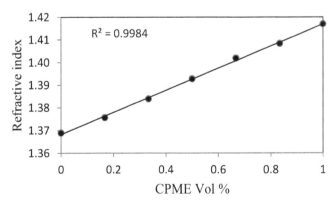

Figure 2.   Refractive index calibration curve for CPME and acetic acid.

3 RESULTS AND DISCUSSION

3.1 *Reliability test of the VLE apparatus*

To test the reliability of the apparatus, a binary system of acetone (1) / benzene (2) was chosen for the experimentation and its VLE data were measured at atmospheric pressure and compared with literature data (Kurihara & Kojima, 1998). The comparison of the experimental x-y data with the literature data is shown in Figure 3. It indicates that the experimental data obtained by the VLE apparatus and the data reported in the literature are matching closely.

3.2 *Vapor-liquid equilibrium measurements*

The vapour-liquid equilibrium data measurement had been carried out for the binary system CPME/Acetic acid and all four variables temperature, pressure and compositions of liquid and vapour phases had been measured. During the experiments, total 26 measurements of VLE data were taken. Out these 26 experimental points, 5 were discarded due to some errors and remaining 21 measurements had been chosen for further assessment which is tabulated (Table 2). The pressure during the experimentation was kept at atmospheric and its value was taken as 101.325 kPa for the calculation purpose.

3.3 *Thermodynamic consistency test for experimentally measured data*

3.3.1 *Herington's consistency test (area test)*
The semi-empirical method reported by Herington was employed to examine the thermodynamic consistency of VLE data for the binary system CPME/Acetic acid. The values of D and J were calculated using Eq. 1 and Eq. 2. The criterion of consistency of this method is that the value of D-J cannot be greater than 10. The value of |D-J| was 4.95 and hence the experimental VLE data for the CPME/Acetic acid system passed the Herington's test.

$$D = 100 \frac{\int_{x_1=0}^{x_1=1} \ln \frac{\gamma_1}{\gamma_2} dx_1}{\int_{x_1=0}^{x_1=1} \left| \ln \frac{\gamma_1}{\gamma_2} \right| dx_1} \tag{1}$$

$$J = 150 \frac{T_{max} - T_{min}}{T_{min}} \tag{2}$$

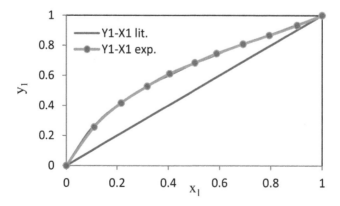

Figure 3.   x-y diagram for acetone (1) / benzene (2) at 101.3 kPa.

Table 2. Experimental P-T-x-y data for CPME/Acetic acid system.

T (K)	$x_{1\,exp}$	$y_{1\,exp}$	γ_1	γ_2
379.15	1.0000	1.0000	1.0000	
379.35	0.9322	0.9408	1.0035	1.2666
379.65	0.8664	0.8844	1.0063	1.2429
380.15	0.7855	0.8211	1.0160	1.1786
380.85	0.6816	0.7309	1.0218	1.1675
381.35	0.6095	0.6761	1.0422	1.1274
381.65	0.5698	0.6438	1.0526	1.1146
382.15	0.5178	0.5981	1.0611	1.1040
382.85	0.4510	0.5378	1.0742	1.0904
383.25	0.4146	0.5121	1.1003	1.0657
383.85	0.3667	0.4668	1.1152	1.0562
384.25	0.3337	0.4453	1.1561	1.0312
384.65	0.3060	0.4131	1.1567	1.0343
385.25	0.2623	0.3663	1.1769	1.0308
386.35	0.2012	0.2889	1.1740	1.0319
386.95	0.1688	0.2530	1.2055	1.0224
387.95	0.1262	0.1932	1.1982	1.0181
388.45	0.1033	0.1648	1.2318	1.0112
389.55	0.0568	0.0957	1.2628	1.0062
390.05	0.0337	0.0582	1.2771	1.0072
391.15	0.0000	0.0000		0.9992

3.3.2 *Van Ness point-to-point test*

In addition to Herington's method, the point to point test of Van Ness (Van Ness, *et al.*, 1973) is used to test the experimental data. The equation used to perform the test is:

$$y_{AAD} = \frac{1}{n} \sum_{i=1}^{n} \left(\frac{|y_{i,\exp} - y_{i,cal}|}{y_{i,\exp}} \right) \qquad (3)$$

where y_{AAD} is the average absolute deviation (AAD) of vapour phase composition; n is the number of experimental points; the subscripts 'cal' and 'exp' present the calculated values and experimental values respectively. The criterion for consistency of this method is requirement of less than 0.01 value of y_{AAD}. The value of y_{AAD} for the system has been tabulated as below.

3.4 *Correlations and predictions for measured VLE data*

The data reduction was carried out by Wilson (Wilson, 1964), NRTL (Renon & Prausnitz, 1968) and UNIQUAC models. The objective function chosen for minimization of error was:

$$\sum_{k=1}^{n} \sum_{i=1}^{2} \left(\frac{\gamma_{ki,\exp} - \gamma_{ki,cal}}{\gamma_{ki,\exp}} \right)^2 \qquad (4)$$

The binary interaction parameters found out by regression for CPME/Acetic acid are reported in Table 5. For regeneration of data using binary interaction parameters, an iterative procedure was used. First of all, the temperature was assumed and vapour pressures of both components were calculated using Antoine equations (Table 6). The activity coefficients γ_1 and γ_2 were calculated using g^E models. Temperature T and liquid phase composition x_1 were taken from the experimental data and pressure P was found out. The correct value of temperature was found out by minimization of the following function:

31

$$\%AAD\sum(\Delta P) = \frac{100}{n}\sum_{i=1}^{n}\frac{\left|P_{i,\exp} - P_{i,cal}\right|}{P_{i,\exp}} \tag{5}$$

The average absolute deviation in temperature and in vapour phase composition were calculated by Eq. 6 and Eq. 7 respectively; and reported along with binary interaction parameters (Table 5).

$$AAD\sum(\Delta T) = \frac{1}{n}\sum_{i=1}^{n}\left|T_{i,\exp} - T_{i,cal}\right| \tag{6}$$

$$AAD\sum(\Delta y) = \frac{1}{n}\sum_{i=1}^{n}\left|y_{i,\exp} - y_{i,cal}\right| \tag{7}$$

The calculated data by various models were compared with the experimentally found data. The x-y diagrams for the binary system have been plotted for Wilson, NRTL and UNIQUAC models and shown in Figure 4 to Figure 6. The calculated data by g^E models are shown as smooth lines and the experimental data are shown by bullets or markers.

Table 3.   Thermodynamic consistency check by Herington's method.

System	D	J	[D-J]	Result
CPME/Acetic acid	9.70	4.75	4.95	Pass

Table 4.   Thermodynamic consistency check by point-to-point test of Van Ness.

System	Model	y_{AAD}	Result
CPME/Acetic acid	NRTL	0.0096	Pass

Table 5.   Binary interaction parameters for CPME/Acetic acid system.

Model	Binary Parameters		AAD (δT)	AAD (δy)
Wilson	Λ_{12} −1687.516	Λ_{21} 2597.678	0.2008	0.0033
NRTL	$g_{12} - g_{22}$ 331.529	$g_{21} - g_{11}$ 567.213	0.1983	0.0033
UNIQUAC	$u_{12} - u_{22}$ 1664.076	$u_{21} - u_{11}$ −971.622	0.2015	0.0034

Table 6.   Antoine equation constants for CPME and Acetic acid.

Compound	Antoine constants			Temperature range/K
	A	B	C	
CPME (Modi, 2014)	15.0255	3798.52	−14.2	357 to 391
Acetic acid (Poling, *et al.*, 2012)	15.0694	3580.79	−48.5	298 to 414

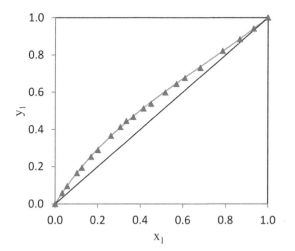

Figure 4.　x-y diagram for Wilson model.

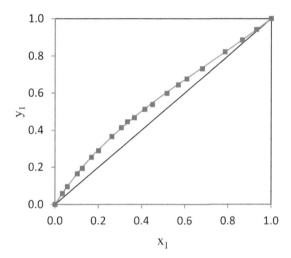

Figure 5.　x-y diagram for NRTL model.

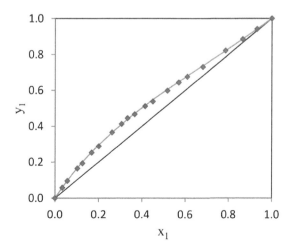

Figure 6.　x-y diagram for UNIQUAC model.

4 CONCLUSION

The VLE data for the binary system comprising green solvent CPME with acetic acid have been measured experimentally by modified Raal dynamic VLE still at atmospheric pressure. The experimental data were found thermodynamically consistent by Herington's area test and Van Ness's point-to-point test. The average absolute deviation in temperature and vapour phase composition for Wilson, NRTL and UNIQUAC models have been reported. It can be concluded from the reported values that although all three models have been successful in fitting the experimentally measured VLE data; NRTL model is the best fit for CPME/Acetic acid system. No azeotrope formation is found for the system.

ACKNOWLEDGMENTS

We acknowledge our gratitude to the management of V.V.P. Engineering College, Rajkot for facilitating the infrastructural support. We are deeply obliged to the Gujarat Council on Science and Technology (GUJCOST), Gandhinagar for providing much needed financial assistance by awarding minor research project grant to this project work.

REFERENCES

Anastas PT. Sustainable (green) chemistry green solvents. Green Chemistry. 45–9. [Online] Available from: www.carloerbareagenti.com/Repository/Download/pdf/Catalogue/EN/catchem100_sez2_green_ en. pdf [Accessed 25th January 2015].

Henderson, R.K., Jiménez-González C., Constable D.J.C., et al. (2011) Expanding GSK's solvent selection guide—embedding sustainability into solvent selection starting at medicinal chemistry. *Green Chemistry*, 13(4), 854–862.

Kirk-Othmer Encyclopedia of Chemical Technology (1998) 4th edition. Vol. 1. New York: John & Wiley Sons Inc. pp. 121–137.

Kurihara K., Hori H. & Kojima K. (1998) Vapor-Liquid Equilibrium Data for Acetone + Methanol + Benzene, Chloroform + Methanol + Benzene, and Constituent Binary Systems at 101.3 kPa. *Journal of Chemical and Engineering Data*, 43, 264–268.

Li H., Xia S., Wu M., et al. (2015) Isobaric (vapour + liquid) equilibria of binary systems containing butylacetate for the separation of methoxy aromatic compounds (anisoleand guaiacol) from biomass fast pyrolysis oil. *Journal of Chemical Thermodynamics*, 87, 141–146.

Malanowski S. (1982) Experimental Methods for Vapour-Liquid Equilibria. Part I. Circulation Methods. *Fluid Phase Equilibria*, 8, 197–219.

Modi C.K. (2014) *Determination of VLE data for system containing CPME*. M.Tech. Thesis, Institute of Technology, Nirma University.

Parsana V.M. & Parikh S.P. (2015) Need for Vapour-Liquid Equilibrium Data Generation of Systems Involving Green Solvents. *Int. Journal of Engineering Research and Applications*, 5, 6(3), 56–62.

Parsana V.M. & Parikh S.P. (2015) Vapor-Liquid Equilibrium Data Prediction by Advanced Group Contribution Methods for a Binary System of Cyclopentyl methyl ether and Acetic acid at Atmospheric Pressure. *Research Journal of Chemical Sciences*, 5(6), 64–72.

Parsana V.M. & Parikh S.P. Isobaric Vapour-Liquid Equilibrium Data Prediction for a Binary System of 2-Methyltetrahydrofuran and Acetic acid using Group Contribution Methods. *Proceedings of the 1st International Conference on Research & Entrepreneurship, 5–6 Jan 2016, RK University, Rajkot, India, ISBN: 978-93-5254-061-7, pp. 1750–1764.*

Poling B.E., Prausnitz J.M., & O'Connell J.P. (2012) *The properties of gases and liquids*. 5th edition. New York: The McGraw-Hill Company limited. pp. 8–23.

Raal J.D., & Muhlbauer A.L. (1998) *Phase Equilibria: Measurement & Computation*. 1st edition. USA: Taylor & Francis.

Renon H., & Prausnitz J.M. (1968) Local Compositions in Thermodynamics Excess Functions for Liquids Mixtures. *AIChE Journal*, 14(1), 135–144.

Sakamoto S. Contribution of cyclopentyl methyl ether (CPME) to green chemistry. *Abstract presentation at Green Chemistry Workshop in Chemspec 2013.*

VanNess H.C., Byer S.M., & Gibbs R.E. (1973) Vapor-Liquid Equilibrium: Part 1. An Appraisal of Data Reduction Methods. *AIChE Journal*, 19(2), 238–244.

Wang X., & Li Y. (2014) Isobaric vapor-liquid equilibrium for binary system of 2-ethylthiophene + n-octane at 101.33 kPa. *Fluid Phase Equilibria*, 378, 113–117.

Wilson G.M. (1964) Vapor-Liquid Equilibrium. XI. A New Expression for the Excess Free Energy of Mixing. *Journal of the American Chemical Society*, 86(2), 127–130.

Zhang H., Liu G., Li C., *et al.* (2012) Liquid–Liquid Equilibria of Water + Acetic Acid + Cyclopentyl methyl ether (CPME) system at Different Temperatures. *Journal of Chemical and Engineering Data*, 57, 2942–2946.

Zhang X., Liu H., Liu Y., *et al.* (2016) Experimental isobaric vapour-liquid equilibrium for the binary and ternary systems with methanol, methyl acetate and dimethyl sulfoxide at 101.3 kPa. *Fluid Phase Equilibria*, 408, 52–57.

Technology Drivers: Engine for Growth – Mahajan, Modi & Patel (Eds)
© 2018 Taylor & Francis Group, London, ISBN 978-1-138-56042-0

A review of greener aromatic/aliphatic separation using ionic liquids

R.C. Gajjar, S.P. Parikh & P.D. Khirsariya
Department of Chemical Engineering, V.V.P Engineering College, Gujarat, India

ABSTRACT: Ionic Liquids (ILs) are a unique class of designer compounds which are salts constituted of cations and anions having a melting point below 100°C. Their properties like negligible vapor pressure, lower melting point, higher ionic conductivity, higher thermal and chemical stability, and tunable physical and chemical properties have made them useful solvents in refineries and the petrochemical sector. This paper precisely reviews the properties and selection of suitable ionic liquids as solvents for greener and efficient separation of various aromatic/aliphatic component systems by liquid–liquid extraction. The conventional sulfolane process employed in the petrochemical industry for the extraction of the aromatics Benzene, Toluene, Ethyl benzene, Xylene (BTEX) from aromatic rich streams of naphtha cracker and reformer, uses the molecular solvent sulfolane which has few physical and chemical limitations when used as an extracting solvent. Replacing it with ionic liquids, which can be tailor-made to achieve desirable solvent characteristics, results in higher selectivity, distribution ratio, purer products, efficient solvent recovery and lower process cost. This paper reports some of the many experimental Liquid–Liquid Extraction (LLE) data collected from the literature regarding IL selectivity and aromatic distribution ratio at particular raffinate compositions, for separation of (aromatic BTEX/aliphatic mixture + IL) systems at various temperatures and atmospheric pressures along with the important conclusions analyzed. This paper focuses mainly on the limitations of sulfolane and highlights the efficiency and greenness of ILs as solvents for aromatic/aliphatic separations. Structural dependency of IL is also reviewed which would help selecting suitable ILs for particular aromatic/aliphatic separations.

Keywords: Ionic liquids, review, aromatic/aliphatic, separation, liquid–liquid extraction

1 INTRODUCTION

Ionic Liquids (ILs), commonly known as molten salts are ionic compounds having a melting point below 100°C, owing to a combination of bulky, asymmetric cations and anions. The commonly explored cations are imidazolium (im), pyridinium (py), pyrrolidinium (pyrr), piperidinium (pip), quaternary ammonium (N_{1111}), quaternary phosphonium (P_{1111}), quinolinium (Qui) and isoquinolinium (iQui). The anions can be halides, triflate (TFO), bistri-flouro amide (NTf_2), tetra flouro borate (BF_4), hexaflouro phosphate (PF_6), acetate (OAc) and so on. Some of the unique and interesting properties of ILs include their negligible vapor pressure, non-flammability, wide liquid range, high thermal stability, high resistance to degradation, good ionic conductivity and an ability to dissolve a variety of compounds (Zhang et al., 2006; Khupse et al., 2010).

The hydrocarbon streams of light naphtha, heavy naphtha and pyrolysis gasoline in refineries are the important sources of Benzene, Toluene, Ethyl benzene, Xylene (BTEX). The separation of aromatics (BTEX) from their mixtures with aliphatic hydrocarbons is carried out by LLE on the basis of their structural differences as the boiling points of these hydrocarbons lie in close range making the separation difficult by distillation. The sulfolane process employed in the petrochemical industry uses the molecular solvent sulfolane for BTEX separation from aromatic rich streams by liquid–liquid extraction as it has the highest selectivity and distribution ratio compared to other volatile organic solvents. For selectively extracting aromatic components from the aromatic/aliphatic stream, the solvent should have higher selectivity, higher aromatic distribution ratio, lower viscosity, lower density, higher thermal and chemical stability and lower reactivity with its process environment along with being environmentally friendly.

2   GREENER AND EFFICIENT AROMATIC/ALIPHATIC SEPARATION WITH ILS

Sulfolane, though it has a higher distribution ratio and selectivity for aromatics, is not chemically and thermally stable at lower temperatures. Its reactive nature with oxygen forms degradation products that corrode the process equipment (Stewart, 2010). ILs are found to exhibit better selectivity, distribution ratio, thermal stability and easy recovery with minimum loss compared to sulfolane (Hossain et al., 2012). The sulfolane process includes extraction of aromatic hydrocarbons by liquid–liquid extraction followed by extractive distillation and a further recovery of sulfolane from the extract and raffinate leading to a costly separation. The immiscibility of ILs with aliphatic hydrocarbons and a greater selectivity for aromatics eliminates the regeneration of IL from raffinate, as well as simple distillation of the extract which would separate IL giving purer components and results being regenerated by usage of recovered IL (Meindersma et al., 2005, 2006). The economic studies conducted on a pilot scale for toluene/n-heptane separation (Graczová et al., 2013) by liquid–liquid extraction proved to have a lesser investment cost than sulfolane (Meindersma et al., 2006, 2007, 2010; Stewart, 2010; Hossain et al., 2012). The negligible vapor pressure of ILs causes no vaporization of IL which avoids air pollution and is a greener alternative to the otherwise conventionally used molecular volatile organic solvents for LLE. The solubility of aromatics in ionic liquids is higher compared to aliphatic hydrocarbons due to the π–π interaction of the ionic liquid aromatic cation with the aromatic hydrocarbon ring, making ionic liquids more selective toward aromatics. COSMO-RS studies have been conducted along with the experimental results generated, to list imidazolium and pyridinium ILs with varying anions which were found to have a greater selectivity and distribution ratio than sulfolane (Meindersma et al., 2005, 2006, 2007). ILs can potentially replace sulfolane and effectively extract aromatics from aromatic lean (20% aromatics) petrochemical streams (Meindersma et al., 2006, 2010). The extract phase is the ionic liquid rich phase and the raffinate phase is the aliphatic rich phase. ILs have shown a higher selectivity ($S = (X_2^{II}/X_2^{I})/(X_1^{II}/X_1^{I})$) and aromatic distribution ratio ($D_2 = X_2^{II}/X_2^{I}$), which ultimately results in lowering the economy of the extraction by a reduction in stages and lowering the solvent requirement for separation. X_2^{II} and X_1^{II} are the mole fractions of the aromatic and aliphatic components in the extract phase respectively. and X_1^{I} are the mole fractions of the aromatic and aliphatic components in the raffinate phase respectively. Replacement of ILs with sulfolane alters the ternary liquid–liquid phase equilibrium from a type I with sulfolane to type II with ILs exhibiting a wide immiscibility range of aromatic/aliphatic mixture with ILs (Meindersma et al., 2006; Ferreira et al., 2011). To search for better solvent properties, binary mixtures of ILs have also been tested (García et al., 2012; García et al., 2012; García et al., 2012; Larriba et al., 2013; Larriba et al., 2014; Larriba et al., 2014).

2.1   *Selection of cations and anions of IL*

The cation family is usually an alkylated nitrogen containing a hetrocyclic ring of imidazolium, pyridinium, pyrrolidinium, isoquinolinium and quinolinium which is aromatic in nature such that it facilitates more selective extraction of BTEX from its mixture with paraffins. A lot of focus has been laid on room temperature imidazolium (Poole & Poole, 2010) ILs on liquid–liquid extraction mainly due to their lower melting point and viscosity compared to other ILs, which makes them easier to work with. The selectivity and distribution ratio of several imidazolium, pyridinium, ammonium and phosphonium based ILs were compared with that of sulfolane (Meindersma et al., 2006, 2007), from which it was concluded that pyridinium based ILs exhibit a higher selectivity and aromatic distribution ratio compared to imidazolium cations due to their more aromatic nature (Corderi et al., 2011). The aromatic distribution ratio was observed to decrease and the aliphatic distribution ratio increased with increasing temperature for these ILs (Meindersma et al., 2006; García et al., 2009; Ferreira et al., 2011; Meindersma et al., 2011; Al-Tuwaim et al., 2011). Ammonium and phosphonium ILs exhibit lesser selectivity than imidazolium and pyridinium ILs due to their aliphatic nature (Pereiro & Rodríguez, 2009; Ferreira et al., 2011). Few aromatic/aliphatic systems have shown comparable or higher selectivity of ammonium based ILs than sulfolane (Domańska et al., 2007) rendering scope for exploring tailor-made ammonium ILs for specific aromatic/aliphatic separations. Pyrrolidinium (Bahadur et al., 2014; Requejo et al., 2015), isoquinolinium (Domańska & Zawadzki, 2011; Domańska et al., 2011; Królikowska et al., 2013; Królikowska et al., 2012; Królikowska & Karpińska, 2013) and quinolinium ILs (Domańska et al., 2011; Domańska et al., 2010, 2011) have shown better extraction properties due to increased π–π interactions between the more polar isoquinolinium and quinolinium cations and the aromatic hydrocarbons. The shorter alkyl chain (ethyl, methyl, butyl) containing N-alkyl-3-methylimidazolium (Cassol et al., 2007) and N-alkyl-3-methylpyridinium cations favor increased selectivity of IL for aromatics, concluding that a shorter alkyl chain length is favored (Arce et al., 2007; Poole & Poole, 2010).

The effect of anions on the solvent characteristics have been analyzed deeply (Ferreira et al., 2011). Aromatics are more soluble in aromatic ionic liquids containing NTf_2 anions (Ferreira et al., 2011), favoring a higher aromatic distribution ratio but a lower IL selectivity for imidazolium based IL (Calvar et al., 2011). NTf_2 (bistriflamide) and OTf (triflate) anions with imidazolium ILs have been reviewed (Visak et al., 2014) with a higher solubility power of NTf_2 anions with aromatics compared to OTf due to the formation of cage like structures enhancing aromatic solubility being noted. Mutually immiscible ILs have been tested with hexane/benzene (Arce et al., 2007) in search for improvement of the IL selectivity and aromatic distribution ratio, the results of which were not satisfactory to encourage the large scale application of the combination of ILs. Cyano containing ILs (Hansmeier et al., 2010; Meindersma et al., 2010, 2011; Larriba et al., 2013) have a higher selectivity and distribution ratio than sulfolane and lower viscosity than pyridinium based ILs thereby enhancing their mass transfer characteristics. Imidazolium based ILs with triiodide anions have a higher selectivity and distribution ratio, but their use is discouraged due to their corrosive nature (Selvan et al., 2000; Meindersma et al., 2005). Cyanide, thiocyanide, and bistriflouro amide anions have proved to be better anion options than halide and fluoride anions.

2.2 *Selection of aromatic/aliphatic system*

The interaction of aromatic hydrocarbons (BTEX) with ionic liquids is by formation of a sandwich like structure between the ionic liquid cation core and the aromatic component. The π–π bonding observed between the aromatic ring structure and the ionic liquid cation core ring favors increased interaction of aromatics with imidazolium, pyridinium, pyrrolidinium, isoquinolinium and quinolinium ionic liquids (Ferreira et al., 2011). The order of decreasing IL selectivity observed experimentally is hexadecane > dodecane > nonane > octane > heptane > hexane (Letcher & Reddy, 2004; González et al., 2009; González et al., 2010; Mokhtarani et al., 2014). The size of the immiscibility region decreases with decreasing size of the n-alkane for a particular IL. As the aromatic nature decreases from aromatic hydrocarbon to n-alkane, the solubility of various hydrocarbons with ionic liquids decreases from aromatics > cyclic hydrocarbons > olefins > n-paraffins (González et al., 2010; Calvar et al., 2011; González et al., 2011; Gutierrez et al., 2011; Corderi et al., 2012; Sakal et al., 2014). This is as per the decrease of the ease of interaction of ionic liquids with the hydrocarbons. The widest area of immiscibility is observed with n-alkanes. As the n-alkane chain length increases, the immiscibility region of the ternary phase diagram increases, irrespective of the ionic liquid used. The tie line points of the raffinate phase lie on the binary mixture line of the aromatic/aliphatic system on the ternary diagram, indicating that the ionic liquid does not enter the raffinate phase, resulting in cost and energy savings for the regeneration of ionic liquids. As the alkyl chain length on the aromatic ring increases, the selectivity and distribution ratio of the IL decreases (Pereiro & Rodríguez, 2009; González et al., 2010; Hansmeier et al., 2010; Al-Tuwaim et al., 2011) because the aromatic character of the hydrocarbon then shifts more toward paraffinic. The selectivity of ILs also decreases as the amount of aromatics in the feed increases (Meindersma et al., 2006) because the interactions between the aromatics and the ionic liquid cation weaken in the presence of more paraffinic hydrocarocarbons (Calvar et al., 2011). The distribution ratio of aromatics with ionic liquids is higher than sulfolane at a low concentration of aromatics in the feed. Thereby, the use of ionic liquids for removal of aromatics from naphtha feed entering the steam cracker to avoid excessive cocking and target energy saving, would be an ideal area for application of ILs as extracting solvents (Meindersma et al., 2005, 2006).

Table 1. Experimental literature LLE data of few aromatic/n-alkane separation using ionic liquids at atmospheric pressure.

System	Ionic liquid	T (k)	X_2^I	S	D_2	Reference
Toluene/n-heptane	[2-bmpy][BF$_4$]	313.2	0.073	60	0.524	García et al., 2010
	[3-bmpy][BF$_4$]		0.067	55	0.666	
	[4-bmpy][BF$_4$]		0.068	57	0.646	
Toluene/n-heptane	[hmim][NTf$_2$]	298.15	0.079	12.49	0.42	Corderi et al., 2012
	[hmim][TfO]		0.135	11.54	0.34	
	[bmim][TfO]		0.06	20.73	0.26	
Toluene/n-heptane	{[bpy][BF$_4$] + [4bmpy][NTf$_2$]}	313.2	0.0343	33.9	0.723	García et al., 2012
Toluene/n-heptane	[emim][NTf$_2$]	298.15	0.104	24.24	0.81	Corderi et al., 2012
	[empy][NTf$_2$]		0.080	13.07	0.92	
Benzene/n-heptane	{[4empy][NTf$_2$] + [emim][DCA]}	313.2	0.0328	73.9	0.659	Larriba et al., 2014
Ethylbenzene/n-heptane			0.0.358	29.1	0.263	
o-xylene/n-heptane			0.0344	38.2	0.317	
m-xylene/n-heptane			0.0368	34.9	0.304	
p-xylene/n-heptane			0.0364	31.4	0.313	
Benzene/hexane	[bmim][MSO$_4$]	298.15	0.108	42.42	0.76	Calvar et al., 2011
			0.053	67.90	0.72	
	[bmim][NTf$_2$]		0.094	18.16	1.72	
Toluene/hexane	[bmim][NTf$_2$]		0.101	12.24	1.20	
Toluene/n-heptane	{[bpy][BF$_4$] + [bpy][NTf$_2$]	313.2	0.0344	34.5	0.529	García et al., 2012
Benzene/heptane	[bmim][MSO$_4$]	298.15	0.053	67.9	0.72	Domíngueza et al., 2012
	[bmim] [NTf$_2$]		0.043	27.91	1.72	
	[pmim][NTf$_2$]		0.043	26.96	1.47	
Benzene/n-hexane	[bmim][FeCl$_4$]	298.15	0.034	13.26	2.27	Sakal et al., 2014
Toluene/n-heptane			0.050	10.69	1.79	
Benzene/hexane	[C$_4$ mim][NTf$_2$]	298.15	0.038	17.39	1.79	Arce et al., 2007
	[C$_8$ mim][NTf$_2$]		0.117	7.35	1.9	
	[C$_{10}$ mim][NTf$_2$]		0.038	5.45	2.21	
	[C$_{12}$ mim][NTf$_2$]		0.092	3.54	1.83	

T: Temperature.
S: Selectivity of IL.
D_2: Distribution ratio of aromatic hydrocarbon.
X_2^I: Mole fraction of aromatic hydrocarbon in the raffinate phase.

3 CONCLUSION

Much research has been conducted in LLE with ILs. Only some of the many literature data are stated in this work as tabulated in Table 1. It can be concluded that the separation of aromatic/aliphatic mixtures using ionic liquids is feasible and is advantageous compared to sulfolane, as the selectivity and distribution coefficients of ionic liquids are higher compared to sulfolane over a wide range of aromatic compositions. The negligible vapor pressure, immiscibility with aliphatics and high solubility with aromatics are the key features that make using ILs a profitable and greener alternative than the conventional molecular solvents. With an increase in temperature the selectivity of ILs decreases. ILs can be designed to tailor make their solvent properties by wise selection of cations and anions in accordance with the aromatic/aliphatic system. The ideal area of application of ILs is for selective extraction of aromatics from aromatic lean hydrocarbon mixtures.

REFERENCES

Alberto, A., Martyn, J.E., Suhas, P.K., Héctor, R. & Kenneth, R.S. (2008). Application of mutually immiscible ionic liquids to the separation of aromatic and aliphatic hydrocarbons by liquid extraction: A preliminary approach. *Physical Chemistry Chemical Physics, 10*, 2538–2542.

Al-Tuwaim, M.S., Alkhaldi, K.H.A.E., Fandary, M.S. & Al-Jimaz, A.S. (2011). Extraction of propylbenzene or butylbenzene from dodecane using 4-methyl-N-butylpyridinium tetrafluoroborate, [mebupy][BF$_4$], as an ionic liquid at different temperatures. *The Journal of Chemical Thermodynamics, 43*, 1804–1809.

Arce, A., Earle, M., Rodríguez, H. & Seddon, K. (2007). Separation of benzene and hexane by solvent extraction with 1-Alkyl-3-methylimidazolium bis{(trifluoromethyl)sulfonyl}amide ionic liquids: Effect of the alkyl-substituent length. *Journal of Physical Chemistry B, 111*, 4732–4736.

Calvar, N., Domínguez, I., Gómez, E. & Domínguez, Á. (2011). Separation of binary mixtures aromatic + aliphatic using ionic liquids: Influence of the structure of the ionic liquid, aromatic and aliphatic. *Chemical Engineering Journal, 175*, 213–221.

Cassol, C.C., Umpierre, A.P., Ebeling, G., Ferrera, B., Sandra S., Chiaro, X. & Dupont, J. (2007). On the extraction of aromatic compounds from hydrocarbons by imidazolium ionic liquids. *International Journal of Molecular Sciences, 8*, 593–605.

Cláudia C., Alexandre P.U., Günter E., Bauer F., Sandra S.X.C. & Jairton D. (2007). On the extraction of aromatic compounds from hydrocarbons by imidazolium ionic liquids. *International Journal of Molecular Sciences, 8*, 593–605.

Corderi, S., Calvar, N., Gomez, E. & Dominguez, A. (2012). Capacity of ionic liquids [EMim][NTf$_2$] and [EMpy][NTf$_2$] for extraction of toluene from mixtures with alkanes: Comparative study of the effect of the cation. *Fluid Phase Equilibria, 315*, 46–52.

Corderi, S., Gonzalez, E., Calvar, N. & Dominguez, A. (2012). Application of [HMim][NTf$_2$], [HMim][TfO] and [BMim][TfO] ionic liquids on the extraction of toluene from alkanes: Effect of the anion and the alkyl chain length of the cation on the LLE. *The Journal of Chemical Thermodynamics, 53*, 60–66.

Domańska, U., Pobudkowska, A. & Królikowski, M. (2007). Separation of aromatic hydrocarbons from alkanes using ammonium ionic liquid C$_2$ NTf$_2$ at T = 298.15 K. *Fluid Phase Equilibria, 259*, 173–179.

Domańska, U., Pobudkowska, A. & Tryznowska, Z. (2007). Effect of an ionic liquid (IL) cation on the ternary system (IL + p-Xylene + Hexane) at T = 298.15 K. *Journal of Chemical Engineering Data, 52*, 2345–2349.

Domíngueza, I., Calvarb, N., Gómeza, E. & Domínguez, A. (2012). Separation of benzene from heptane using tree ionic liquids: BMimMSO$_4$, BMimNTf$_2$, and PMimNTf$_2$. *Procedia Engineering, 42*, 1597–1605.

Farshad, F., Iravaninia, M., Kasiri, N., Mohammadi, T. & Ivakpour, J. (2011). Separation of toluene/n-heptane mixtures experimental, modeling and optimization. *The Chemical Engineering Journal, 173*, 11–18.

Ferreira, A.R., Freire, M.G., Ribeiro, J.C., Lopes, F.M., Crespo, J.G. & Coutinho, A.P. (2011). An overview of the liquid–liquid equilibria of (ionic liquid þ hydrocarbon) binary systems and their modeling by the conductor-like screening model for real solvents. *Industrial & Engineering Chemical Research, 50*, 5279–5294.

Ferreira, A.R., Freire, M.G., Ribeiro, J.C., Lopes, F.M., Crespo, J.G. & Coutinho, J.A.P. (2011). Overview of the liquid–liquid equilibria of ternary systems composed of ionic liquid and aromatic and aliphatic hydrocarbons, and their modeling by COSMO-RS. *Industrial & Engineering Chemical Research, 50*, 5279–5294.

García, J., Fernández, A., Torrecilla, J.S., Oliet, M. & Rodríguez, F. (2009). Liquid–liquid equilibria for {hexane + benzene + 1-ethyl-3-methylimidazolium ethylsulfate} at (298.2, 313.2 and 328.2) K. *Fluid Phase Equilibria, 282*, 117–120.

García, J., Fernández, A., Torrecilla, J.S., Oliet, M. & Rodríguez, F. (2010). Ternary liquid–liquid equilibria measurement for hexane and benzene with the ionic liquid 1-butyl-3-methylimidazolium methylsulfate at T = (298.2, 313.2, and 328.2) K. *Journal of Chemical Engineering Data, 55*, 258–261.

García, S., Larriba, M., Casas, A., García, J. & Rodríguez, F. (2012). Separation of toluene and heptane by liquid–liquid extraction using binary mixtures of the ionic liquids 1-Butyl-4- methylpyridinium bis(trifluoromethylsulfonyl)imide and 1-Ethyl-3- methylimidazolium ethylsulfate. *Journal of Chemical Engineering Data, 57*, 2472–2478.

García, S., Larriba, M., García, J., Torrecilla, J. & Rodríguez, F. (2012). Separation of toluene from n-heptane by liquid–liquid extraction using binary mixtures of [bpy][BF$_4$] and [4bmpy][NTf$_2$] ionic liquids as solvent. *The Journal of Chemical Thermodynamics, 53*, 119–124.

García, S., Larriba, M., García, J., Torrecilla, J. & Rodríguez, F. (2012). Liquid–liquid extraction of toluene from n-heptane using binary mixtures of N-butylpyridinium tetrafluoroborate and N-butylpyridinium bis (trifluoromethylsulfonyl) imide ionic liquids. *Chemical Engineering Journal, 180*, 210–215.

González, E.J., Calvar, N., Gómez, E. & Domínguez, Á. (2010). Separation of benzene from alkanes using 1-ethyl-3-methylpyridinium ethylsulfate ionic liquid at several temperatures and atmospheric pressure: Effect of the size of the aliphatic hydrocarbons. *The Journal of Chemical Thermodynamics, 42*, 104–109.

González, E.J., Calvar, N., Gómez, E. & Domínguez, Á. (2011). Extraction of toluene from aliphatic compounds using an ionic liquid as solvent: Influence of the alkane on the (liquid + liquid) equilibrium. *The Journal of Chemical Thermodynamics, 43*, 562–568.

González, E.J., Calvar, N., González, B. & Domínguez, A. (2010). Liquid–liquid equilibrium for ternary mixtures of hexane + aromatic compounds + [EMpy][ESO$_4$] at T = 298.15 K. *Journal of Chemical Engineering Data, 55*, 633–638.

González, E.J., Calvar, N., González, B. & Domínguez, Á. (2010). Separation of toluene from alkanes using 1-ethyl-3-methylpyridinium ethylsulfate ionic liquid at T = 298.15 K and atmospheric pressure. *The Journal of Chemical Thermodynamics, 42*, 752–757.

Graczová, E., Steltenpohl, P., Šoltýs, M. & Katriňák, T. (2013). Design calculations of an extractor for aromatic and aliphatic hydrocarbons separation using ionic liquids. *Chemical Papers, 12*, 1548–1559.

Hansmeier, A.R., Jongmans, M., Meindersma, G.W. & de Haan, A.B. (2010). LLE data for the ionic liquid 3-methyl-N-butyl pyridinium dicyanamide with several aromatic and aliphatic hydrocarbons. *The Journal of Chemical Thermodynamics, 42*, 484–490.

Hansmeier, A.R., Ruiz, M., Meindersma, G.W. & de Haan, A.B. (2010). Liquid–liquid equilibria for the three ternary systems (3-methyl-N-butylpyridinium dicyanamide + toluene + heptane), (1-butyl-3-methylimidazolium dicyanamide + toluene + heptane) and (1-butyl-3-methylimidazolium thiocyanate + toluene + heptane) at T = (313.15 and 348.15) K and p = 0.1 MPa. *Journal of Chemical Engineering Data, 55*, 708–713.

Hossain, M.A., Dai Hyun Kim, J.L., Nguyen, D.Q., Cheong, M. & Kim, H.S. (2012.) Ionic liquids as benign solvents for the extraction of aromatics. *Bulletin of the Korean Chemical Society, 33*, 3241–3247.

Khupse, N.D. & Kumar, A. (2010). Ionic liquids: New materials with wide applications. *Indian Journal of Chemistry, 49*, 635–648.

Larriba, M., Navarro, P., García, J. & Rodríguez, F. (2013). Liquid–liquid extraction of toluene from heptane using [emim][DCA], [bmim][DCA], and [emim][TCM] ionic liquids. *Industrial & Engineering Chemical Research, 52*, 2714–2720.

Larriba, M., Navarro, P., García, J. & Rodríguez, F. (2014). Extraction of benzene, ethylbenzene, and xylenes from n-heptane using binary mixtures of [4empy][Tf$_2$ N] and [emim][DCA] ionic liquids. *Fluid Phase Equilibria, 380*, 1–10.

Letcher, T.M. & Reddy, P. (2005). Ternary (liquid + liquid) equilibria for mixtures of 1-hexyl-3-methylimidazolium (tetrafluoroborate or hexafluorophosphate) + benzene + an alkane at T = 298:2 K and p ¼ = 0.1 MPa. *The Journal of Chemical Thermodynamics, 37*, 415–421.

Meindersma, G.W., Acker, T.V. & de Haan, A.B. (2011). Physical properties of 3-methyl-N-butylpyridinium tricyanomethanide and ternary LLE data with an aromatic and an aliphatic hydrocarbon at T = (303.2 and 328.2) K and p = 0.1 MPa. *Fluid Phase Equilibria, 307*, 30–38.

42

Meindersma, G.W., Galán Sánchez, L.M., Hansmeier, A.R. & de Haan, A.B. (2007). Application of task-specific ionic liquids for intensified separations. *Monatshefte für Chemie, 138*, 1125–1136.

Meindersma, G.W., Hansmeier, A.R. & de Haan, A.B. (2010). Ionic liquids for aromatics extraction. Present status and future outlook. *Industrial & Engineering Chemical Research, 49*, 7530–7540.

Meindersma, G.W., Podt, A.J.G. & de Haan, A.B. (2005). Selection of ionic liquids for the extraction of aromatic hydrocarbons from aromatic/aliphatic mixtures. *Fuel Processing Technology, 87*, 59–70.

Meindersma, G.W., Podt, A.J.G. & de Haan, A.B. (2006). Ternary liquid–liquid equilibria for mixtures of, toluene + n-heptane + an ionic liquid. *Fluid Phase Equilibria, 247*, 158–168.

Meindersma, G.W., Podt, A.J.G., Klaren M. & de Haan, A.B. (2006). Separation of aromatic and aliphatic hydrocarbons with ionic liquids. *Chemical Engineering Communications, 193*, 1384–1396.

Meindersma, G.W., Simons, T.J. & de Haan, A.B. (2011). Physical properties of 3-methyl-N-butylpyridinium tetracyanoborate and 1-butyl-1-methylpyrrolidinium tetracyanoborate and ternary LLE data of [3-mebupy]B(CN)$_4$ with an aromatic and an aliphatic hydrocarbon at T = 303.2 K and 328.2 K and p = 0.1 MPa. *The Journal of Chemical Thermodynamics, 43*, 1628–1640.

Meindersma G.W., Podt A.J.G. and de Haan A.B. (2006). Ternary liquid–liquid equilibria for mixtures of an aromatic + an aliphatic hydrocarbon + 4-methyl-N-butylpyridinium tetrafluoroborate. *Journal of Chemical Engineering Data, 51*, 1814–1819.

Min, G., Yim, T., Lee, H.Y., Huh, D.H., Lee, E., Mun, J., Oh, S.M. & Kim, Y.G. (2006). Synthesis and properties of ionic liquids: Imidazolium tetrafluoroborates with unsaturated side chains. *Bulletin of the Korean Chemical Society, 27*, 847–852.

Pereiro, A.B. & Rodríguez, A. (2009). Application of the ionic liquid Ammoeng 102 for aromatic/aliphatic hydrocarbon separation. *The Journal of Chem. Thermodynamics, 41*, 951–956.

Pereiro, A.B. & Rodríguez, A. (2010). An ionic liquid proposed as solvent in aromatic hydrocarbon separation by liquid extraction. *AIChE Journal, 56*, 381–386.

Sakal, S., Lu, Y., Jiang, X., Shen, C. & Li, C. (2014). A promising ionic liquid [BMIM][FeCl$_4$] for the extractive separation of aromatic and aliphatic hydrocarbons. *Journal of Chemical Engineering Data, 59*, 533–539.

Stewart, O. (2010). *Sulfolane technical assistance and evaluation report*. Alaska Department of Environmental Conservation

Visak, Z.P., Calado, M.S., Vuksanovic, J.M., Ivanis, G.R., Branco, A.S.H., Grozdanic, N.D., Kijevcanin, M. & Serbanovic, S.P. (2014). Solutions of ionic liquids with diverse aliphatic and aromatic solutes—Phase behavior and potentials for applications: A review article. *Arabian Journal of Chemistry*, 1–13.

Zhang, S., Sun, N., He, X., Lu, X. & Zhang, X. (2006). Physical properties of ionic liquids: Database and evaluation. *Journal of Physical and Chemical Reference Data*, 35.

Zoran, P.V., Marta, S.C., Jelena, M.V., Gorica, R.I., Adriana, S.H.B., Nikola, D.G., Mirjana, L.K. & Serbanovic, S.P. (2014). Solutions of ionic liquids with diverse aliphatic and aromatic solutes—Phase behavior and potentials for applications: A review article *Arabian Journal of Chemistry*, [http://dx.doi.org/10.1016/j.arabjc.2014.10.003].

Civil engineering track

Technology Drivers: Engine for Growth – Mahajan, Modi & Patel (Eds)
© 2018 Taylor & Francis Group, London, ISBN 978-1-138-56042-0

Damage detection in structures using particle swarm optimization method

Aakash Mohan & Sudhirkumar V. Barai
Department of Civil Engineering, Indian Institute of Technology Kharagpur, Kharagpur, West Bengal, India

ABSTRACT: The present study focuses on identifying the damage and finding the location and extent of damage in a structure through the use of particle swarm optimization. The proposed technique assesses the damage through static nodal displacements of the damaged structure. The presence of damage is confirmed through sensor readings on selected nodes where sensors are placed. The objective function is defined as summation of the square of difference of actual nodal displacements in the damaged state and the simulated nodal displacement with the damage parameters. The objective function is minimized by particle swarm optimization. The proposed method of structural damage detection was tested on a 25-bar space truss with different damage scenarios. The extensive study of the proposed method with different cases gives reliable results with significant accuracy.

1   INTRODUCTION

Any change in material or geometric property, such as change in boundary conditions or connectivity, which severely affect asystem's performance is defined as damage. Damage may occur due to anyone or combination of these causes, for example: exceeding the load limits, error in design or construction, structure ageing (end of serviceability period) or any other environmental or physical factors. Structural Health Monitoring is a non-destructive technique to predict the structure's performance. Damage detection is an important component of structural health monitoring. There are different components of structural health monitoring; among them, the most important is the detection of damage at an early stage, which reduces the loss of life and property.

On a broad scale, damage detection methods can be mostly classified into three categories, namely visual inspection methods, local methods, and global methods. Visual inspection can only detect the external damage, and moreover it can't give an idea of the severity of damage. Among the local methods, the most famous techniques are Ultrasonic Thickness Testing (UTT), Infrared Thermography (IRT), Acoustic Emission (AE), Magnetic Thickness Gauging (MTG), Strain Gage Application (SGA) and X-ray radiography. The limitations of the local methods are that they can only reveal information about the local condition of the structure, which may not be enough to assess performance of the entire structure. However, the global methods can predict the behavior of the whole structure, by evaluating the response of structures at one or more locations. Hence, a number of methodologies have been adopted for detecting structural damage using vibration-based global methods in the past decade. These methods can be categorized into optimization-based and signal processing-based damage detection methods. Soft computing methods are an alternate approach, which has been very popular in recent times.

Particle Swarm Optimization (PSO) is a population-based algorithm and it has been applied to many complex engineering optimization problems because of various advantages including simplicity and convergence speed. PSO has been widely used in engineering, such as for control design, in logic circuit design, for topology optimization, in power systems design, and for shape optimization. Fallahian and Seyedpoor (2010) proposed a two-stage structural damage assessment method using PSO and an adaptive neuro-fuzzy inference system. A hybrid particle swarm optimization–simplex algorithm (PSOS) using frequency domain data was suggested by Begambre and Laier (2009) for structural damage assessment. Perara et al. (2010) implemented PSO and GA for multi-objective inverse damage detection problems. An improved particle swarm optimization algorithm was recommended by Yu and Chen (2010) based on macro-economic strategies.

The present work demonstrates a static method of damage detection using the damaged nodal displacements through the use of an evolutionary algorithm. The structure focused is space truss and the evolutionary algorithm used is particle swarm optimization.

2 PROBLEM FORMULATION

In this paper, damage in a member is considered as the reduction in area of members and hence reduction in stiffness, assuming that the length of member (L) and modulus of elasticity (E) is constant during the serviceability period. The present study aims to detect damage through the damage displacements, which are recorded by sensors placed at selected nodes.

2.1 *Sensor placement strategy*

It has been shown that damage in any structure occurs at the locations that are sensitive to that particular loading. A part of a structure is called sensitive if it is more prone to displacements and rotations. The structure in this study is a truss structure hence there are no rotations. The technique applied is that the nodal displacements are located in the undamaged state and sensors are placed only at those nodes at which displacements are significant. Whether a displacement is significant or not, depends on the type of structure and the allowable limits of displacements; for example, for a normal building, a displacement of 0.5 m at the tip is not acceptable, but for a 100-storey building, it is acceptable. The maximum displacement was determined from the undamaged displacements of all the nodes. Values of displacement that are greater than 10% of the maximum displacement are considered to be significant displacements. Hence, a sensor is placed at a node if displacements in all the orthogonal directions are greater than 10% of the maximum displacement or displacement in at least one direction is greater than 50% of the maximum displacement.

The primary variable here is displacement, hence the technique of sensor placement based on nodal displacement values is reliable. However, this technique may not hold true if the primary variable is not displacement and this sensor placement method therefore cannot be generalized for any structure and problem type.

2.2 *Damage identification, locations and extent*

Once the sensors are placed, the damage can be identified in a structure through sensor readings. Any change in the sensor reading that is more than 5% of that of the undamaged state indicates a damage in structure. Damage locations (members in which damage is present), as well as damage extent, are found out through particle swarm optimization. The technique is as follows:

Initially it is assumed that all the members are damaged. Damage is induced in a member by multiplying the stiffness of that member by a scalar value 'Damage fraction' (D), which lies between 0 and 1. These damage fractions are the unknown variables in the PSO. The value of damage fractions for each member is simulated by PSO in each iteration. So, in any particular iteration, damage in member 'i' can be given as $K_i^{damaged} = D_i * K_i^{undamaged}$, where 'i' will vary from 1 to the number of members in that structure. After obtaining the damage stiffness matrix for the whole structure, the corresponding damage displacements are obtained.

Let the displacement vector in each iteration with 'n' degrees of freedom be given as shown in Equation 1.

$$U' = [U'_1 \; U'_2 \; U'_3 \; U'_4 \; U'_5 \; U'_6 \; U'_7 \ldots\ldots \; U'_i \; U'_j \; U'_k \ldots\ldots U'_t \; \ldots\ldots\ldots\ldots \; U'_n] \tag{1}$$

Let the nodal displacement vector for the actual damaged state, obtained from the sensor reading be as given as shown in Equation 2.

$$U = [U_i \; U_j \; U_k \ldots\ldots U_t \;] \tag{2}$$

where, i, j, k…t are the degrees of freedom of the node at which sensors are placed.

The fitness function is defined as shown in Equation 3.

$$F(U') = (U'_i - U_i)^2 + (U'_j - U_j)^2 + (U'_k - U_k)^2 + \ldots\ldots + (U'_t - U_t)^2 \tag{3}$$

2.3 *Formulation of the optimization problem*

The fitness function will only contain the degree of freedom at which we know the sensor reading. The fitness function will be a function of the unknown damage fractions. The fitness function is minimized using particle swarm optimization. In every iteration, the PSO will simulate different values of damage fractions until it completely minimizes the fitness function. When the fitness function gets minimized, each of its terms also get minimized, which implies that the simulated nodal displacements are almost the same as the actual damaged nodal displacements. Hence, the values of damage fractions obtained at this state are the actual damage extent in each member. The idea behind is that we are finding that particular combinations of damage fractions, which is giving exactly the same deflections as the damaged deflections and hence we are getting the exact damage extent. Since fitness function is quadratic, it will have only one unique minimum, so we will get a unique convergent solution. When the algorithm terminates, the value of the unknown variable 'D' is the final damage fraction for each member. Thus, by looking at the damage fraction for each member, we can find the member that is damaged and the extent of damage. In this way, both the damage locations and the damage extent are found out by particle swarm optimization.

The optimization problem for the damage detection technique can be formulated as: 'Find the values of the damage fractions (D) for each member, such that the fitness function F(U') is minimized'. The damage fractions obtained give the member that is damaged and the extent of damage.

2.4 *Implementation of the proposed method*

This study is only a proposed method; it is not conducted in context of a real time structure. So to test the proposed method, different structural damage scenarios were taken and damages were found in them. The general truss code which works on the stiffness method was written in MATLAB. The undamaged displacements were found through the MATLAB truss code and the sensor locations were decided as per the sensor placement strategy. Different damage scenarios were taken. The different damage configurations were produced in MATLAB through the reduction of member areas. The nodal damage displacement for each configuration was found out from the MATLAB truss code. The flow chart of the proposed method is shown in Fig. 1. Since sensors are not at every node, only those damaged nodal displacements that were recorded through the sensors were considered in the fitness function. The fitness function was then minimized through the PSO, keeping the damage fractions as the unknown variables. The damage fractions obtained at the termination of the algorithm will give the exact damage locations and damage extent, which should match with the damage scenario taken. The values of various PSO parameters taken in this method are as follows:

1. Self-adjustment factor: 1.49
2. Social adjustment factor: 1.49
3. Maximum iterations: '200*number of variables'

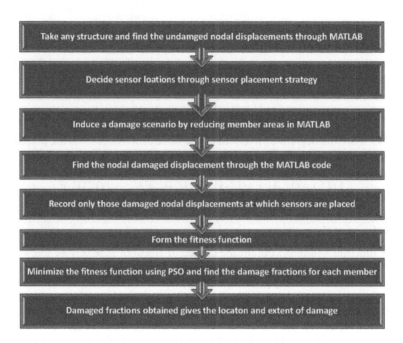

Figure 1. Flowchart for PSO.

4. Max time: Infinity
5. Minimum fraction neighbors: 0.25
6. Stall iterations limit: 150
7. Swarm size: 150
8. Tolerance function limit: 1e-14

3 NUMERICAL EXAMPLES

In order to test the capabilities of the proposed method for identifying the damaged members and finding the locations and extent of damage within these members, one illustrative example of a 25-bar space truss is considered.

3.1 *25-bar space truss*

A 25-bar 3D truss is shown in Fig. 1. The elastic modulus of each member is 6.89E10 MPa and undamaged area of members is 9.7E-4 m^2. The node number and the dimensions are shown in Fig. 2. The locations and extent of damage are found out by sensor reading using particle swarm optimization. The member numbers are as per the MATLAB truss code and are given below. Each row of the matrix represents the member number serially; for example, the member connecting node 1 and node 2 represents member number 1. The loading applied is shown in Table 1. The damage scenarios considered are shown in Table 2.

Connectivity = [1 2;1 4;2 3;1 5;2 6;2 5;2 4;1 3;1 6;3 6;4 5;3 4;5 6;3 10;6 7;4 9;5 8;3 8;4 7;6 9;5 10;3 7;4 8;5 9;6 10].

As per the sensor placement strategy, three sensors were required for this particular loading configuration. Sensors were placed at nodes 1, 2 and 5. A typical result for the case of four-member damage is compared in Table 3 for case 12.

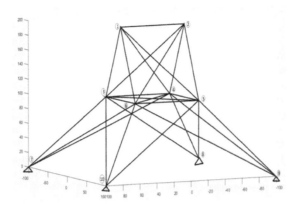

Figure 2.   25-bar space truss. (All dimensions are in m.).

Table 1.   Loading for 25-bar space truss.

Node	F_X (KN)	F_Y (KN)	F_Z (KN)
1	44.482	−44.482	−44.482
2	0	−44.482	−44.482
3	2.224	0	0
6	2.669	0	0

Table 2.   Different damage cases for space truss.

One-member damage	Two-member damage	Three-member damage	Four-member damage
20% in 9 (case 1)	30% in 8, 40% in 24 (case 4)	30% in 3, 40% in 7, 50% in 9 (case 7)	20% in 4, 30% in 5, 40% in 15, 50% in 18 (case 10)
40% in 21 (case 2)	30% in 2, 40% in 4 (case 5)	20% in 4, 40% in 8, 50% in 20 (case 8)	30% in 3, 40% in 9, 45% in 14, 50% in 25 (case 11)
50% in 1 (case 3)	40% in 22, 50% in 25 (case 6)	50% in 3, 40% in 13, 50% in 23 (case 9)	30% in 1, 20% in 7, 40% in 11, 50% in 22 (case 12)

Table 3.   Comparison of results of space truss for case 12.

Case No.	Theoretical results	Results obtained from PSO	Percentage error
12	30% in 1, 20% in 7, 40% in 11, 50% in 22	23.72% in 1, 17.3% in 7, 32.09% in 11, 48.75% in 22	20.93% in 1, 13.5% in 7, 19.775% in 11, 2.5% in 22

The damage fraction obtained in other members was less than 10%, so they can be neglected. Results are shown only for case 12 because it is the worst damage scenario, with 4 members damaged. In all other cases, one-member damage, two-member damage and three-member damage, the results obtained were better with fewer errors. Hence, the results for other cases are not shown here; only the result for the worst scenario is shown. The percentage error values obtained are the maximum values of the error.

4 CONCLUSION

In this study, particle swarm optimization is applied to damage detection in space truss. Presence of damage was identified by sensor readings. This method of sensor placement is able to reduce the number of sensors to a significant extent. Only three sensors were used for a 25-bar space truss. The primary variable used in this method is displacement, hence this technique of sensor placement based on nodal displacement values is reliable. However, this technique may not hold true if the primary variables are different. This sensor placement method therefore cannot be generalized for any structure and problem type. The robustness of the proposed method can be tested by applying it to different structures with different loadings so that it can be validated. The locations and extent of damage are found through particle swarm optimization. The results obtained match the theoretical results to a significant extent for both plane trusses.

REFERENCES

Begambre, O. & Laier, J.E. (2009). A hybrid Particle Swarm Optimization–Simplex algorithm (PSOS) for structural damage identification. *Advances in Engineering Software*, *40*(9), 883–891.

Fallahian, S. & Seyedpoor, S.M. (2010). A two stage method for structural damage identification using an adaptive neuro-fuzzy inference system and particle swarm optimization. *Asian Journal of Civil Engineering (Building and Housing)*, *11*(6), 797–810.

Perera, R., Fang, S.E. & Ruiz, A. (2010). Application of particle swarm optimization and genetic algorithms to multi objective damage identification inverse problems with modelling errors. *Meccanica*, *45*(5), 723–734.

Yu, L. & Chen, X. (2010). Bridge damage identification by combining modal flexibility and PSO methods. In *Prognostics and Health Management Conference, 2010. PHM'10.* (pp. 1–6). IEEE.

Technology Drivers: Engine for Growth – Mahajan, Modi & Patel (Eds)
© *2018 Taylor & Francis Group, London, ISBN 978-1-138-56042-0*

Evaluation of mechanical properties and light transmission of light-transmitting concrete

Eesh Kumar Taneja, Tejas M. Joshi & Urmil Dave
Institute of Technology, Nirma University, Ahmedabad, Gujarat, India

ABSTRACT: Sustainable development is a requirement of society. Energy saving and safe evaluation of engineering structures play a large role in this. Light-Transmitting Concrete (LTC) is a building material that has a light transmitting property due to embedded optical fiber. Light-transmitting concrete is made of optical fibers, cement, water, and sand. The main function of LTC is to use sunlight as a light source to reduce power consumption and this type of concrete is used for architectural and interior design purposes in structures. The experimental results show that LTC has good transmissive, mechanical, and self-sensing properties. Factors such as quality of fiber, fraction of fiber, and the variety of light sources that affect the optical power of LTC are studied in this paper.

1 INTRODUCTION

Hungarian architect Aron Losonczi developed light-transmitting concrete (LTC) at the Technical University of Budapest. LTC comes in precast blocks of different sizes. Optical fibers are used to transfer light through the concrete. It is used for architectural and interior design purposes and also in load-bearing members. Fibers are placed parallel or perpendicular to the direction of load as per the requirement. Optical fibers transmit light efficiently at any angle without loss of light conducted through the fiber. LTC is made of mortar and optical fibers. Coarse aggregates are not used in its manufacture. LTC can be made in any size and with an option of embedded heat isolation. Light-transmitting building blocks can also be created in the desired texture and color. It is expensive due to the high cost of fibers and the requirement for skilled laborers.

1.1 *Literature review*

LiTracon (2001) was the first company to make light-transmitting concrete. The patent protected by LiTracon (2001) products presents the phenomenon of light-transmitting concrete in the form of widely applicable new building materials. Luccon (2004) is the main company focused on strength. Luccon design is a precast concrete product. In a special process, fiber optic cables are inserted in a fine-grain concrete. It provides furniture, shop interiors, kitchens, partition walls, sanitary equipment, stairs among other things. A further company,

Table 1. Standard values of illuminance.

Particular space	Illuminance, lux	Particular space	Illuminance, lux
Study room	150	Toilet	100
Bedroom	100	Stair	100
Kitchen	200	Corridor	70
Dining room	100	Garage	70
Bath room	100		

Lucem (2003), provides concrete with architectural properties as well as strength. Diffuse natural light and sunlight provide the full spectrum of colors shining through the LUCEM (2003) panels. Central Building Research Institute, India (1947) provides the standard values of illumination. Table 1 shows the illumination requirements of different spaces in building.

2 MATERIAL USED IN INVESTIGATION

2.1 Ordinary Portland Cement (OPC 53 grade)

For the experimental investigation, OPC 53 grade cement was used. Table 2 shows the properties of OPC 53 grade cement.

2.2 Fine aggregate

For the experimental investigation, annore sand was used. Table 3 shows the properties of annore sand.

Table 2. Properties of cement.

Particulars		Test results	Requirement of IS:12269–2013
Fineness (m²/kg)		314	Min 225
Standard Consistency (%)		30	
Setting time	Initial	180	Min 30
	Final	280	Max 600
Soundness (mm)		0.8	Max 10
Compressive strength (N/mm²)	3 days	37.6	Min 27
	7 days	50.6	Min 37
	28 days	55.4	Min 53

Table 3. Properties of fine aggregate.

Material	Sand
Zone	I, II, III
Silt (% by volume)	2.50
Specific gravity	2.65
Water absorption (%)	1.10
Chloride content (%)	0.003
Sulfate content (%)	0.0239

Table 4. Mix proportion in different trials.

Material		Quantity
Cement		200 g
Sand	Zone I	200 g
	Zone II	200 g
	Zone III	200 g
Optical fiber	Trial I	10 fibers/hole
	Trial II	15 fibers/hole
	Trial III	20 fibers/hole
Water		[(P/4)+3]*(% of combined weight of cement and sand)

2.3 *Optical fiber*

In this experimental program step index fibers were used, which are very low-quality fibers compared to other fibers available in the market. The efficiency of light transmission in these fibers is low but they are economical.

2.4 *Mix proportion*

In this experimental investigation, cement mortar was used. The ratio of cement: sand was taken as 1:3. Coarse aggregates are not used for making light-transmitting concrete. For making light-transmitting concrete, optical fibers are inserted with different numbers in each cube. Table 4 shows the mix proportions used for the investigation.

3   EXPERIMENTAL PROGRAMME

3.1 *Principle of operation*

Light concrete is made of mortar and optical fibers. It cannot be casted at site so it is provided as prefabricated building blocks and panels. Small sizes are preferred for better results. Fibers are used in concrete like small pieces of aggregate. It works based on a 'nano-optics' principle. When tiny slits are placed directly on top of each other, fibers pass light efficiently. Therefore, optical fibers behave like slits in concrete and carry light through the concrete.

3.2 *Fabrication and analysis of light-transmitting concrete*

Fabrication of light-transmitting concrete is important because the spacing between fibers and the quantity of fibers affect the strength and light emission. For the present study, low-quality optical fiber (step index fiber) was used. In this study, the dimensions of the produced specimen were 150 mm × 150 mm × 150 mm. A lux meter was used to measure the intensity of light passing through the concrete.

3.2.1 *Mold and specimen fabrication*
The fabricated samples containing optical fibers were cuboids with dimensions of 150 mm × 150 mm × 150 mm. The mold was made up of plywood. The optical fiber strands, batched by volume (or fiber to cement ratio), were placed through the holes individually. The cement paste was then prepared in 1.0: 3.0: 0.4 proportions and poured into the mold, and vibration was carried out with the help of vibrator to avoid void formation. As Figure 1 shows, the holes are spaced at 1 cm in both horizontal and vertical directions.

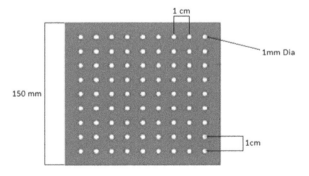

Figure 1.   Side view of mold.

4 RESULTS AND DISCUSSION

4.1 *Light transmittance test on specimen*

A lux meter was used to measure the transmission of light through specimen. Two readings were taken—one without the sample (A_1) and one with the sample (A_2). Equation (1) shows the light transmittance of different samples. As per the results shown in Table 5, as the amount of optical fiber increased, the intensity of passing light also increased. Figure 2 shows the light transmission through a sample.

$$\text{Light transmittance} = \frac{A2}{A1} \times 100 \tag{1}$$

4.2 *Compressive strength*

The compressive strength of samples was determined after measuring the light transmitted by compressive testing machines; the test results are shown in Table 6. As can be seen from Table 6, the compressive strength of light-transmitting concrete increased as the quantity of fibers increased from 10 to 20.

4.3 *Density*

Density is weight per unit volume. The density of light-transmitting concrete should be between 22 and 24 kN/m³. As Table 7 shows, the average density lies within this range.

Table 5. Percentage of passing light through concrete.

Sr. No.	No. of fibers/hole	Passing light intensity (lux) (A_2)	Light intensity of source (lux) (A_1)	% light passing
1.	10	398		3.24
2.	15	570	12,300	4.63
3.	20	650		5.28

Figure 2. Light transmission through a sample of concrete.

56

Table 6.   Compressive strength of light-transmitting concrete.

Sr. No	No. of fibers/hole	Load (kN)	Compressive strength (MPa)
1.	10	560	25
2.	15	780	34.67
3.	20	900	40

Table 7.   Density of light-transmitting concrete.

Sr. No.	No. of fibers/hole	Mass (N)	Density (kN/m^3)
1.	10	78.87	23.37
2.	15	77.30	22.90
3.	20	77.50	22.96

5 CONCLUSIONS

The following conclusions can be drawn from the present investigation:

– Light-transmitting concrete can be developed by any type of optical fiber or large diameter glass fiber in the concrete mixture.
– It has good light transmissive properties and the ratio of optical fiber to concrete is proportional to transmission of light.
– It doesn't lose the strength parameter up to a limit of quantity of optical fibers when compared to regular concrete and also it has essential properties for architectural and interior design purposes.
– This new kind of building material can integrate the concepts of energy saving and sustainable development.

REFERENCES

Central Building Research Institute. (1947). CBRI. Retrieved from: http://cbri.res.in/ [Accessed 5th March 2016].
Litracon. (2001) LTC: Light Transmitting Concrete, [Online] Available from: http://litracon.hu/ [Accessed 3rd January 2016].
Luccon. (2004) LTC: Light Translucent concrete, [Online] Available from: http://www.luccon.com/ [Accessed 3rd January 2016].
Lucem (2003). Light Concrete, [Online] Available from: http://www.lucem.com/de/ [Accessed 3rd January 2016].
Shanmugavadivu, P.M., Scinduja, V., Sarathivelan, T. & Shudesamithronn, C.V. (2014). An experimental studyon light transmitting concrete. *IJRET* [Online], 2319–1163.

Technology Drivers: Engine for Growth – Mahajan, Modi & Patel (Eds)
© 2018 Taylor & Francis Group, London, ISBN 978-1-138-56042-0

Comparative study of compressive strength and durability properties of binary, ternary and quaternary mortar

R.S. Anjaria & U.V. Dave
Department of Civil Engineering, Institute of Technology, Nirma University, Ahmedabad, Gujarat, India

ABSTRACT: The present study uses supplementary cementitious materials such as fly ash (FA), ground granulated blast furnace slag (GGBS) and metakaolin (MK). Eight mortar mixes consisting of control (M1), binary (M2), ternary (M3) and five quaternary mixes (M4 to M8) are prepared. M2 is made up using 65% OPC and 35% FA. M3 consists of 60% OPC, 35% FA and 5% MK. M4 is made up using 50% OPC, 35% FA, 5% MK and 10% GGBS. M5 consists of 40% OPC, 35% FA, 5% MK and 20% GGBS. 40% OPC, 25% FA, 5% MK and 30% GGBS are used to prepare M6. M7 is composed of 30% OPC, 25% FA, 5% MK and 40% GGBS. M8 is made up of 30% OPC, 35% FA, 5% MK and 30% GGBS. Binder to sand ratio used is 1:3 for all mortar mixes. Compressive strength test sat 3, 7, 28, and 90 days are conducted for all mortar mixes. Rapid chloride penetration tests (RCPT) at 28 and 90 days and drying shrinkage test sat 35 days are conducted for all mortar mixes. The compressive strength of M2, M3, M4, and M5 is at par compared to that of control mortar at 90 days. Lesser chloride ion penetration is observed for binary, ternary, and all quaternary mortars as compared to that of control mortar at 28 and 90 days. Drying shrinkage of binary, ternary, and all quaternary mixes is lower than that of control mortar at 35 days. Thus, it has been observed that compressive strength and RCPT, as well as drying shrinkage test, have been found to be better in quaternary mortar as compared to OPC, binary, and ternary mortars.

1 INTRODUCTION

Binary and ternary blends prepared using OPC and Supplementary Cementitious Materials (SCMs) show improvements in early and later age strength, durability, reduction in the heat of hydration and economy as compared to that of control mortar and concrete (Ahmed et al., 2008; Khatib et al., 2009; Dhinakaran et al., 2017; Erdem & Kirca, 2008). Ternary blends showed better results in terms of mechanical and durability properties as compared to binary and OPC mixes. As the results improve from binary to ternary mixes, studies were carried out on quaternary mix. Quaternary binder mixes showed further improvement in durability and later age strength as compared to that of control, binary and ternary mixes (Dave et al., 2016; Ghrici et al., 2007; Mackhloufi et al., 2014). Moreover, SCMs such as fly ash (FA), ground granulated blast furnace slag (GGBS) and metakaolin (MK) are available as by-products from various industrial processes. These SCMs can be used in quaternary mixes to enhance theirdurability and mechanical properties. By increasing the use of SCMS, carbon footprints can be reduced as cement consumption is reduced.

2 MATERIALS AND MIX PROPORTIONING

2.1 *Materials*

Ordinary Portland cement (53 Grade) (C), FA, GGBS and MK were used in the present study. The physical and chemical properties of OPC, FA, GGBS and MK are presented in Table 1. Natural sand confirming to IS:650 (Bureau of Indian Standard, 1991) was used.

Table 1. Properties of materials.

	OPC	Fly ash	GGBS	MK	Water
Physical properties					
Specific gravity	3.20	2.90	2.91	2.68	1.00
Blaine's fineness (m²/kg)	309	333	379	10090	–
Chemical properties					
Calcium oxide, CaO%	0.92	0.96	34.4	0.32	–
Aluminum oxide, Al_2O_3%	1.25	30.90	17.12	32.30	–
Silicon dioxide, SiO_2%	16.70	61.4	36.80	61.70	–
Ferric oxide, Fe_2O_3%	2.23	0.40	0.92	0.10	–
Magnesium oxide, MgO%	1.29	1.42	3.92	0.16	–
Loss of ignition %	1.81	1.05	0.60	0.55	–

Table 2. Proportioning of materials in different mixes.

Mix	Binder proportions
M1	100C
M2	65C + 35FA
M3	60C + 35FA + 5MK
M4	50C + 35FA + 5MK + 10GGBS
M5	40C + 35FA + 5MK + 20GGBS
M6	40C + 25FA + 5MK + 30GGBS
M7	30C + 25FA + 5MK + 40GGBS
M8	30C + 35FA + 5MK + 30GGBS

2.2 *Preparation of specimen*

For preparation of mortar, the binder to sand ratio used was 1:3. Zone I, zone II and zone III sand was used in equal proportions in the form of 3 parts of sand.M1 is the control mix, M2 is the binary mix, M3 is the ternary mix and M4 to M8 are quaternary mixes (Table 2).

The compressive strength of mortar mixes was evaluated using the provisions of IS:516 (Bureau of Indian Standard, 1959). Mortar cubes of size 70.6 mm × 70.6 mm × 70.6 mm were prepared and tested on a universal testing machine. Rapid chloride penetration testing of mortar mixes was carried out using the provisions of ASTM 1202 C (1997). Cylinders of 100 mm diameter and 50 mm height were prepared for all mortar mixes. A voltage of 60 V DC was maintained across the ends of the sample throughout the test. One lead was immersed in a 3.0% salt (NaCl) solution and another in a 0.3 N sodium hydroxide (NaOH) solution. Drying shrinkage tests on mortar mixes were carried out using the provisions of IS: 4031 (Part-10) (Bureau of Indian Standard, 1988). Beams of sizes 25 mm × 25 mm × 280 mm were prepared for evaluating drying shrinkage for all mortar mixes. Initial readings of drying shrinkage were taken at 7 days, with length comparator and final reading 28 days after the initial reading, for all mortars.

3 MECHANICAL PROPERTIES

3.1 *Compressive strength*

The compressive strength results of OPC, binary, ternary, and quaternary mortars are shown in Figure 1. The compressive strengths of binary and ternary mortars were lower than that of the control mortar at 90 days. The compressive strength of quaternary mixes M4 and M5 was similar to that of the control mortar at 90 days. Quaternary mixes M6, M7, and M8 had

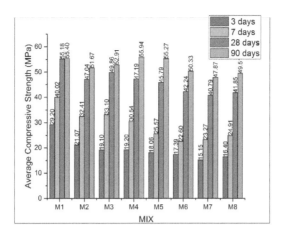

Figure 1.   Compressive strength of mortars.

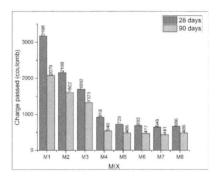

Figure 2.   Charge passing through mortars.

lower compressive strength than the control mortar at 90 days. The early age compressive strength of binary, ternary, and quaternary mixes was lower than the later age compressive strength, as FA and GGBS provide the strength at a later age. With increase in age of mortar from 7 to 28 days and 28 to 90 days, higher increments in the compressive strength of binary, ternary, and quaternary mixes were observed as compared to that of the control mortar.

4   DURABILITY PROPERTIES

4.1   Rapid chloride penetration test

Rapid chloride penetration test results of all mortar mixes are shown in Figure 2. The chloride ion penetration was decided from the charge passed (Table 3).

The control mortar had moderate chloride ion penetrability at 28 and 90 days. Binary mortar had low chloride ion penetrability at 90 days. At 28 and 90 days, ternary mortar had low chloride ion penetrability. Quaternary mixes M4 to M8 had very low chloride ion penetrability at 28 and 90 days. In binary mortar, two materials with different fineness were used: OPC and FA. Therefore, the pore spaces are reduced as compared to the control mortar. In ternary mortar, the pore spaces are further reduced as compared to that of binary mortar because MK, which is a finer material, is used. With use of four different materials of different fineness, fewer pores are likely to be present in case of quaternary mortars. The fewer the pore spaces in the mortar, the lower the charge passing through the mortar mixes. Hence,

Table 3. Chloride ion penetrability.

Mix	Chloride ion penetrability (ASTM 1202C (1997))		Mix	Chloride ion penetrability (ASTM 1202C (1997))	
	28 days	90 days		28 days	90 days
M1	Moderate	Moderate	M5	Very Low	Very Low
M2	Moderate	Moderate	M6	Very Low	Very Low
M3	Low	Low	M7	Very Low	Very Low
M4	Very Low	Very Low	M8	Very Low	Very Low

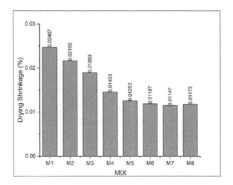

Figure 3. Drying shrinkage of mortars.

quaternary mixes exhibit very low chloride ion penetrability as compared to that of other mortar mixes.

4.2 *Drying shrinkage*

Drying shrinkage test results of OPC, binary, ternary, and quaternary mortars are shown in Figure 3. Binary mortar M2 had less drying shrinkage compared to the control mortar. Drying shrinkage of ternary mortar was further reduced compared to drying shrinkage of binary and control mortar mixes. The drying shrinkage of quaternary mortar was observed to be further reduced as compared to that of control, binary, and ternary mortar mixes. It was observed that with the use of FA, MK and GGBS, there was a reduction in the drying shrinkage of mortars. Thus, with higher amounts of SCMs, the drying shrinkage of mortar mixes reduces. The lower the shrinkage, the lower the crack formation in the case of mortars. Thus, it is postulated that due to less shrinkage there are fewer chances of the formation of cracks in quaternary mixes as compared to that of other types of mortar mixes.

5 CONCLUDING REMARKS

From the study on OPC, binary, ternary, and quaternary mortars, the following conclusions have been drawn:

– 50–70% replacement of OPC by means of various SCMs resulted in the development of better compressive strength of mortar mixes at 90 days as compared to that of OPC in comparison to that of 28 days. Here it is anticipated that use of SCMs is likely to result in further increase in compressive strength of quaternary mortar mixes due to continuous hydration.

- Permeability was reduced for binary, ternary, and quaternary mortars as compared to that of the control mortar. The reduction in permeability is due to the reduction in pore spaces in mortar due to incorporation of SCMs in mortars.
- Drying shrinkage of binary, ternary, and quaternary mortars was lower than that of control mortar. As the hydration process is delayed, the shrinkage in the quaternary mortar mixes is further reduced.
- All quaternary mortar mixes exhibited better durability as compared to that of the control, binary, and ternary mortars. Hence, quaternary mixes can be used to produce quaternary mortar and concrete where higher durability is required, such as for infrastructure and other important construction projects during exposure to severe environments.

REFERENCES

Ahmed, M.S., Kayali, O. & Anderson, W. (2008). Chloride penetration in binary and ternary blended cement concretes as measured by two different rapid methods. *Cement and Concrete Composites*, *30*(7), 576–582.

American Society for Testing and Materials. (1997). *ASTM 1202C-1997 Standard Test Method for Electrical Indication of Concrete's Ability to Resist Chloride Ion Penetration.*

Bureau of Indian Standard. (1959). *IS: 516−1959. Methods of tests for strength of concrete.* New Delhi.

Bureau of Indian Standard. (1988). *IS: 4031(Part-10)-1988. Methods of Physical Test for Hydraulic Cement (Determination of drying shrinkage)*, New Delhi.

Bureau of Indian Standard. (1991). *IS: 650–1991. Specification of Standard Sand for Testing of Cement.* New Delhi.

Dave, N., Misra, A.K., Srivastava, A. & Kaushik, S.K. (2016). Experimental analysis of strength and durability properties of quaternary cement binder and mortar. *Construction and Building Materials*, *107*, 117–124.

Dhinakaran, G., Kumar, K.R., Vijayarakhavan, S. & Avinash, M. (2017). Strength and durability characteristics of ternary blend and lightweight HPC. *Construction and Building Materials*, *134*, 727–736.

Erdem, T.K. & Kırca, Ö. (2008). Use of binary and ternary blends in high strength concrete. *Construction and Building Materials*, *22*(7), 1477–1483.

Ghrici, M., Kenai, S. & Said-Mansour, M. (2007). Mechanical properties and durability of mortar and concrete containing natural pozzolana and limestone blended cements. *Cement and Concrete Composites*, *29*(7), 542–549.

Khatib, J.M., Kayali, O. & Siddique, R. (2009). Dimensional change and strength of mortars containing fly ash and metakaolin. *Journal of Materials in Civil Engineering*, *21*(9), 523–528.

Makhloufi, Z., Bederina, M., Bouhicha, M. & Kadri, E.H. (2014). Effect of mineral admixtures on resistance to sulfuric acid solution of mortars with quaternary binders. *Physics Procedia*, *55*, 329–335.

Technology Drivers: Engine for Growth – Mahajan, Modi & Patel (Eds)
© 2018 Taylor & Francis Group, London, ISBN 978-1-138-56042-0

Effect of freeze-thaw on K-based Geopolymer Concrete (GPC) and Portland Cement Concrete (PCC)

Fernanda Belforti
Department of Civil Engineering, University of Estadual Paulista, Brazil

Peiman Azarsa & Rishi Gupta
Department of Civil Engineering, University of Victoria, Victoria, Canada

Urmil Dave
Department of Civil Engineering, Nirma University, Gujarat, India

ABSTRACT: In this paper, the durability of K-based Geopolymer Concrete (GPC) and Portland Cement Concrete (PCC) against freeze-thaw cycles subjected to 0, 17 and 27 freeze-thaw cycles have been investigated. Sub-zero temperatures are common in cold weather countries and it can cause surface deterioration, such as spalling and scaling, exposure of aggregates and induced internal crack growth in the concrete. The mass loss and relative dynamic modulus of elasticity of specimens, by means of resonant frequency test, were measured before and after the freeze-thaw cycles. Results showed that dynamic modulus of elasticity fluctuated for the Geopolymer Concrete and increased for Portland Cement Concrete, and the compressive strength decreased as freeze-thaw cycles were repeated for both types of concrete. Moreover, a fluctuation in the mass occurred due to the amount of water that permeated through existing cracks in the specimens.

1 INTRODUCTION

The concern about environment has become paramount regarding protecting worldwide. Concrete is considered as one of the most used engineering materials in the world. However, to produce cement, the main component of Portland Cement Concrete (PCC), a high environmental footprint is originated. Hence, alternative binder materials have been applied to produce a cement-less concrete, named Geopolymer Concrete (GPC). Such type of concrete is formed by alkali activation of industrial aluminosilicate waste materials, for instance fly ash, a by-product obtained from coal burning in power generation industries. Therefore, GPC is an innovative and eco-friendly construction material whose manufacturing process emits less greenhouse gases and requires less energy, making this a feasible and potential green alternative to PCC.

Concrete is likely to be damaged when subjected to freeze-thaw cycles, which may occur in countries that have sub-zero temperature conditions. The deterioration caused by frost damage is progressive and it can start with scaling or spalling of the surface and end in complete collapse. Deteriorations succeed after continued freeze and thaw cycles, and the material gradually loses its structural strength and durability.

The deterioration in the concrete caused by freeze-thaw actions occurs when it is critically saturated, which happens when concrete pores are filled with water about 90% or more. After temperature drops below freezing and water turns to ice, its volume increases by 9% compared to liquid phase. After this volume expansion, the water has no extra space to occupy, subsequently an internal stress develops, causing cracking [Portland cement Association. N.P. 2002].

The damage caused by frost has been investigated at University of Victoria Materials Laboratory by accelerated freeze and thaw cycles. The resistance against the cycles is commonly assessed and classified according to the type of damage, whether external or internal. Internal damage in concrete can be evaluated by the fundamental longitudinal frequency measurements. Weight change and loss of strength can also be used as one of the criteria to evaluate the frost resistance.

2   LITERATURE REVIEW

M Al Bakri et al, (2001) have found that fly ash based geopolymer is better than normal concrete in many aspects such as compressive strength, exposure to aggressive environment, workability and exposure to high temperature.

Al Bakri et al, (2013) have discovered that for 1, 7 and 28 days the fly ash-based geopolymer concrete produced higher compressive strength compared to PCC, the density of geopolymer concrete is denser than PCC, but still comparable to the properties as PCC, and the water absorption and porosity of geopolymer concrete was lower than PCC.

In another study, P. Sun and H.-C. Wu, (2013) investigated the freeze-thaw resistance of fly ash-based alkali activated mortars at ambient conditions. It was observed that the initial compressive strength of PCC mortars was higher than the fly ash-based alkali activated (FA) mortars, but the second one gained strength with time. The dynamic modulus loss of PCC was higher. Thus, the freeze-thaw resistance of FA was superior compared with PCC. In addition, the use of air-entraining agent improved both specimens, against freeze-thaw cycles.

Y. Fu et al, (2001) studied the performance of Alkali-activated slag concrete (ASC) against freeze-thaw cycles. It was stated that ASC has a high compressive strength of 90 MPa, excellent apparent performance with few denudation of concrete's surface, great corrosion resistance of materials, no transition strips and compact structure. Hence, ASC showed an excellent freeze-thaw durability.

Cai, L. et al, (2013) studied alkali-activated slag concrete (ASC) against freeze-thaw cycles and observed that the air-void structure is the crucial factor. The freeze–thaw resistance tends to be better with smaller air bubble space coefficient and bigger specific surface area.

Huai-Shuai Shang and Ting-Hua, (2013) have analyzed the durability of air-entrained concrete specimens with varied strength and subjected a freeze-thaw cycles. It was noted that the dynamic modulus of elasticity and weight decreased as the freeze-thaw cycles were repeated and the deterioration of freeze-thaw durability for air-entrained concrete is slower than that of plain concrete.

Shang, H. et al, (1991) investigated the behavior of air entrained concrete after different freeze-thaw cycles were completed, and the experimental results showed that the-dynamic modulus of elasticity and strength decreased as the freeze-thaw was repeated. The improvement was also noticed in the frost resistance of concrete using air entraining admixture.

E. Douglas et al, (2015) investigated the compressive strength development of alkali activated ground granulated blast-furnace slag concretes with sodium silicate and hydrate lime in the concrete mixture. It showed that the development happens but at a lower rate than in Portland Cement Concrete. The specimens have a good compressive strength that could reach 59.6 MPa at 28 days. Moreover, air entraining admixture can improve the workability and durability of the concrete.

Limited researches related to freeze-thaw using bottom ash as a precursor with K-based in geopolymer mix design could be identified. Hence, the goal of this study is to investigate the efficiency of bottom ash-based geopolymer concrete under the freeze-thaw cycles in order to obtain a durable and sustainable construction material.

3 EXPERIMENTAL WORK

3.1 *Material*

To produce K-based GPC, fly ash was used as the cementitious material. Two classes of fly ash are defined by ASTM C618: Class F fly ash and Class C fly ash [ASTM Int. 2015]. Fly ash used in this study is low calcium class F and obtained from Lafarge Company, Canada. Table 1 shows the composition of the material used in this study reported by manufacture.

The alkaline liquid used was a combination of Potassium hydroxide and Potassium silicate solution. Potassium hydroxide (KOH) in solid pellets form (85% pure) and purchased from Sigma-Aldrich was used. Potassium Silicate ($K2SiO3$) powder obtained from PQ Corporation (USA) supplier has been used. The ratio of $SiO2/K2O$ is 1.62 and the amount of water is 14.8. Sand and coarse aggregates obtained from a quarry in British Columbia, Canada, have been used. Sand presented 2.685 Relative Dry Density, 1.42% Water absorption and coarse aggregates size of 10 mm presented 2.767 Relative Dry Density, and 0.954% Water absorption.

To produce the PCC, General Use (GU) Cement was utilized, as well as aggregate with maximum size of 7 mm. The sand used was the same as for GPC. 80 ml of Water Reducing Admixture (WRA) and tap water from the laboratory was also employed. The air content measured was 2.5%. Due to the rapid setting time of GPC, air content could not be measured.

3.2 *Sample preparation*

A set of six GPC specimens were prepared. First, the alkaline solution, Potassium Hydroxide (KOH) solids were dissolved in water before the mixing to achieve a molarity of 12 M. Then, the dry materials, including the $K2SiO3$, were mixed for 4 minutes. During the mixing, potassium hydroxide solution and extra water were added until the GPC presented uniform consistency. Immediately after the mixing, the GPC mixture was casted using cylinder molds of size (100 × 200) mm in three layers and vibrated for 60–90 seconds. The specimens were left into the oven at a temperature of 60°C for 24 h. Table 2 shows the mix design of the GPC.

A set of six cylinders (100 × 200) mm of PCC were prepared. The basic mix design used for the plain concrete samples is shown in Table 3.

Table 1. Physical and chemical compositions of fly ash.

Properties	SiO_2	Al_2O_3	Fe_2O_3	CaO	MgO	SO_3	LOI
Values (%)	47.1	17.4	5.7	14	5.4	0.8	0.19

Table 2. Details of mix proportion of GPC.

Material	Fly ash	Coarse aggregates	Sand	KOH (12 M)	K_2SiO_3	Extra water
Mix proportions (kg/m³)	388	1170	630	85.16	125.74	38.71

Table 3. Details of mix proportion of PCC.

Material	PC	Gravel	Sand	Water
Mix proportions (kg/m³)	340	1120	820	181

3.3 Testing method for resistance of concrete to rapid freezing and thawing

Based on ASTM C666 [ASTM C666/C666M, 2003], concrete specimens, widths and depths from 75 mm to 125 mm and lengths between 275 mm to 405 mm, are subjected to temperature cycling from −18°C to 4°C and from 4°C to −18°C. Cycles should last from 2 to 5 hours. In the present study, the tested specimens were cylinders (100 × 200) mm, subjected to the approximately specified temperature cycling, from 3°C to −17°C and from −17°C to 3°C and with a cycle duration of 3 hours for Procedure B, Rapid Freezing in Air and Thawing in Water. The GPC and PCC specimens were placed in a conventional freezer. Temperature inside the freezer was measured with a laser thermometer temperature gun. The cycles were controlled manually by a timer that had been set for specific time intervals and temperature based on freeze-thaw standards. The cylinders were placed out of the freezer in a bucket filled with water, two times per day, for a period of 1.5 hours, designed for the thaw part of the cycle.

3.4 Test method for fundamental longitudinal resonant frequencies of concrete specimens

This method [ASTM C215, 2014] was used to evaluate the resonant frequency of specimens that is undergoing degradation after freeze-thaw cycles, and it was used to measure the longitudinal frequency, following the setup show in Figure 1.

The GPC and PCC cylinders (100 × 200) mm were placed horizontally in a support located underneath the middle of specimens.

The procedure was executed by removing the cylinder from the conventional freezer after 17 and 27 cycles. The specimens were weighed, then placed in the support, and an accelerometer (output signal) was attached to the centre of one end by using a tape. Using the impact hammer, the cylinder was tapped perpendicularly at the opposite end at the middle of its edge, and the resonant frequency indicated by the waveform analyzer was recorded. The fundamental longitudinal resonant frequency is the frequency with the highest peak in the amplitude spectrum. Testing were performed at 0, 17 and 27 cycles. The Relative Dynamic Modulus of Elasticity (RDME), was calculated as follows:

$$P_c = \frac{n_c^2}{n^2} \times 100 \tag{1}$$

where c is the number of cycles of freeze-thaw, n_c is the resonant frequency after c cycles, and n is the initial resonant frequency (at zero cycles).

Figure1.   Fundamental longitudinal resonant frequency test.

The Durability Factor (DF) is defined as:

$$DF = \frac{N}{M} \times P_c \tag{2}$$

where P_c is the relative dynamic modulus, N is the number of cycles completed, and M is the planned duration of testing, which usually ends when the relative dynamic modulus falls below 50–60 percent of its initial value or when 300 cycles are completed.

4   RESULTS AND DISCUSSION

In the end of freeze-thaw cycles, the specimens were tested for compressive strength, resonant frequency and unit mass.

4.1   *Slump test*

This test is performed to check the consistency of fresh paste of concrete and to determine the workability of concrete. The workability of fresh concrete was measured by the slump test with average value of 22 mm for GPC and 60 mm for PCC.

4.2   *Compression test*

The compression test [ASTM C39/C39M-17, 2017] shows the strength of specimens in the hardened state. The obtained experimental results were studied to compare the compressive strength of GPC and PCC. In comparison, PCC was found to be better than GPC. The compressive strength of PCC, attaining 7 days, were found to be 48% and 79% more than GPC, before and after 27 freeze-thaw cycles, respectively.

4.3   *Relative Dynamic Modulus of Elasticity (RDME) and weight loss*

The experimental results of the Resonant Frequency, RDME and weight loss of GPC and PCC in the freeze-thaw cycling tests are given in Table 5. It is possible to observe that the highest Resonant Frequency of GPC is 6602 Hz and for PCC it is 9766 Hz.

Analysing the results, from Table 6, for GPC the RDME first decreased then increased during the cycles of freeze-thaw, while for PCC, it increased after freeze-thaw cycles, showing different behaviour of both. Table 6 shows the number of cycles (N) at which the RDME reaches the specified minimum value for discontinuing the test or the specified number of cycles at which the exposure is to be terminated, whichever is less. So, 27 days was the time frame that maximum specimens could be tested in this project.

Thereby, from Equation 1, how the Durability Factor (DF) is N divided by M times the RDME, the DF is the proper RDME and all the values are over 95 once the modulus of

Table 4.   Compressive strength of GPC and PCC.

Compressive strength (MPa)		
Specimens	Cycle = 0	Cycle = 27
GPC1	32	25
GPC2	31	25
GPC3	31	23
PCC1	46	44
PCC2	45	43
PCC3	46	43

Table 5. Longitudinal resonant frequency.

Longitudinal resonant frequency (Hz)

Specimens	Cycle = 0	Cycle = 17	Cycle = 27
GPC1	6601.56	6074.22	6464.84
GPC2	6562.50	5996.09	6406.25
GPC3	6230.47	5820.31	6210.94
PCC1	9433.59	9707.03	9765.63
PCC2	9335.94	9628.91	9687.50
PCC3	9355.47	9628.91	9687.50

Table 6. Relative dynamic modulus of elasticity and durability factor.

	No cycles	RDME	N = M (number of cycles to be terminated)	RDME	DF
GPC1	17	84.7	27	113.3	113.3
GPC2	17	83.5	27	114.2	114.2
GPC3	17	87.3	27	113.9	113.9
PCC1	17	105.9	27	101.2	101.2
PCC2	17	106.4	27	101.2	101.2
PCC3	17	105.93	27	101.2	101.2

Table 7. Standard deviations for Procedure A and B and the limits of results between the specimens [5].

Range of average durability	Procedure A		Procedure B	
	Standard deviation[A]	Acceptable range of two results[A]	Standard deviation[A]	Acceptable range of two results[A]
0 to 5	0.8	2.2	1.1	3.0
5 to 10	1.5	4.4	4.0	11.4
10 to 20	5.9	16.7	8.1	22.9
20 to 30	8.4	23.6	10.5	29.8
30 to 50	12.7	35.9	15.4	43.5
50 to 70	15.3	43.2	20.1	56.9
70 to 80	11.6	32.7	17.1	48.3
80 to 90	5.7	16.0	8.8	24.9
90 to 90	2.1	6.0	3.9	11.0
Over 95	1.1	3.1	2.0	5.7

[A]These numbers represent the (1S) and (D2S) limits as described in Practice C670.

elasticity increased after 27 cycles. As can be seen in Table 7, from Standard C666, the Standard deviation for Procedure A and Procedure B should be 1.1 and 2.0 between two specimens respectively; and comparing the specimens GPC2 and PCC1 that have the higher difference between their values, the derivation is 1.12. So, all the other specimens will have less than 2.0 of derivation, which follows the standard.

The dynamic modulus of elasticity can be measured by means of longitudinal vibration. It reflects the elasticity performance of the material. Thus, the loss of the dynamic modulus of elasticity with freeze-thaw cycles means the loss of the elasticity performance.

The weight variation during freeze-thaw cycles is due to water permeating through the specimen and cracking. As soon as micro-cracking occurs, the deteriorated zones filled with surrounding water will cause an increase in the weight of the specimen. The weight loss caused by the desquamation of concrete surface was also observed with freeze-thaw cycles.

Table 8. Mass of GPC and PCC before and after freeze-thaw cycles.

Cycles	0	17	27	0–17	17–27
Specimens	Mass (g)	Mass (g)	Mass (g)	Mass change (%)	Mass change (%)
GPC1	3795	3825	3820	0.01007	0.0099
GPC2	3765	3845	3790	0.01021	0.0098
GPC3	3785	3820	3815	0.01009	0.0099
PCC1	3785	3825	3835	0.01010	0.0100
PCC2	3800	3805	3850	0.01001	0.0101
PCC3	3780	3815	3830	0.01009	0.0100

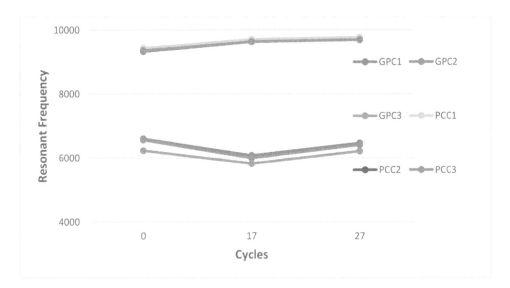

Figure 2. Resonant frequency by the number of freeze thaw cycles.

Moreover, as it can be seen from Table 8, mass of both GPC and PCC first increased after 17 cycles. However, after 27 cycles, mass decreased for GPC and continued increasing for PCC. As the weight of specimens was measured after the thaw part of the cycle, when specimens were immersed in the bucket filled with water, mass of both type of specimens increased after 17 cycles due to water inside their pores. Even though water was still in the pores for 27 cycles, the further decrease in mass for GPC can be related to the amount of concrete deterioration being superior to the amount of water that entered the concrete specimens.

5 CONCLUSION

In this study, the durability of Geopolymer concrete and Portland cement concrete subjected to 0, 17 and 27 of freeze-thaw was explored. The mass loss and dynamic elastic properties of Geopolymer concrete and Portland cement concrete specimens, including the resonant frequency and dynamic modulus of elasticity, were measured before and after the freeze-thaw cycles. The main conclusions and comments regarding this work are:

– Results showed that dynamic modulus of elasticity suffered a drop for the GPC and increased for PCC, and the compressive strength decreased as the freeze-thaw cycles were repeated for both types of concrete; however, this decrease was significant in geopolymer.

- Elastic modulus of PCC found increased, possibly due to the continuation of the process of cement hydration;
- Fluctuation in the mass occurred due the amount of water that went in and out of the specimen and cracks;
- The freeze-thaw durability of concrete should be taken into consideration in structural design and maintenance;
- For future studies, air entrained admixture can be added to improve the resistance of concrete, once it will increase the air content of the specimens.
- Fluctuation in the results were observed, therefore more cycles are necessary to have a better understanding of the phenomenon and more repeatability of data.

REFERENCES

A. C618–15 (2015), Standard Specification for Coal Fly Ash and Raw or Calcined Natural Pozzolan for Use in Concrete, *ASTM Int. West Conshohocken, PA.*

Al Bakri, A. M. M., H. Kamarudin, M. Binhussain, I. K. Nizar, A. R. Rafiza, and Y. Zarina (2013) Comparison of geopolymer fly ash and ordinary portland cement to the strength of concrete, *Adv. Sci. Lett., vol. 19, no. 12, pp. 3592–3595.*

ASTM C215 (2014), Standard Test Method for Fundamental Transverse, Longitudinal, and Torsional Resonant Frequencies of Concrete Specimens, *Am. Soc. Test. Mater., pp. 1–7.*

ASTM C39/C39M-17 (2017), Standard Test Method for Compressive Strength of Cylindrical Concrete Specimens, *ASTM Int. West Conshohocken, PA.*

ASTM C666/C666M (2003), Standard Test Method for Resistance of Concrete to Rapid Freezing and Thawing ASTM Int. *West Conshohocken, PA, vol. 3, no. Reapproved, pp. 1–6.*

Cai, L., H. Wang, and Y. Fu (2013) Freeze—thaw resistance of alkali—slag concrete based on response surface Methodology, *vol. 49, pp. 70–76.*

Douglas, V. M., E., Bilodeau, A., Brandstetr, J., & Malhotra (1991) Alkali activated ground granulated blast-furnace Slag concrete: *preliminary investigation,Cem. Concr. Res., vol. 21, pp. 101–108.*

Fu, Y., L. Cai, and W. Yonggen (2011) Freeze-thaw cycle test and damage mechanics models of alkali-activated slag Concrete, *Constr. Build. Mater. vol. 25, no. 7, pp. 3144–3148.*

Mustafa, M., A. Bakri, H. Mohammed, H. Kamarudin, I. K. Niza, and Y. Zarina (2011) Review on fly ash-based geopolymer concrete without Portland Cement, *J. Eng. Technol. Res., vol. 3, no. 1, pp. 1–4.*

Portland cement Association. N.P. (2002), Types and Cause of Concrete Deterioration, *Web Serial No. 2617.*

Shang H. and T. Yi (2013) Freeze-Thaw Durability of Air-Entrained Concrete, *the Scientific World Journal vol. 2013, Article ID 650791, 6 pages.*

Shuai Shang H., (2013), Triaxial T-C-C behavior of air-entrained concrete after freeze-thaw cycles, *Cold Reg. Sci. Technol., vol. 89, no. 3, and pp. 1–6.*

Sun, P., and H. C. Wu (2013), Chemical and freeze-thaw resistance of fly ash-based inorganic mortars Fuel, *vol. 111, pp. 740–745.*

Project risk management by fuzzy EVM for elevated metro rail corridor projects

Manvinder Singh & Debasis Sarkar
Department of Civil Engineering, School of Technology, PDPU, Gujarat, India

ABSTRACT: This paper is an attempt to develop a new project risk management method for elevated metro corridor projects. Mega infrastructure projects are exposed to grander risks due to numerous phases like feasibility, land acquisition, design, development, implementation, execution and final commissioning. So, a methodical technique of risk analysis is required to identify, collect, compute and analyse these risks, and accordingly articulate risk response plans. Expected Value Method is a well-organised method that helps in identifying the risk severity by computing Composite Likelihood factor and Composite Impact factor for each activity. But inter-relationship between the input terms of EVM i.e. Likelihood and impact, is not possible within the scope of EVM. So, fuzzy logic is assimilated within EVM to map the interrelationship between input terms. Case study of Bangalore metro rail project construction in numerous phases are presented to validate the concept of this technique. The use of this approach helps to establish the degree of compatibility between the quantitative results obtained from questionnaires and the qualitative terms supporting the same. Ranking of major activities of transportation project of Southern India is done by using Fuzzy EVM method.

1   INTRODUCTION

Over the last 50 years, many complex mega projects like construction of underground and elevated corridors for metro rail, tunnels, bridges, airports, skyscrapers etc. have experienced large variations in cost and time causing massive amount of delays in the execution and final commissioning of these projects. Due to increase in project complications and magnitudes, high levels of risk and uncertainties are encountered through all the phases of the project starting from feasibility, land acquirement, tendering and contracts, design, development, execution and final commissioning. These risks and uncertainties causes' terrific time and cost overrun which ultimately affects successful accomplishment of the project. Mass Rapid Transit System projects like execution of underground and elevated metro rail project sen compass high degree of risks during the process of land acquisition, tendering and contracts, design, piling, pier, casting and erection of segments, launching girder pre-stressing operations.

The objective of this paper is an attempt to develop new project risk management method (Fuzzy EVM) for elevated metro corridor projects. This method would help to classify the risks according to priority, thus enabling the project experts to articulate risk mitigation actions accordingly.

2   LITERATURE REVIEW

Subramanyam et al. (2012) have focused on research work to identify factors that influence the smooth completion of a project and develop a risk assessment model. They have made an attempt to analyze the present risk condition in the construction industry by meeting experts in the field to collect first-hand information as a first step towards risk assessment and suggest a risk response strategy. A total of 93 risk factors were identified. Sarkar and Dutta (2011) had developed a risk management model for underground metro construction projects. The identification of the risks was carried out by interacting with the experts associated with similar projects. They had assessed these risks in terms of likelihood of failure and impact of risks. Subramanyam et al. (2012), endeavoured to identify factors that influence the smooth completion of the project. Kuo and Lu (2013), have expressed that, construction projects in metropolitan areas is very risky and requires very reliable risk assessment model for proper planning and execution. Abdelgawad and Fayek (2010) developed a model by combining fuzzy logic with both FMEA and AHP in a comprehensive framework that provides a practical and thorough approach for assessing the level of criticality of risk events in the construction projects. Subramanyam et al. (2012) had developed quantifiable model which is based upon the likelihood of occurrence of a risk and its level of importance. Choi et al. (2004) developed fuzzy based software for risk analysis. Chan et al. (2009) had reviewed and published the fuzzy literature of 10 years in8topmost quality journals. Eom and Paek (2009) established an environmental risk index model for contractors to minimize third-party environmental disputes at construction sites.

3 METHODOLOGY

3.1 *Expected Value Method (EVM)*

Expected Value Method is a well-organised method that helps in identifying the risk severity by multiplication of Composite Likelihood factor and Composite Impact factor for each activity. The questionnaire's data collected from experts to be analysed using Microsoft Excel sheet and EVM. The data from 55 experts in 23 major risk groups/activities to be entered in the excel sheet to generate an average risk weights, probability of failures of the identified risks and impact values ranging from 0 to 1 for each attribute. After that composite likelihood factor and composite impact factor to be computed for each attribute using equation (2) and (3). Sarkar and Dutta (2011) assume a network of deterministic time and cost. They had extended the work of Nicholas (2007). They define the variables as follows:

L_{ij}: Likelihood of ith risk source for jth activity
W_{ij}: Weightage of ith risk source for jth activity
I_{ij}: Impact of ith risk source for jth activity
CLF_j: Composite Likelihood Factor for jth activity
CIF_j: Composite Impact Factor for jth activity

An activity may have several risk sources each having it sown likelihood of occurrence. The value of likelihood should range between 0 to 1. The likelihood of failure (L_{ij}) defined above, of the identified risk sources of each activity were obtained through a questionnaire survey. The corresponding weight age (W_{ij}) of each activity has also been obtained from the feedback of the questionnaire survey circulated among experts.

$$\sum_{i=1}^{M} W_{ij} = 1 \, for \, all \, j \, (j=1.......N) \tag{1}$$

The weightages can be based on local priority (LP) where the weight ages of all the sub-activities of a particular activity equal 1. The mean of all the responses should desirably be considered for analysis. Inconsistent responses can be modified using a second round questionnaire survey using the Delphi technique. The likelihood (L_{ij}) of all risk sources for each activity j can be combined and expressed as a single composite likelihood factor (CLF)$_j$. The weight ages (W_{ij}) of the risk sources of the activities are multiplied with their respective likelihoods to obtain the CLF for the activity. The relationship of computing the CLF as a weighted average is given below:

$$Composite\ Likelihood\ Factor\ CLF_j = \sum_{i=1}^{M} L_{ij}W_{ij}\ for\ all\ j \tag{2}$$

$$Composite\ Likelihood\ Factor\ CIF_i = \sum_{i=1}^{M} I_{ij}W_{ij,} \tag{3}$$

$$0 \leq I_{ij} \leq 1\ and \sum_{i=1}^{M} W_{ij} = 1\ for\ all\ j$$

3.2 *Application of fuzzy in EVM (fuzzy EVM)*

One membership function for all risk factors has to be defined in order to convert each qualitative variable into fuzzy values. The membership functions are to be defined with triangular fuzzy numbers. The membership functions are defined with values "Very Low", "Low", "Medium", "High", and "Critical" according to their probability (Likelihood of failure) and their impact of risk. The values of the linguistic scale are chosen from 0 to 5 at an increment of 0.5. The values of likelihood and their impact of risk ranges from 0 to 1. The severity of a risk is to be described in both qualitative and quantitative.

4 CASE STUDY

The case study considered for this work is Bangalore Elevated Metro Rail Corridor construction. The construction starting from the Nayandahalli to RV college station is being executed by Infrastructure Leasing & Financial Services Company Limited. The length of the section under study is 4.0 km. The elevated stations are Nayandahalli, Rajarajeshwari Nagar, Gnanabharathi and RV College. Total 133 piers and 540 piles would be constructed. The weight of each segment is 14 tons. To build one span between two adjacent piers, 11 segments are required. For this section, Infrastructure Leasing & Financial Services Company will produce, erect and launch 1180 segments for the viaduct from Nayandahalli to RV college station. Infrastructure Leasing & Financial Services Company Limited had started pier construction on Nayandahalli and Rajarajeshwari Nagar. The methodology of research work was formulation of questionnaire (firstly to identify the major activities involved from beginning to handing over the project) and survey where the replies of the 55 experts from the metro rail corridor projects were used as inputs for articulating the risk severity ranks. Hence three to four rounds of discussions with metro higher officials and staff (including machine operators, foreman, supervisors, engineers, site in charges, managers, Deputy General Managers), consultants and chief engineer were carried out for the finalisation of questionnaire, rating scales and their responses. The risk assessment is carried out in terms of risk severity level and their ranks.

5 RESULTS OF THE RESEARCH

5.1 *Final risk severity and ranking by fuzzy EVM method*

The composite likelihood factor and composite impact factor values computed from expected value method for all 23 major risk groups were used as inputs for Fuzzy method by using Matlab. Twenty five rules and 5 Membership functions were formed in Matlab. The outputs attained for all 23 major risk groups are as risk severity value of each risk activity. All values are tabulated in Table 1. For example: Final Risk severity in feasibility and Detailed Project Report activity is shown in Figure 1.

Table 1. Final values of composite likelihood factor and composite impact factor for all major risk categories calculated. Sample of 3 major risk categories/activities given below.

S.No	Description of project risk category	Composite likelihood factor	Composite impact factor
1	Risks in traffic and utility diversion	0.575	0.872
2	Sub structure work to super structure work risks (pile, pile cap, pier and pier cap)	0.404	0.773
3	Risks in launching girder	0.465	0.783

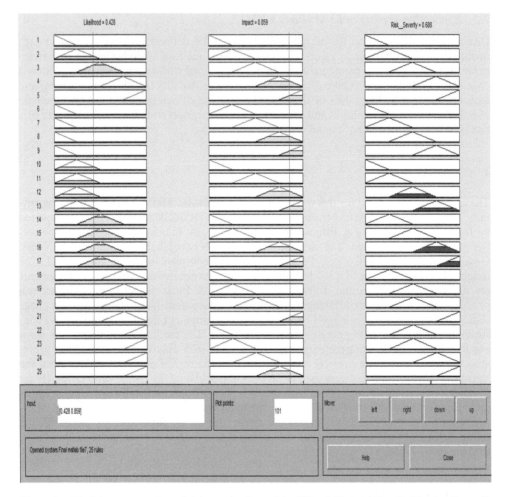

Figure 1. Matlab representation of risk severity for activity "Feasibility and Detailed Project Report".

6   DISCUSSION AND INTERPRETATION

The final fuzzy rankings and risk severity values for all 23 major risk categories of an elevated metro project attained from Fuzzy EVM (Fuzzy method is incorporated in EVM method) are calculated and sample of 6 major risk categories are tabulated in Table 2.

By application of Fuzzy expected value method, Risks in traffic and utility diversion activity for an elevated metro corridor project are having highest quantitative risk severity value of

Table 2. Final risk severity values (quantitative & qualitative) and ranking using the concept of Fuzzy EVM for all 23 major risk categories computed. Sample of 6 major risk categories are given below.

S.No	Description of project risk category	Final fuzzy risk severity		FEVM ranking
		Quantitative	Qualitative	
1	Risks in traffic and utility diversion	0.781	Critical risk	1
2	Risks in Land acquisition and handover	0.777	Critical risk	2
3	Risks in segment erection	0.769	Critical risk	3
4	Risks in launching girder	0.708	Critical risk	4
5	Risks in segment casting	0.700	Critical risk	5
6	Risks in Parapet erection	0.689	Critical risk	6

0.781 and are having critical risks and FEVM ranking is one. Land acquisition and handover activity is having critical risk severity and quantitative risk severity value is 0.777 and ranking is 2. Risks in segment erection activity stood at third rank with quantitative risk severity value of 0.769. Risks in launching girder activity are of critical nature in both quantitative (0.708) and qualitative and ranking is 4 by Fuzzy EVM.

Risks in segment casting (risk severity rank – 5), parapet erections (risk severity rank – 6), feasibility and detailed project report (risk severity rank – 7), drawings preparation, submission and approvals (risk severity rank – 8), tendering and award of contracts(risk severity rank – 9), risks in sub structure work to super structure work like pile and pier etc, come under critical risk severity category and all these works are to be executed very carefully. Risks in casting yard setup, cable tray, parapet casting, risks in pile test and barricading works are come under high risk severity category. Risks in Expansion joint and road widening activities come under medium risk severity category and project site office set up come under low risk severity category.

There is good range of risk severity quantitative values from 0.310 to 0.781 obtained by Fuzzy EVM method for all 23 major risk activities of an elevated metro corridor project. There is clear risk ranking for all 23 major risk activities of an elevated metro corridor project by Fuzzy EVM method.

7 CONCLUSION

After discussions with field experts and interpretations of results, it is concluded that although EVM is being widely used for determination of risk severity values by using the well-known concept of Composite Likelihood factor and Composite Impact factor for each activity. But interrelationship between the input terms of EVM i.e. Likelihood and, impact, is not possible within the scope of EVM. So, fuzzy logic is integrated within EVM to map the interrelationship between input terms and better results in terms of practical applications for risk mitigations actions as rankings are not overlapped and there is wide range of risk severity quantitative values. The fuzziness in results of Expected value method is eliminated by integrating fuzzy logic. Hence Fuzzy EVM method is better for defining risk ranking/index.

REFERENCES

Abdelgawad, M. and Fayek, A.R. (2010). "Risk management in the Construction Industry Using Combined Fuzzy FMEA and Fuzzy AHP." Journal of Construction and Management, vol. 136(9), pp.. 1028–1036.

Chan, A.P.C., Chan, D.W.M. and Yeung, J.F.Y. (2009). "Overview of the Application of "Fuzzy Techniques" in Construction Management." *Journal of Construction Engineering and Management,* vol. 135(11), pp. 1231–1252.

Choi, H., Cho, H. and Seo, J.W. (2004) "Risk Assessment Methodology for underground Construction Projects" *Journal of Construction Engineering and Management*, vol. 130(2), pp. 258–272.

Eom, C.S.J. and Paek, J.H. (2009) "Risk Index Model for Minimizing Environmental Disputes in Construction." *Journal of management in Engineering*, vol. 135(1), pp. 34–41.

Kuo, Y.C. and Lu, S.T. (2013) "Using fuzzy multiple criteria decision making approach to enhance risk assessment for metropolitan construction projects". *International Journal of Project Management*, vol.31, pp. 602–614.

Nicholas, J.M. (2007) Project Management for Business and Technology: Principles and Practice, Second Edition, Pearson Prentice Hall, New Delhi.

Sarkar, D and Dutta, G. (2011) "A Framework of Project Risk Management for the Underground Corridor Construction of Metro Rail." *International Journal of Construction project Management*, vol.4(1), pp. 21–38.

Subramanyam, H., Sawant, P.H. and Bhatt, V. (2012) "Construction Project Risk Assessment: Development of Model Based on Investigation of opinion of Construction Project Experts from India." *Journal of Construction Engineering and Management*, vol.138 (3), pp. 409–421.

Annexure A

Table 3.   Risk severity classification (For fuzzy expected value method).

S.No.	Ranges	Classification
I	0.00–0.249	V. Low Risk
Ii	0.250–0.449	Low Risk
Iii	0.450–0.549	Medium Risk
Iv	0.550–0.649	High Risk
V	0.650–1.000	Critical Risk

Annexure B

Table 4.   Questionnaire of tenders and award of contracts risk.

Risk description	Likelihood (L_{ij})	Weightage $(LP) (W_{ij})$	Impact (I_{ij})
Delay in submission of drawings by detailed design consultant (DDC)			
Lack of accuracy in internal detailed estimate			
Two packet system (Technical and financial evaluation) is not implemented			
Delay in preparation and approval of tender document			
Delay in issuing NIT (Notice invite tender)			
Inadequate or insufficient site information (soil test and survey report)			
Improper evaluation of contractor			
Delay in pre-bid meeting			
Delay in award of contract			
Delay in mobilization of resources by contractor			
Variations by the client			

Technology Drivers: Engine for Growth – Mahajan, Modi & Patel (Eds)
© 2018 Taylor & Francis Group, London, ISBN 978-1-138-56042-0

Investigation on mechanical properties of crushed sand concrete

A.A. Jadeja & U.V. Dave
Department of Civil Engineering, Institute of Technology, Nirma University, Gujarat, India

ABSTRACT: Fine aggregate occupies considerable volume in concrete. Conventionally, natural sand is used as the fine aggregate in concrete. The scarcity of good quality natural sand and strict government laws on its mining have made it necessary to look for an alternative for the fine aggregate to be used in the concrete. Crushed sand is one of the most considerable alternative of natural sand nowadays for its use in the concrete. Natural sand is replaced by the crushed sand in proportion of 50%, 75% and 100% volume of the fine aggregate for M25 concrete grade in the present investigation. Mechanical properties like compressive strength, flexure strength, split tensile strength, bond strength and modulus of elasticity for four concrete mixes are investigated. It is observed that 50% replacement of natural sand by the crushed sand in the concrete mixes helped to improve the mechanical properties. The reason behind maximum improvement in properties at 50% replacement of natural sand by the crushed sand may be considered as the presence of lesser voids in the concrete due to better particle packing.

1 INTRODUCTION

In a developing country like India, demand of the new infrastructure is going to increase with time. Hence, the concrete production will be more which will lead to higher demand of fine aggregate material i.e. natural sand. In the scenario of scarcity of good quality natural sand, it is required to look for another material which can be termed as more suitable as fine aggregate in concrete.

Rocks undergo natural weathering over years and river sand is formed. The sand is mined from river beds and used to make concrete. On the other hand, the crushed sand is produced by crushing the parent rock to the required smaller sizes. It is also known as Manufactured Sand or M-Sand.

Effect of the crushed sand and fly ash on the compressive strength and flexure strength properties of M25 grade of the concrete was studied. Natural sand was replaced by the crushed sand in the proportions of 30%, 50%, 70% and 100%, respectively with 35% replacement of cement by fly ash for all the concrete mixes. Considerable reduction was observed for the compressive strength and the flexure strength of the concrete mixes wherein the natural sand was replaced by the crushed sand in the proportion of 70% and 100%, respectively (Mundra et al. 2010). Compressive strength, flexural strength and bond strength of the M20 & M30 grade of concrete mixes were marginally higher with M-sand as compared to that of the concrete with natural sand as the fine aggregate. Stress-strain behavior of concrete with M-sand and natural sand as fine aggregate was similar (Reddy, 2012). High Strength Concrete mix made with the crushed sand obtained the higher compressive strength as compared to that of the natural sand concrete, while its elastic modulus was lower (Donza et al. 2002). Improvement in the compressive strength, flexure strength and split tensile strength of M30 grade of the concrete was observed when 60% natural sand was replaced by the crushed sand (Vijay et al. 2015). M20 grade of concrete with 60% replacement of the natural sand by the crushed sand performed better as compared to that of the control concrete in terms of the compressive strength, flexure strength and split tensile strength (Jadhav & Kulkarni, 2012).

The main objective of this investigation is to study the change in behavior of the concrete in terms of mechanical properties when natural sand is replaced by the crushed sand in different proportions. For that purpose, concrete grade of M25 is selected and natural sand is replaced by the crushed sand by 50%, 75% and 100% proportion respectively. The change in terms of mechanical properties is to be studied.

2 EXPERIMENTAL PROGRAMME

This section provides information about properties of aggregates, concrete mix proportioning and testing of concrete.

2.1 Properties of aggregates

2.1.1 Coarse aggregate
Crushed angular aggregate of 20 mm and 10 mm down size are used as the coarse aggregate. Properties of the coarse aggregate are presented (Table 1).

2.1.2 Fine aggregate
Locally available natural river sand is used as fine aggregate for control concrete and other three mixes. It is replaced by the crushed sand in the concrete mixes except control concrete. Properties of fine aggregate including results of sieve analysis are presented (Table 1).

2.2 Mixture proportioning

The concrete mixes are designed as per the provisions of IS 10262. In the control concrete mix M25-CC of M25 grade is with 100% natural sand. The natural sand is replaced by the crushed sand by 50%, 75% and 100% and the mixes are designated as M25-50, M25-75 and M25-100, respectively (Table 2). All other ingredients are kept constant for all the four mixes.

Table 1. Properties of aggregate (Jadeja, 2017).

Material	Loose bulk density (kg/m^3)	Compacted bulk density (kg/m^3)	Specific gravity	Fineness modulus	Zone
Crushed Sand	1477	1706	2.73	2.88	II
Natural Sand	1546	1681	2.75	2.79	II
Coarse Aggregate (20 mm)	1353	1517	2.79	7.28	–
Coarse Aggregate (10 mm)	1523	1618	2.82	6.03	–

Table 2. Concrete mixture composition (Jadeja, 2017).

Material		Unit	M25-CC	M25-50	M25-75	M25-100
Cement		kg/m^3	334	334	334	334
Water		lit/m^3	167	167	167	167
Coarse	20 mm	kg/m^3	690	690	690	690
Aggregate	10 mm	kg/m^3	456	456	456	456
Fine	Natural sand	kg/m^3	883	442	221	–
Aggregate	Crushed sand	kg/m^3	–	438	658	877
Chemical admixture		%	0.9	1.1	1.3	1.3
W/C		–	0.50	0.50	0.50	0.50

3 RESULTS AND DISCUSSION OF MECHANICAL PROPERTIES

3.1 *Density of hardened concrete*

The density observed during the experiment for the M25-CC, M25-50, M25-75 and M25-100 is 2684 kg/m^3, 2694 kg/m^3, 2677 kg/m^3 and 2672 kg/m^3, respectively. The percentage difference between the theoretical and experimental density of M25-CC, M25-50, M25-75 and M25-100 is 6.0%, 6.5%, 5.8% and 5.7%, respectively.

3.2 *Compressive strength*

The compressive strength is evaluated by testing the cube specimen of 150 mm × 150 mm × 150 mm size for concrete mixes at 28 days as per the provision of IS 516. Average results of the three specimens are considered as the final result of the compressive strength for each concrete mix. Results of the compressive strength of the concrete mixes made with the crushed sand are compared with the results of control concrete as shown in Figure 1. Compressive strength is improved by 7.53% for M25-50 mix as compared to that of M25-CC mix. Here, the percentage difference observed between the compressive strength of individual specimen and the average compressive strength for M25-CC concrete mix is in the range of 3.6% to −3.8% and for M25-50 concrete mix is in the range of 2.2% to −2.1% which is less than 7.53%. Therefore, 7.53% increment in compressive strength for M25-50 concrete mix as compared to that of M25-CC is satisfactory. It is observed that the compressive strength is reduced by 9.16% and 7.97% respectively for M25-75 and M25-100 mixes as compared to that of M25-CC mix.

3.3 *Flexure strength*

Flexure strength is evaluated by testing the beam specimen of 100 mm × 100 mm × 500 mm size for all concrete mixes at 28 days as per the provision of IS 516. Average results of the three specimens are considered as the final result of the flexure strength for each concrete mix. Results of the flexure strength of the concrete mixes made with the crushed sand are compared with the results of the control concrete as shown in Figure 2. 8.01% improvement in flexure strength is observed for M25-50 mix as compared to that of M25-CC mix. Here, the percentage difference observed between the flexure strength of individual specimen and the average flexure strength for M25-CC concrete mix is in the range of 4.0% to −3.3% and for M25-50 concrete mix is in the range of 2.3% to −1.5% which is less than 8.01%. Therefore, 8.01% increment in flexure strength for M25-50 concrete mix as compared to that of M25-CC is satisfactory. Flexure strength is reduced by 6.46% for M25-75 mix and 3.1% for M25-100 mix respectively as compared to that of M25-CC mix.

3.4 *Split tensile strength*

Split tensile strength is evaluated by testing the cylindrical specimen of 150 mm × 300 mm size for all concrete mixes at 28 days as per the provision of IS 5816. Average results of the

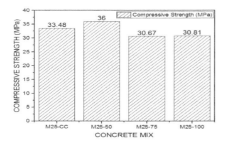

Figure 1.   Variation in compressive strength of concrete due to addition of the crushed sand.

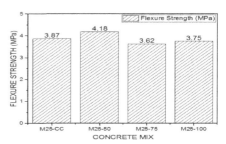

Figure 2.   Variation in flexure strength of concrete due to addition of the crushed sand.

three specimens are considered as the final result of the split tensile strength for each concrete mix. Results of the split tensile strength of the concrete mixes made with the crushed sand are compared with the results of the control concrete as shown in Figure 3. 2.07% improvement in split tensile strength is observed for M25-50 mix as compared to that of M25-CC mix. Split tensile strength is reduced by 2.41% for M25-75 mix and 3.10% for M25-100 mix respectively as compared to that of M25-CC mix.

3.5   Bond strength

Bond strength is evaluated by testing the cube specimen of 150 mm × 300 mm × 150 mm size with embedded reinforcement of 12 mm diameter for all concrete mixes at 28 days as per the provision of IS 2770. Average results of the three specimens are considered as the final result of the bond strength for each concrete mix. Results of the bond strength of the concrete mixes made with the crushed sand are compared with the results of the control concrete as shown in Figure 4. 5.08% improvement in bond strength is observed for M25-50 mix as compared to that of M25-CC mix. Bond strength is reduced by 3.45% for M25-75 mix and 8.88% for M25-100 mix respectively as compared to that of M25-CC mix.

3.6   Modulus of elasticity

Modulus of elasticity is evaluated by testing the cylindrical specimen of 150 mm × 300 mm size for all concrete mixes at 28 days as per the provision of IS 516. Average results of the three specimens are considered as the final result of the modulus of elasticity for each concrete mix. Tangent modulus method has been used to evaluate the modulus of elasticity for all the concrete mixes. Slope of stress-strain curve gives the value of the modulus of elasticity for corresponding concrete mix. Sample calculation of modulus of elasticity for a specimen of M25-CC and the stress-strain graph for the same is presented in Figure 5.

Results of the modulus of elasticity of the concrete mixes made with the crushed sand are compared with the results of the control concrete as shown in Figure 6. 5.32% improvement

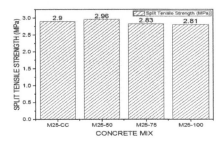

Figure 3.   Variation in split tensile strength of concrete due to addition of the crushed sand.

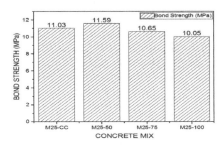

Figure 4.   Variation in bond strength of concrete due to addition of the crushed sand.

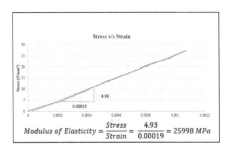

Figure 5.   Stress-strain curve for a specimen of M25-CC mix.

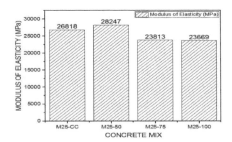

Figure 6.   Variation in modulus of elasticity of concrete due to addition of the crushed sand.

in modulus of elasticity is observed for M25-50 mix as compared to that of M25-CC mix. Modulus of elasticity is reduced by 11.20% for M25-75 mix and 11.74% for M25-100 mix respectively as compared to that of M25-CC mix.

Similar trend in behavior is observed for all the mechanical properties considered in the present investigation in context to all four concrete mixes. The reason behind the maximum improvement in properties for M25-50 mix may be considered as the presence of lesser voids in concrete due to better particle packing. When the natural sand is replaced by the crushed sand by more than 50% proportion, the excessive fine particles are likely to weaken the bond between the cement paste and the coarse aggregate and therefore it may have resulted into the inferior performance of M25-75 and M25-100 mixes as compared to that of M25-50 mix.

4 CONCLUSIONS

Based on the experimental results presented herewith, the following conclusions are drawn:

– Compressive strength of crushed sand concrete is higher at 50% replacement of the natural sand by the crushed sand as compared to that of control concrete. With increase in the crushed sand content in the concrete mix, the compressive strength of the concrete gets reduced.
– Flexure strength followed the same trend as has been observed in case of the compressive strength for all the concrete mixes. 50% replacement of natural sand with the crushed sand further enhanced the flexure strength of the concrete.
– Improvement in the split tensile strength of the concrete is observed at 50% replacement of natural sand as compared to that of control concrete.
– Bond strength ofthe crushed sand concrete and control concrete is observed at par. Improvement in the bond strength is observed when crushed sand is replaced by 50% of the natural fine aggregate.
– Modulus of Elasticity is improved at 50% proportion of the crushed sand. Beyond 50% replacement of the crushed sand, the modulus of elasticity of concrete mixes reduced as compared to that of control concrete.

Thus, replacing 50% natural sand by the crushed sand results into improvement in the mechanical properties of the concrete.

REFERENCES

Bureau of Indian Standards (1959) IS 516:1959. *Methods of Tests for Strength of Concrete.* Delhi, BIS.
Bureau of Indian Standards (1965) IS 2770-1:1965. *Methods of Testing Bond in Reinforced Concrete.* Delhi, BIS.
Bureau of Indian Standards (1999) IS 5816:1999. *Splitting Tensile Strength of Concrete—Method of Test.*Delhi, BIS.
Bureau of Indian Standards (2009) IS 10262:2009. *Concrete Mix Proportion Guidelines.* Delhi, BIS.
Donza, H., Cabrera, O. & Irassar, E.F. (2002) "High-strength concrete with different fine aggregate". *Cement and Concrete Research.* [Online] 32, 1755–1761.
Jadeja, A.A. (2017) *Suitability of Crushed Sand as Fine Aggregate in Concrete.* M. Tech Major Project. Institute of Technology. Nirma University. Ahmedabad.
Jadhav, P.A., Kulkarni, D.K. (2012) "Experimental Investigation on the Properties of Concrete Containing Manufactured Sand". *International Journal of Advanced Engineering Technology.* [Online] 2, 101–104.
Mundra, S.,Sindhi, P.R., Chandwani, V., Nagar, R. & Agrawal, V (2016). "Crushed Rock Sand-An Economical and Ecological Alternative to Natural Sand to Optimize Concrete Mix". *Perspective in Science.* [Online] 8, 345–347.
Reddy, B.V.V. (2012) *Suitability of Manufactured Sand (M-Sand) as Fine Aggregate in Mortars and Concrete.* [Report] Indian Institute of Science. Bangalore.
Vijay, B., Selvan, S.S., Felix, K.T. & Annadurai, R. (2015). "Experimental Investigation on the Strength Characteristics of Concrete Using Manufactured Sand". *International Journal for Innovative Research in Science & Technology.* [Online] 1, 174–178.

Technology Drivers: Engine for Growth – Mahajan, Modi & Patel (Eds)
© 2018 Taylor & Francis Group, London, ISBN 978-1-138-56042-0

Size effect and non-linear fracture behavior of fiber-reinforced self-compacting concrete

B.G. Patel
Faculty of Engineering Technology and Research, Bardoli, Surat, India

A.K. Desai
Department of Applied Mechanics, S.V. National Institute of Technology, Surat, India

S.G. Shah
Department of Civil Engineering, Babaria Institute of Technology, BITS Edu Campus, Vadodara, India

ABSTRACT: This paper comprises an experimental study to characterize the size effect and fracture behavior of plain and fiber-reinforced self-compacting concrete. To understand fracture behavior, geometrically similar plain and fiber-reinforced SCC specimens were casted and tested in a three-point bending setup under a closed loop servo-controlled testing machine. The non-linear fracture mechanics parameters such as fracture energy, length of process zone, brittleness number, critical mode I stress intensity factor, critical crack-tip opening displacement, size of fracture process zone for infinitely large specimens, and the crack growth resistance curve (R-curve) were determined using Bazant's size effect method. The resistance against crack growth was high for fiber-reinforced specimens compared to that of plain specimens. It is concluded that fiber-reinforced concrete specimens show better behavior under crack propagation as compared to plain specimens.

1 INTRODUCTION

Self-compacting concrete (SCC) was defined by Okamura (1997) as concrete that is able to flow in the interior of the formwork, filling it in a natural manner and passing through the reinforcing bars and other obstacles, flowing and consolidating under the action of its own weight. When the structure is loaded, the micro-cracks open up and propagate. The development of such micro-cracks, results in inelastic deformation in concrete. Fiber-reinforced concrete (FRC) is a cementing concrete mixture reinforced with more or less randomly distributed small fibers.

The size effect law proposed by Bazant (1984) approximately describes the transition from the strength criterion, for which there is no size effect, to the linear elastic fracture mechanics (LEFM) criterion, for which the size effect is the strongest possible. The main advantage of the size effect method is that by only measuring the maximum load values of geometrically similar specimens of different sizes, one can obtain fracture parameters that are size-independent. Furthermore, from the size effect method the effective length of the fracture process zone can also be determined, from which one can further obtain the R-curve and the critical effective crack-tip opening displacement, which are very useful parameters in non-linear fracture mechanics (Gettu et al., 1990; Bazant et al., 1991).

2 EXPERIMENTAL PROGRAM

2.1 *Materials and mix proportions*

Standard 53 grade ordinary Portland cement, class F fly ash, polycarboxylate ether-based super plasticizer, locally available river sand, and two different sizes of coarse aggregates (maximum sizes 10 mm and 20 mm) were used for preparation of the SCC. Two types of fibers, namely glass fiber (Cem-FIL) and steel fiber (Dramix RCBN 35/65 hooked end) were used to develop glass-fiber-reinforced self-compacting concrete (GFRSCC) and steel-fiber-reinforced self-compacting concrete (SFRSCC) respectively. The details of the mix proportion are given in Table 1.

2.2 Specimen size

In order to study the size effect and other parameters, geometrically similar specimens, as per the draft recommendation of RILEM (RILEM, 1990), were prepared.

2.3 Testing of specimens

In order to qualify as 'self-compacting concrete', the passing and filling abilities of all the mixes were tested by workability tests, namely slump flow, T_{50} slump flow, J-Ring, V-Funnel, L-Box, and U-Box, as per the EFNARC (EFNARC, 2005) standards. The results of the fresh properties of plain self-compacting concrete (PSCC), glass-fiber-reinforced self-compacting concrete (GFRSCC) and steel-fiber-reinforced self-compacting concrete (SFRSCC) satisfied the EFNARC requirement for self-compacting concrete. The compressive, split tensile, and flexural tests were done as per the IS standard (IS 516, 2013). Compressive strengths varied in the range 62 MPa to 66 MPa, spilt tensile strengths varied in the range 4.5 MPa to 6.5 MPa, while flexural strengths varied in the range 4.3 MPa to 5.9 MPa for plain self-compacting concrete and fiber-reinforced self-compacting concrete.

Three-point bending tests were performed on the geometrically similar beam specimens. All the specimens were tested in a closed loop servo-controlled testing machine with a capacity of 50 kN, as shown in Figure 1. A specially calibrated 30 kN load cell was used for measuring the load. The load-point displacement was measured using a built-in linear variable displacement transducer (LVDT). The crack mouth opening displacement (CMOD) was measured using a clip gage. All the tests were performed in CMOD control with a rate of opening of 0.0005 mm/sec. The results of load, displacement, CMOD, and time were simultaneously acquired through a data acquisition system.

Table 1. Details of mix proportion.

Constituent	PSCC	GFRSCC	SFRSCC
Cement (kg/m³)	354	354	354
Fly ash (kg/m³)	96	96	96
Fine aggregate (kg/m³)	634	634	634
Coarse aggregates <10 mm size (kg/m³)	224	224	224
Coarse aggregates <20 mm size (kg/m³)	332	332	332
Water (liter/m³)	160	160	160
Superplasticizer (liter/m³)	2.225	2.225	2.225
Glass fiber (gm/m³)	–	350	–
Steel fiber (kg/m³)	–	–	32

Table 2. Details of the dimensions of beams.

Beam designation	Depth d (mm)	Span S (mm)	Length L (mm)	Thickness b (mm)	Notch depth a_0 (mm)
Small	76	190	241	50	15.2
Medium	152	380	431	50	30.4
Large	304	760	810	50	60.8

Figure 1.   Testing of a specimen on a closed loop servo control testing machine.

Table 3.   Area under load-displacement curves (kN.mm).

	PSCC	GFRSCC	SFRSCC
Small	0.53	1.58	6.34
Medium	1.17	4.86	12.21
Large	5.06	10.34	25.69

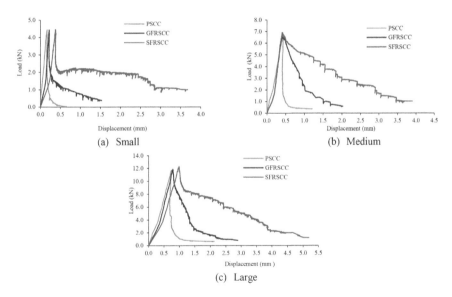

Figure 2.   Load versus displacement curves for small, medium and large specimens.

3   RESULTS AND DISCUSSION

The combined results of load versus displacement for PSCC, GFRSCC, and SFRSCC specimens are shown in Figure 2(a), 2(b) and 2(c) for small, medium, and large specimens respectively. The area under the load-displacement curve is a measure of the fracture energy and is presented in Table 3 for all types and sizes of SCC.

The following observations are made from these plots:

1. The area under the load-displacement curves, which is a measure of the fracture energy, was high for SFRSCC specimens compared to GFRSCC and PSCC specimens. The fracture energy is quantified in Table 3.

2. Energy absorption capacity increased by 198% and 1096% in small specimens due to the addition of glass fibers and steel fibers respectively. Similarly, in medium specimens, energy absorption capacity increased by 184% and 614% due to addition of glass fibers and steel fibers respectively. Energy absorption capacity also increased in large specimens by 104% and 407% due to the addition of glass fibers and steel fibers respectively.
3. It can be concluded that the addition of fibers to specimens greatly increases the energy absorption capacity in all sizes of specimens. It can also be concluded that fiber-reinforced materials behave in a better way than material without fibers.

Figure 3(a), 3(b) and 3(c), show the load versus CMOD curves for small, medium and large specimens respectively, for all PSCC, GFRSCC and SFRSCC specimens.

The following observations are made from these plots:

1. The post-peak behavior, and hence the area under the load-CMOD curve, was much higher in SFRSCC specimens followed by GFRSCC specimens and PSCC specimens. Once again this is due to the fact that a large fracture process zone is developed in the case of an SFRSCC specimen. Furthermore, the post-peak load-CMOD response of SFR-SCC specimens shows a softening type of behavior, indicating the presence of a fracture process zone.
2. The size of this fracture process zone is small due to the absence of a fiber interlock mechanism, which is present in FRSCC specimens.

To characterize the brittleness of the structural response quantitatively, various definitions of the so-called brittleness numbers have been proposed by Carpinteri (1982), Bazant (1984), and Hillerborg (1985). However, only Bazant's brittleness number β is independent of the geometrical shape of the specimen, which was justified experimentally in Bazant and Pfeiffer (1987). The brittleness number is capable of characterizing the nature of failure regardless of structure geometry. The Bazant's brittleness number is given by $\beta = d/d_0$, where d is depth of specimen and d_0 is an empirical constant. Table 4 shows the brittleness number for PSCC, GFRSCC and SFRSCC specimens.

The following observations are made from this plot:

1. As expected, the data points corresponding to the large size specimens fell near the vicinity of the LEFM curve, which shows that as the structure size increased its brittleness increased.

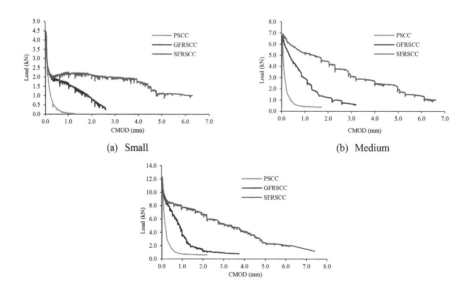

Figure 3.   Load versus CMOD curves for small, medium and large specimens.

Table 4. Brittleness number (β).

Specimen size	Type of SCC		
	PSCC	GFRSCC	SFRSCC
Small	2.317	1.153	1.046
Medium	4.635	2.305	2.091
Large	9.269	4.610	4.182

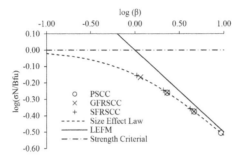

Figure 4. Experimental data following size effect law.

Table 5. Non-linear fracture parameters from size effect law.

Types of SCC	G_f (N.m/m^2)	c_f (mm)	K_{Ic} MPa$\sqrt{m}$	CTOD$_c$ δ_c (μm)	Size of FPZ l_0 (mm)	Average G_F (WOF method) N.mm/mm^2
PSCC	55.26	6.25	1.120	12.45	102.43	0.261
GFRSCC	62.16	12.57	1.214	18.31	120.45	0.723
SFRSCC	63.80	13.86	1.265	18.95	130.63	2.069

2. The specimens without fibers as well as specimens with fibers followed the trend of Bazant's size effect law. Hence both for SCC and FRSCC specimens, the size independent non-linear fracture parameters obtained from the size effect method can describe the fracture behavior better than other models.
3. For the plain specimens, the data points fell near to the LEFM criterion, which shows the increase in the brittleness. The addition of fibers to concrete reduced the brittleness number, so the concrete became ductile.

The non-linear fracture mechanics parameters, such as fracture energy, length of the process zone, critical mode I stress intensity factor, critical crack-tip opening displacement, size of fracture process zone for infinitely large specimens, and the crack growth resistance curve (R-curve) were determined using Bazant's size effect method (Shah and Chandra Kishen, 2010) and are presented in Table 5:
From these results, the following observations are made:

1. Addition of fibers to self-compacting concrete increased the size independent fracture energy (G_f). The size independent fracture energy was higher for SFRSCC specimen than for the GFRSCC and PSCC specimens.
2. The dimensions of the fracture process zone (c_f and l_0) were smaller for PSCC compared to GFRSCC and SFRSCC specimens. This is because in the case of GFRSCC and SFRSCC specimens, a fracture process zone is formed ahead of the crack-tip, and mechanisms

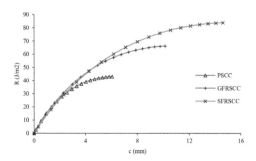

Figure 5. R-curve for all the specimens.

like micro-cracking, mortar interlocking with fibers, crack-branching, and crack-bridging take place. These mechanisms are either very small or absent in case of PSCC specimens. Furthermore, the addition of fibers in self-compacting concrete increased the dimensions of the fracture process zone, which is the reason for increase in the load-carrying capacity.

(3) Trends similar to fracture energy (G_f) and dimensions of fracture process zone c_f and l_o were observed for fracture toughness parameters (K_{Ic}) as well as crack-tip opening displacement ($CTOD_c$).

(4) Fracture energy is computed using the 'work of fracture' method as per the recommendation of RILEM, which is based on Hillerborg's fictitious crack model (Hillerborg et al., 1976). It is seen that average fracture energy by work of fracture (WOF) method of SFR-SCC is more, as compared to GFRSCC and PSCC specimens. It shows that more energy is required to break fiber-reinforced specimens.

The *R*-curve was computed for PSCC, GFRSCC and SFRSCC specimens as shown in Figure 5. Since the *R*-curve is the plot of resistance to crack propagation, it can be seen that for SFRSCC specimens, the resistance to crack propagation was greater than for GFRSCC and PSCC specimens. The addition of fibers to SCC increased the resistance to crack propagation of the specimens.

4   CONCLUSIONS

The following conclusions can be made from this study:

1. Insertion of glass and steel fibers into SCC resulted in better behavior of the concrete under crack propagation. Among the two fibers, the steel fibers showed better performance due to higher material strength as compared to glass fibers.
2. Fiber-reinforced self-compacting concrete has higher residual load-carrying capacity as compared to plain self-compacting concrete. It can be concluded that fiber-reinforced materials behave in a better way than materials without fibers.
3. The brittleness number (β) for plain specimens was higher than for the fiber-reinforced SCC specimens. It is also observed that the brittleness number of SFRSCC specimens was less than GFRSCC specimens because steel fibers have better mechanical properties compared to glass fibers. All the specimens have a brittleness number of between 0.1 and 10, which validates the applicability of the non-linear fracture mechanics parameters.
4. The non-linear fracture parameters, namely fracture energy (G_f), length of process zone (c_f), fracture toughness (K_{Ic}), and crack-tip opening displacement $CTOD_c$ (δ_c) increased due to the addition of fibers to self-compacting concrete. It is concluded that SFRSCC exhibits better crack resistance parameters than GFRSCC; however, both FRSCC show better performance than PSCC under cracking.
5. The crack growth resistance curve (R – *curve*) was developed for all the specimens. The resistance against crack growth (R) was higher for SFRSCC than for GFRSCC and PSCC specimens.

REFERENCES

Bazant, Z.P. & Pfeiffer, P.A. (1987). Determination of fracture energy from size effect and brittleness number. *ACI Materials Journal*, 463–480.

Bazant, Z.P. (1984.) Size effect in blunt fracture: Concrete, rock, metal. *Journal of Engineering Mechanics, 110*(4), 518–535.

Bazant, Z.P., Gettu, R. & Kazemi, M. (1991). Identification of nonlinear fracture properties from size effects tests and structural analysis based on geometry dependent R-curves. *International Journal of Rock Mechanics and Mining Sciences, 28*(1):43–51.

Carpinteri, A. (1982). Notch sensitivity in fracture testing of aggregative materials. *Engineering Fracture Mechanics, 16*(4), 467–481.

EFNARC. (2005). *The European guidelines for self-compacting concrete. Specification, production and use.* Retrieved from EFNARC website: http://www.efnarc.org/pdf/SCCGuidelinesMay2005.pdf

Gettu, R., Bazant, Z.P. & Karr, M.E. (1990). Fracture properties and brittleness of high-strength concrete. *ACI Materials Journal, 87*(6), 608–618.

Hillerborg, A. (1985). A theoretical basis of a method to determine the fracture energy GF of concrete. *Materials and Structures, 18*(106), 291–296.

Hillerborg, A., Moder, M. & Petersson, P.-E. (1976). Analysis of crack formation and crack growth in concrete by means of fracture mechanics and finite elements. *Cement and Concrete Research, 6,* 773–782.

Okamura, H. (1997). Self-compacting high-performance concrete. *Concrete International, 19*(7), 50–54.

RILEM. (1990). Draft recommendations—size-effect method for determining fracture energy and process zone size of concrete. *Materials and Structures, 23,* 461–465.

Shah, S.G. & Chandra Kishen, J.M. (2010). Nonlinear fracture properties of concrete-concrete interfaces. *Mechanics of Materials, 42,* 916–931.

Technology Drivers: Engine for Growth – Mahajan, Modi & Patel (Eds)
© *2018 Taylor & Francis Group, London, ISBN 978-1-138-56042-0*

Farm level land suitability assessment for wheat and mustard crops using geomatics: A case study of Badipur village in Patan district, Gujarat

Gautam Dadhich & Parul R. Patel
Institute of Technology, Nirma University, Ahmedabad, India

M.H. Kaluberne
Bhaskaracharya Institute for Space Applications and Geo-Informatics (BISAG), Gandhinagar, Gujarat, India

ABSTRACT: Agriculture is the major livelihood of the majority of the rural population in India. Farmers are forced to produce more food resources with limited land and in adverse climatic conditions in order to cater for the exponentially increasing demand for food. The assessment of the suitability of the crop land can provide a suitable solution for sustainable agricultural production. In this study, efforts have been made to carry out land suitability assessments at farm level using easily available evolution parameters. A methodology has been developed for land suitability assessment using soil health card data. The aim of this study is to determine the physical suitability of the land for producing major crops using a Spatial Multi-Criteria Evaluation (SMCE) approach, and to compare present land use against potential land use. The study was carried out in Badipur village in the Patan district of Gujarat state. The relevant parameters listed in the soil suitability manual of the National Bureau of Soil Survey and Land Use Planning (NBSS & LUP), such as Soil Texture, Soil pH, Soil Salinity, Soil Sodicity, Soil Depth, Soil Electrical Conductivity, Ground Water Quality, Soil Nutrients [Nitrogen (N), Phosphorous (P), Potassium (K)] and Organic Carbon, are considered for suitability analysis. The land evaluation criteria were developed based on farmers' knowledge, the suggestions of agriculture experts, literature, and available crop land suitability criteria. All data were stored in a Geographical Information System (GIS) environment and the maps were generated for each and every parameter. For Multi-Criteria Evaluation (MCE), the pairwise comparison matrix known as the Analytical Hierarchy Process (AHP) was applied, and the areas suitable for major crops were identified. Each farm in Badipur village was classified and mapped into four categories of suitability (Highly suitable, Moderately suitable, Marginally suitable, and Unsuitable) as per the Food and Agriculture Organization (FAO) (1976) qualitative evaluation. The study demonstrates that the spatial Multi-Criteria Decision Making Method (MCDM) technique is a powerful support system for agriculture resource management at farm level. The methodology used in this study can be utilized for implementing crop land suitability at farm level.

1 INTRODUCTION

India is a country of villages. The rural civilization in the country encompasses the core of Indian culture and also symbolizes the real India. The Indian government agencies have initiated many schemes intended to improve the living standards of the rural masses. Nearly 75 percent of the population are living in villages and are still reliant on agriculture. Around 43 percent of India's land territory is utilized for farming activity. The rural masses have several issues with regards to rural poverty, as most farmers in the villages are small land holders with diverse farming methods, which are highly risk susceptible. In addition, these farmers have limited access to advanced agricultural tools and techniques. Various studies have been made by researchers regarding agricultural development using advanced technologies. There is a need to penetrate advanced and efficient agricultural improvement techniques, such as precision farming, organic farming, and so on, in order to improve the current conditions of the rural masses. Precision farming may be adopted by incorporating land suitability assessment. Land suitability assessment is carried out to evaluate the suitability of farmland for a particular use, such as the cultivation or farming of a specific crop. The Food and Agricultural Organization (FAO, 1976) outlined a well-defined method for suitability analysis by considering various parameters. These parameters included Land Use Type, Soil Texture, Soil pH, Soil Salinity, Soil Sodicity, Soil Depth, Soil Drainage, Ground Water Quality and Soil Nutrients [Nitrogen (N), Phosphorus (P), and Potassium (K); NPK]. Many climatic and farming factors, such as Rainfall, Temperature, Humidity, Irrigation Type, and so on, also affect crop development (Dadhich et al., 2017). These factors can be utilized to assess any farmland for the suitability of a specific crop. The Geographical Information System (GIS) is a beneficial tool for the effective mapping of land suitability and evaluation (Collins et al., 2001). Using GIS tools, it is possible to generate a crop land suitability map for specific land by considering a thematic map of the various parameters affecting land suitability (Rossiter, 1996). In this study, the Analytical Hierarchy Process (AHP), which is a Multi-Criteria Decision Making (MCDM) method, is assimilated to obtain efficient results by analyzing various suitability parameters. AHP is used to determine the weight of every parameter based on their relative importance. The weightage analysis has been combined with the GIS technology to generate an accurate and reliable crop land suitability map. FAO guidelines for the land suitability evaluation (FAO, 1976) are widely used for such types of evaluation. These guidelines classify the land into different classes, which vary from Suitable (S) to Not suitable (N), where S denotes land suitable for the cultivation of a selected crop, and N denotes that the land qualities do not meet the required criteria for the cultivation of a selected crop (FAO, 1976). Various attempts have been made to carry out land suitability evaluation at district level or regional level, but very few efforts have been made to evaluate land suitability at a micro level. This study aims to carry out land suitability analysis for a wheat and mustard crop for Badipur village. Maddahi et al. (2014) and Agidew (2015) emphasized carrying out land suitability for economically important crops or cash crops for the sustainable management of agricultural resources. Wheat is one of the major food crops and mustard is a major cash crop, and both are vital parts of the food industry. Earlier studies (Mendas et al., 2012; Mustafa et al., 2011; Shahbazi et al., 2009; Wang et al., 2011) on crop land suitability evaluation depict that wheat and mustard have been taken as a primary crop for similar studies. Major factors affecting the suitability of wheat and mustard crops are land quality factors, such as Land Use Type, Soil Texture, Soil pH, Soil Salinity, Soil Sodicity, Soil Depth, Soil Electrical Conductivity (EC), Ground Water Quality and Soil Nutrients [Nitrogen (N), Phosphorus (P), and Potassium (K); NPK] (Ashraf & Normohammadan, 2011; Ashraf, 2010; Bidadi et al., 2015; Haldar, 2013; Mokarram et al., 2010; Rabia e& Terribile, 2013). The objective of this study is to analyze the spatial distribution of wheat and mustard crops in Badipur village in the Patan district in order for farmers, decision makers and policy makers to cater for the needs of precision farming and sustainable agricultural development.

2 STUDY AREA

Badipur is a village panchayat located in the Patan district of Gujarat state, India. Badipur village lies at a central coordinate of 23.821 N latitude, 72.013 E longitude, covering an area of 629.71 Ha. It is situated near the State Highway (SH) 10 and is about 16 km 13 KM west from District headquarters, Patan. The total village population is 949 and the total number of households is 194 (Census, 2011). The total literacy of the village is 66.50 percent. Figure 1 shows a survey number map of Badipur village. Badipur village is comprised of 488 farms along with gametal and a river.

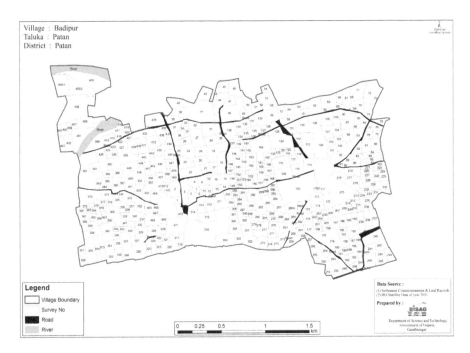

Figure 1. Survey number map of Badipur village in Patan district.

3 METHODOLOGY

To achieve the objectives of the study, both primary and secondary raster and vector data were used. The methodology that was used to evaluate the land suitability for different Land Utilization Types (LUTs) was based on FAO guidelines (FAO, 1976). In the present study, there were nine major parameters that affected crops, such as Land Use Type, Soil Texture, Soil pH, Soil Salinity, Soil Sodicity, Soil Depth, Soil Electrical Conductivity (EC), Ground Water Quality and Soil Nutrients [Nitrogen (N), Phosphorus (P), and Potassium (K); NPK], and thematic maps were prepared for every selected parameter. The methodology adopted for this study is presented in Figure 2.

A Spatial Multi-Criteria Decision Making (SMCDM) technique has been used as the decision making technique (Perveen et al., 2013). SMCDM involves data input, user priorities, manipulation and interpretation of information using well-defined decision making criteria (Jankowski, 1995). A weighted overlay technique (Saaty's method) is one of the approaches that is usually adopted for applying a standardized measurement scale to diverse and dissimilar inputs.

In this study, a weighted overlay technique has been used for crop land suitability analysis. The weight for each and every parameter is derived from the analytical hierarchy process (AHP). The suitable factors for cultivation and the respective favorable conditions and criteria are adopted from NBSS LUP, Nagpur (India) Manual (Naidu et al., 2006). The most favorable condition is termed as "Highly suitable" and presented as "S1". The "Moderately suitable" conditions are termed as "S2", which are areas that are potentially suitable for wheat crops with some improvements and changes in conditions. The "Marginally suitable" conditions are termed as "S3", which are areas that are less suitable for wheat crops and with unfavorable conditions. The areas that are not at all suitable for wheat crop cultivation are termed as "N".

The various parameters used in this study have different levels of relative importance, so it is necessary to assign proper weightages to the parameters. In this study, the relative importance of each and every parameter is considered by determining the weightage factor using Saaty's (AHP) method (Saaty, 1977). Table 1 shows the pair wise comparison matrix of

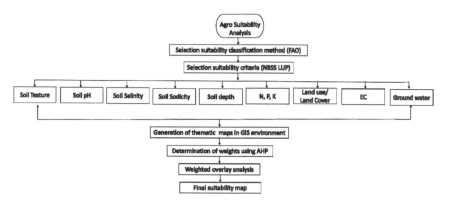

Figure 2. Methodology flow chart for crop land suitability.

Table 1. Comparison matrix of wheat land suitability.

Parameters	Soil nutrients	Soil salinity	Soil pH	Soil texture	Soil depth	Soil EC	OC	Soil sodicity	Ground water quality
Weights	0.301	0.219	0.155	0.109	0.078	0.053	0.037	0.025	0.018

wheat land suitability. The final suitability map is derived in a GIS environment by obtaining weight from Saaty's method. The final suitability in the weighted overlay technique is derived by the following equation.

4 RESULTS AND DISCUSSION

Soil PH is responsible for the availability or solubility of phytotoxity elements for crops and consequently specifies the soil suitability for a particular crop (Halder 2013). The soil PH of the study area ranged from 6.48 to 8.65. The reclassified soil PH map shows that 49.17 percent (pH 6.5 to 7.0) and 7.51 percent (pH 7.0 to 7.5) of the study area has a highly suitable soil PH. The reclassified soil depth map reveals that 82.61 percent of the study area has very deep soil. Most of the physical characteristics of the soil depend upon the texture class (Mustafa et al., 2011). There are three textural classes in the study area, namely, coarse loamy (75.89 percent), fine (8.16 percent) and fine loamy (9.83 percent). Soil fertility is the most important soil characteristic and it has a great impact on crop productivity. The status of Phosphorus, Nitrogen and Potassium is Low (74.22 percent), Medium (44.24 percent) and High (56.69 percent) respectively in Badipur village. Soil nutrients are classified in terms of Low, Medium and High. These maps show that the study area has low phosphorus, medium nitrogen and high potassium. The soil salinity is a major constraint for the cultivation of wheat and mustard crops. Soil salinity is measured by Electric Conductivity (EC), which indicates the total concentration of soluble salts in the soil. In the root zone, the presence of soil with substantial amounts of natural salt leads to a reduction of soil water, which is extracted by plants and may cause a nutrient imbalance. Ashraf (2010) and Kalogirou (2002) used salinity as an important criterion to carry out crop land suitability analysis. The percentage distribution of land that is Highly-Saline, Moderately-Saline and other areas is 18.30 percent, 35.83 percent and 5.80 percent respectively. Exchangeable Sodium Percentage (ESP) or sodicity affects the productivity of wheat and mustard crops by reducing the water that is available to the plant root. Sodicity also disrupts soil structure, which tends to decrease pore space and reduce permeability and water availability in the root zone (Nelson, 2001). Soils

Figure 3. Overall suitability map for wheat and mustard crops in Badipur village.

having an ESP of greater than six are designated as sodic soils. Highly sodic, Marginally sodic and other types are 4.82 percent, 89.37 percent, and 5.80 percent respectively. The type and characteristics of these salts are influenced by the source of the groundwater and the soil strata through which it flows. A higher quantity of soluble salts may be harmful for many crops. The results of the water samples were analyzed for pH, TDS, EC (electrical conductivity) and Residual Alkalinity (RSC) in 2010 (July), and were collected from the State Water Data Centre (SWDC), Gandhinagar. Four ground water quality zones are categorized based on the value of EC and RSC, as follows (Kalubarme et al., 2012). The ground water quality for wheat cultivation is classified as Good (EC < 2 dS/m, RSC < 2 me/L), Marginal saline (EC 2–4 dS/m, RSC < 2.5 me/L), Marginal sodic (EC < 2–4 dS/m, RSC < 2.5–4 me/L) and Poor (EC > 4 dS/m, RSC > 4 me/L) (Goosen & Shayya, 1999).

The weighted maps/layers were combined by performing the weighted overlay in a GIS environment. Finally, the crop suitability map for wheat and mustard was prepared as shown in Figure 3. The distribution of wheat acreage under various suitability classes was, Highly suitable: 15.74 percent (99.13 Ha), Moderately suitable: 7.78 percent (48.97 Ha), Marginally suitable: 24.58 percent (154.76 Ha) and Non-suitable: 51.90 percent (326.84 Ha). The distribution of mustard acreage under various suitability classes was, Highly suitable: 26.41 percent (166.31 Ha), Moderately suitable: 44.24 percent (278.58 Ha), Marginally suitable: 21.92 percent (138.04 Ha) and Non-suitable: 7.43 percent (46.77 Ha). Results indicate that mustard crops are relatively more suitable than wheat crops in Badipur village.

5 CONCLUSIONS

– The land suitability assessment for wheat and mustard crops has been conducted in Badipur village, Patan district, Gujarat state, in order to help the decision makers and farmers, as well as the agricultural development planners.
– The study focuses on this issue because land suitability analysis plays a vital role in achieving the best utilization of the available and limited land resources.
– The various factors used to analyze the suitability of the wheat crop in the study area were N, P, K, EC, Organic Carbon (OC), pH, Soil Texture, Soil Salinity, Soil Sodicity, Soil Depth and Ground Water Quality.
– These factors were ranked based on the FAO land evaluation system using FAO (1976) guidelines. The weighted overlay technique has been used for crop land suitability analysis. The weight for each and every parameter is derived from the AHP process.
– In this study, the relative importance of each and every parameter is considered by determining the weightage factor using Saaty's method and is based on the weight derived from Saaty's method. The final suitability map is derived in a GIS environment.
– This study provides information at a local level that may be utilized by planners and farmers for improving staple food production.

- The distribution of wheat acreage under various suitability classes was, Highly suitable: 15.74 percent (99.13 Ha), Moderately suitable: 7.78 percent (48.97 Ha), Marginally suitable: 24.58 percent (154.76 Ha) and Non-suitable: 51.90 percent (326.84 Ha).
- The distribution of mustard acreage under various suitability classes was, Highly suitable: 26.41 percent (166.31 Ha), Moderately suitable: 44.24 percent (278.58 Ha), Marginally suitable: 21.92 percent (138.04 Ha) and Non-suitable: 7.43 percent (46.77 Ha).
- Results indicate that mustard crops are relatively more suitable than wheat crops in Badipur village.

REFERENCES

Agidew, A.A. (2015). Land suitability evaluation for sorghum and barley crops in South Wollo Zone of Ethiopia. *Journal of Economics and Sustainable Development, 6*(1), 14–25.

Ashraf, S. & Normohammadan, B. (2011). Qualitative evaluation of land suitability for wheat in Northeast-Iran using FAO methods. *Indian Journal of Science and Technology, 4*(6), 703–707.

Ashraf, S. (2010). Land suitability analysis for wheat using multicriteria evaluation and GIS method. *Research Journal of Biological Sciences, 5*(9), 601–605.

Baniya, M.S.N. (2008). *Land suitability evaluation using GIS for vegetable crops in Kathmandu valley/ Nepal* (Doctoral dissertation). Humboldt-Universität zu Berlin.

FAO. (1976). *A frame work for land evaluation* (Soils Bulletin No. 32). Food and Agriculture Organisation of the United Nations, Rome.

Goosen, M.F. & Shayya, W.H. (1999). *Water management, purification, and conservation in arid climates: Water management (Vol. 1).* CRC Press.

Halder, J.C. (2013). Land suitability assessment for crop cultivation by using remote sensing and GIS. *Journal of Geography and Geology, 5*(3), 65.

Kalubarme, M.H., Chauhan, D.S. & Saroha, G.P. (2013). Cotton suitability analysis using geoinformatics in Punjab State, India. *Asian Journal of Geoinformatics, 12*(4), 27–36.

Maddahi, Z., Jalalian, A., Zarkesh, M.M.K. & Honarjo, N. (2014). Land suitability analysis for rice cultivation using multi criteria evaluation approach and GIS. *European Journal of Experimental Biology, 4*(3), 639–648.

Mendas, A. & Delali, A. (2012). Integration of multicriteria decision analysis in GIS to develop land suitability for agriculture: Application to durum wheat cultivation in the region of Mleta in Algeria. *Computers and Electronics in Agriculture, 83*, 117–126.

Mokarram, M., Rangzan, K., Moezzi, A. & Baninemeh, J. (2010). Land suitability evaluation for wheat cultivation by fuzzy theory approach as compared with parametric method. *Proceedings of the international archives of the photogrametry, remote sensing and spatial information sciences, 38(part II)* (pp.1440–145).

Mustafa, A.A., Singh, M., Sahoo, R.N., Ahmed, N., Khanna, M., Sarangi, A. & Mishra, A.K. (2011). Land suitability analysis for different crops: A multi criteria decision making approach using remote sensing and GIS. *Researcher, 3*(12), 1–24.

Naidu, L.G.K., Ramamurthy V., Challa O., Hegde R. & Krishnan P. (2006). *Manual, soil-site suitability criteria for major crops.* National Bureau of Soil Survey and Land Use Planning, ICAR.

Patan District Punchayat. [Online] patandp.gujarat.gov.in/patan/images/krushi-mahotsav–2013.pdf (Accessed 16 November 2016).

Perveen, S., Arsalan, M.H., Siddiqui, M.F., Khan, I.A., Anjum, S. & Abid, M. (2013). GIS-based multi-criteria model for cotton crop land suitability: A perspective from Sindh Province of Pakistan. *FUUAST Journal of Biology, 3*(1), 31.

Rabia, A.H. & Terribile, F. (2013). Introducing a new parametric concept for land suitability assessment. *International Journal of Environmental Science and Development, 4*(1), 15.

Rossiter, D.G. (1996). A theoretical framework for land evaluation. *Geoderma, 72*(3), 165–190.

Saaty, T.L. (1977). A scaling method for priorities in hierarchical structures. *Journal of Mathematical Psychology, 15*(3), 234–281.

Sys, C., Van Ranst, E. & Debaveye, J. (1991). *Land evaluation. Part II: Methods in land evaluation.* Agricultural Publications, GADC, Brussels, Belgium.

Experimental investigation of reinforced concrete infill frames under pseudo dynamic loading

J.M. Popat, T.H. Bhoraniya, S.P. Purohit & S.A. Sharma
Department of Civil Engineering, Institute of Technology, Nirma University, Gujarat, India

ABSTRACT: Reinforced Concrete (RC) bare frames show flexible behavior when subjected to a lateral load. The addition of infill into the RC frame alters its behavior, and further complexity is added when different materials are used for the infill. During a seismic event, RC frames with infill are subjected to lateral cyclic loading. Thus, it is important to study the behavior of RC frames with different types of infill when subjected to pseudo dynamic loading. The present paper considers three scaled RC frames, comprising of a bare RC frame and RC frames with two types of infills, namely, Autoclaved Aerated Concrete blocks (AAC) and Fly Ash Bricks (FAB). All of the scaled RC frames are subjected to half cyclic pseudo dynamic loading through a lateral load test assembly. The behavior of the bare RC frame and the RC frames with infills is studied in terms of Deformation Capabilities, Lateral Stiffness, Strain Capabilities, Energy Dissipation, and Types of Failures. It has been found that the RC frame with the FAB infill yields the maximum lateral load, followed by the RC frame with AAC blocks infill and the bare RC frame, and inversely for the lateral displacement. Energy dissipation is highest for the RC frame with FAB infill, while it is least for the RC frame with AAC blocks infill. In addition, the maximum stiffness degradation is observed in the RC frame with AAC blocks.

1 INTRODUCTION

In India, most of the multistory buildings consist of Reinforced Concrete (RC) moment resisting frames. Gravity loads (Dead and Live) do not cause much of a problem, but the lateral loads, due to wind or earthquake loads, are important and they need special consideration during the analysis and design of RC buildings. In design practices generally, the infill walls are considered to be secondary load carrying members, while the primary loads are carried by slabs, beams and columns. Thus, the design of infill walls is neglected because of the complex interaction between the frame and the infill. Other factors that affect the behavior of infill frames without infill are local skills, workmanship and materials. In the present study, the behavior of a bare RC frame and RC frames infilled with AAC blocks and FAB are studied under pseudo dynamic lateral loading to simulate a seismic event scenario. Several parameters, such as load, lateral deformation, lateral stiffness, energy dissipation and types of failure, are given primary weightage in order to study the behavior of the RC frame, both with and without the infills of different materials, and parameters such as deflection, stiffness and energy dissipation are carried out.

2 LITERATURE REVIEW

Masonry infill walls remarkably increase the initial stiffness of RC frames. Infill walls, being a stiffer component of the RC frame, attract the major portion of the lateral seismic shear forces and thereby reduce the demand on the RC frame members (Kaushik et al., 2005). Masonry infill adds lateral stiffness to the RC building and the structural load transfer mechanism is changed from frame action to predominant truss action for the RC frame (Murthy & Jain, 2000). The experimental results indicated that infill panels can significantly improve the performance of RC frames. The lateral loads developed by the infilled RC frame specimens were always higher than that of the bare RC frame (Mehrabi et al., 1996). Infilled RC frames exhibit significantly higher ultimate strength, residual strength, and initial stiffness when compared to a bare RC frame (Al-Chaar et al., 2002). Full scale single room masonry buildings of different typologies, such as unreinforced, reinforced and confined masonry, that were tested under cyclic quasi static loads showed failure modes, such as: sliding of brick at the mortar-unit interface; discrete shear cracks in masonry walls; crushing of masonry units; and bending of reinforcement in tie columns (Chourasia et al., 2016). A new finite element technique to determine the influence of brick infill panels for RC frames under lateral loading has also been developed. An analytical study shows that an infill panel helps to decrease the shear force for the column of the RC frame when the infill is fully in contact with the RC frame (Asteris, 2003). The structural behavior of the RC frame with the masonry wall when subjected to in-plane monotonic loading was also investigated. The investigation confirms that the presence of infill increases the stiffness of the RC frame and that infill masonry is highly influenced by the failure of mortar (Tzeng et al., 1999).

3 PROBLEM FORMULATION

A ground plus two story (G+2) RC building of plan dimensions 16 m × 16 m, and having 3.2 m story height, is considered for the present study (Sharma et al., 2015). The RC building is analyzed and designed using prevailing codal stipulation (IS: 456–2000).

The prototype building is scaled down by 1/3rd scale in order to conduct the experimental study, in line with the literature review. The 1/3rd scale RC building is analyzed to identify the maximum bending moment occurring due to governing load combinations following codal stipulation (IS:456–2000). The sections are designed for the RC frame using analysis results, which were found to be governed by the minimum reinforcement requirement of codal stipulation

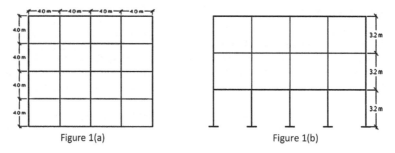

Figure 1(a)   Figure 1(b)

Figure 1. (a) Plan of prototype G+2 RC building. (b) Elevation of prototype G+2 RC building.

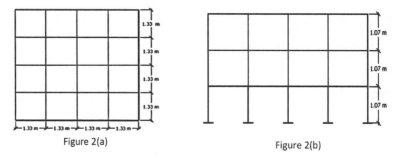

Figure 2(a)   Figure 2(b)

Figure 2. (a) Plan of scaled down G+2 RC building. (b) Elevation of scaled down G+2 RC building.

100

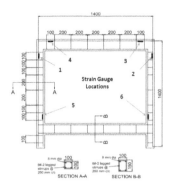

Figure 3. Reinforcement cage with strain gauges.

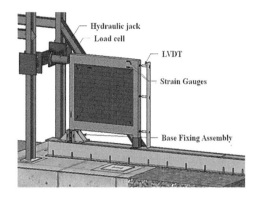

Figure 4. Test setup.

(IS: 456–2000). Figure 1(a) and Figure 1(b) show the plan of a prototype G+2 story RC building and the elevation of the RC building, respectively. The plan and elevation of the scaled down G+2 story RC building are shown in Figure 2(a) and Figure 2(b). The dimensions of the scaled down RC frame components, columns and beams are 100 mm × 100 mm and 100 mm × 150 mm, respectively. The scaled down RC frame without any infill (bare) and with infill walls are cast using the M25 grade of concrete in order to conduct the experimental studies.

4 EXPERIMENTAL PROGRAM

An experimental program was developed, which included characterization of the constituent materials, experimental test setup and instrumentation, and testing of the RC frames, in order to study the behavior of the RC frames, both with and without infills, under a mortar cube yields average compressive strength of 11.17 N/mm². The average concrete cube strength for concrete is 30.32 N/mm² and the yield strength of reinforcing steel is 453.79 N/mm².

4.1 *Test setup and instrumentation*

All of the RC frames are subjected to lateral pseudo dynamic half cyclic loading through a hydraulic jack (250 kN capacity) and a load cell attached to the top of the RC frame, as shown in Figure 3. The lateral displacement of the RC frame at three locations, top, middle and bottom, are measured through LVDT (200 mm & 100 mm capacity). In addition, six strain gauges are attached on the reinforcing bars of the RC frame at the points of critical zones. A 16 channel data analyzer system is used to capture load and displacement, as well as strain. Figure 3 shows the reinforcement cage with the locations of the strain gauges. Figure 4 shows the complete experimental setup of the RC frame subjected to the lateral load.

5 RESULTS AND DISCUSSIONS

All of the RC frames are tested under the pseudo dynamic half cyclic lateral load. The lateral load, lateral displacement, and strain are measured for each of the RC frame specimens. The initial stiffness, stiffness degradation, energy dissipation, maximum lateral displacement, strain, first crack load, and ultimate load for each of the RC frame specimens are recorded. Failure modes for each of the RC frame specimens are also critically observed and recorded. All of the RC frames are subjected to a pseudo dynamic half cycle load in different steps of loading.

5.1 *Testing of bare RC frame (RCF-1)*

RCF-1 is tested with a first cycle of 0 → 4 kN, wherein a first shear crack is observed at 3.5 kN load at the beam-column junction at the bottom of the loading end. The lateral

101

displacement of the RC frame at the top end is found to be 28.84 mm. An offset of 9.72 mm for displacement on the unloading of the RC frame confirms the non-linear behavior in the first cyclic loading itself. RCF-1 is applied to a second cycle of 0 → 6 kN loading, wherein further widening of the shear cracks are observed. The spalling of concrete due to compression at the top of the windward column and at the bottom of the loaded column is also observed. The lateral displacement of RCF-1 at the end of the second cycle is found to be 72.88 mm. A permanent offset of displacement of about 40.04 mm is observed. RCF-1 is applied to a third cycle of 0 → 6.6 kN loading, wherein a lateral displacement of 157.32 mm is observed. A permanent offset of displacement of 99.80 mm is observed after unloading the RC frame. Figure 5(a), Figure 5(b) and Figure 5(c) show the testing of the bare RC frame, lateral load vs. lateral displacement plot, and lateral load vs. strain plot for the strain gauge locations 1 & 2 of the RC frame, respectively. The total energy dissipated up to the ultimate failure of the bare RC frame is found to be 667.22 N-m (J). The stiffness degradation is calculated, from Figure 5(b), as 69.75% for the third cycle, using secant stiffness. Figure 5(c) shows the strain in the steel due to compression and tension.

5.2   Testing of the RC frame with AAC blocks infill (RCF-2)

RCF-2 is tested with a first cycle of 0 → 9 kN. At the application of the first cycle there is a spalling of mortar from the infill and RC frame joints. The lateral displacement of the RC frame at the top end is found to be 3.05 mm. RCF-2 is applied to a second cycle of 0 → 12 kN loading, wherein debonding of the mortar from the infill is observed. The lateral displacement of RCF-2 at the end of the second cycle is found to be 6.0 mm. A permanent offset of displacement of about 1.6 mm is observed. RCF-2 is applied to a third cycle of 0 → 15 kN loading, wherein a lateral displacement of 18.56 mm is observed. RCF-2 is applied to a fourth cycle of 0 → 20 kN loading, wherein a lateral displacement of 45.92 mm is observed. Figure 6(a) shows the testing of the RC frame with AAC block infill, while Figure 6(b) shows the lateral load vs. lateral displacement plot, and Figure 6(c) displays the lateral load vs. strain plot for the strain gauge locations 1 & 2 of the RC frame. The total energy dissipated up to the ultimate failure of the bare RC frame is calculated to be 540.635 N-m (J). The lateral stiffness degradation is calculated from Figure 6(b) for each cycle using secant stiffness and shows a maximum lateral stiffness degradation of 85.22%, which is calculated after the final cycle of loading. Figure 6(c) shows the strain in the steel due to compression and tension.

5.3   Testing of the RC frame with fly ash bricks infill (RCF-3)

RCF-3 is tested with a first cycle of 0 → 10 kN. The lateral displacement of the RC frame at the top end is found to be 4.3 mm. RCF-3 is applied to a second cycle of 0 → 20 kN loading, wherein the spalling of mortar from the infill is observed. The lateral displacement of RCF-3 at the end of the second cycle is found to be 15.0 mm. A permanent offset of displacement of about 5.06 mm is observed. RCF-3 is applied to a third cycle of 0 → 25 kN loading, wherein

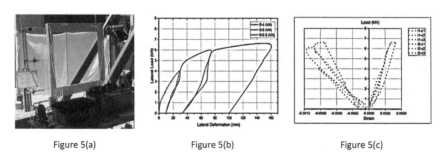

Figure 5(a)   Figure 5(b)   Figure 5(c)

Figure 5.   (a) Testing of bare RC frame. (b) Lateral load vs. lateral deformation plot of bare RC frame. (c) Lateral load vs. strain at locations 1 & 2 for bare RC frame.

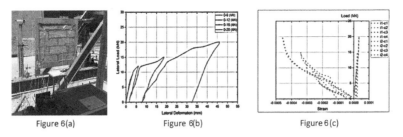

| Figure 6(a) | Figure 6(b) | Figure 6(c) |

Figure 6.   (a) RC frame with AAC blocks infill. (b) Lateral load vs. lateral deformation for RCF-2. (c) Lateral load vs. strain for RCF-2.

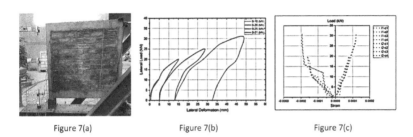

| Figure 7(a) | Figure 7(b) | Figure 7(c) |

Figure 7.   (a) RC frame with fly ash bricks infill. (b) Lateral load vs. lateral deformation for RCF-3. (c) Lateral load vs. strain for RCF-3.

Table 1.   Comparison of parameters for bare and infill RC frames.

Sr. no.	Specimen	Maximum lateral displacement (mm)	Maximum failure load (kN)	Energy dissipation (N·m) (After final cycle of loading)	Total stiffness degradation (%) (At the failure load w.r.t. initial stiffness)
1	RCF-1	157.32	6.60	667.22	69.75
2	RCF-2	45.92	20.00	540.63	85.22
3	RCF-3	48.52	31.00	825.75	72.52

a lateral displacement of 28.50 mm is observed. RCF-3 is applied to a fourth cycle of 0 → 31 kN loading, wherein a lateral displacement of 48.52 mm is observed. Figure 7(a) shows the testing of the RC frame with fly ash bricks infill, while Figure 7(b) shows the lateral load vs. lateral displacement plot, and Figure 7(c) displays the lateral load vs. strain plot for the strain gauge locations 1 & 2 of the RC frame. The total energy dissipated up to the ultimate failure of the bare RC frame is calculated to be 825.75 N-m (J). The lateral stiffness degradation is calculated from Figure 7(b) for each cycle using secant stiffness and shows a maximum lateral stiffness degradation of 72.52%, which is calculated after the final cycle of loading. Figure 7(c) shows the strain in the steel due to compression and tension.

   The comparisons of various parameters, such as maximum lateral displacement, maximum failure load, energy dissipation, and total stiffness degradation, are given in Table 1. It is evident from Table 1 that the RC frame with FAB infill yields the highest lateral load and higher energy dissipation. The highest lateral stiffness degradation is found to be in the RC frame with AAC blocks.

6   SUMMARY AND CONCLUSIONS

A bare RC frame and RC frames with different types of infill are considered for the present study. The RC frames are subjected to half cyclic pseudo dynamic loading and tested on a

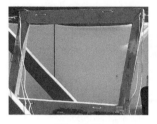

| Figure 8 (a) | Figure 8 (b) | Figure 8 (c) |

Figure 8.   Failure pattern of (a) RCF1 (b) RCF2 and (c) RCF3.

lateral load assembly test setup. The behavior of the bare and infilled RC frames is studied in terms of lateral stiffness, lateral displacement, failure load, energy dissipation, and stiffness degradation. The following conclusions are derived based on the study.

– The RC frames with fly ash bricks infill and with AAC blocks infill yield 7 times and 3.03 times higher lateral load, respectively, when compared to the bare RC frame. This proves the contribution of the infill material toward load sharing.
– The reduction in lateral displacement for the RC frame with FAB infill is 70.81%, while for the RC frame with AAC blocks it is 69.16%, vis-à-vis the bare RC frame. This is due to the stiffness contribution by the infill material.
– The energy dissipation for the RC frame with AAC blocks is the least, followed by the bare RC frame and the RC frame with fly ash bricks infill. This is due to the fact that the RC frame with FAB and the bare RC frame undergo major cracking under lateral load. This may help in increasing the damping value of the RC frame with infill.
– The total stiffness degradation for the RC frame with AAC blocks and the RC frame with fly ash bricks infill is 22.18% and 3.98% higher, respectively, when compared to the bare RC frame. This indicates that the RC frame with AAC blocks degrade stiffness heavily on set of cracking.
– While the bare RC frame shows conventional failures at each tension and compression zone, the RC frame with AAC blocks shows a failure of the mortar joint and a block unit failure. However, the RC frame with fly ash bricks undergoes a mortar joint failure followed by a stepped type of failure.

REFERENCES

Al-Chaar, G., Issa, M. & Sweeney, S. (2002). Behaviour of masonry-infilled nonductile reinforced concrete frames. *Journal of Structural Engineering*, *128*(8), 1055–1063.
Asteris, P.G. (2003). Lateral stiffness of brick masonry infilled plane frames. *Journal of Structural Engineering*, *129*(8), 1071–1079.
Chourasia, A., Bhattacharyya, S.K., Bhandari, N.M. & Bhargava, P. (2016). Seismic performance of different masonry buildings: Full-scale experimental study. *Journal of Performance of Constructed Facilities*, *30*(5), 1–12.
Indian Standard. (2000). IS:456–2000. *Plain and reinforced concrete—Code of practice*. New Delhi, BIS.
Kaushik, H.B., Rai, D.C. & Jain, S.K. (2006). Code approaches to seismic design of masonry-infilled reinforced concrete frames: A-state-of-the-art review. *Earthquake Spectra*, *22*(4), 961–983.
Mehrabi, A.B., Shing, P.B., Schuller, M.P. & Noland, J.L. (1996). Experimental evaluation of masonry-infilled RC frames. *Journal of Structural Engineering*, *122*(3), 228–237.
Murthy, C.V.R. & Jain, S.K. (2000). Beneficial influence of masonry infill walls on seismic performance of RC frame buildings. *Proceedings of the 12th World Conference on Earthquake Engineering*, *12th WCEE, 30 January–4 February 2000, Auckland, New Zealand* (pp. 190–196). New Zealand: Upper Hutt.
Sharma, S., Purohit, S.P., Patel, P.V. & Bhoraniya, T.H. (2015). Behaviour of reinforced concrete infill panels under lateral load. *International Journal of Research in Engineering and Technology*, *4*(13), 183–194.
Tzeng, J.C., Liou, Y.W. & Chiou, Y.J. (1999). Experimental and analytical study of masonry infilled frames. *Journal of Structural Engineering*, *125*(10), 1109–1117.

Technology Drivers: Engine for Growth – Mahajan, Modi & Patel (Eds)
© *2018 Taylor & Francis Group, London, ISBN 978-1-138-56042-0*

Effect of randomly distributed jute fibers on design of soil subgrade

H.M. Rangwala, H.M. Kamplimath, A. Kanara, M. Ratlami, M. Kothari & M. Khatri
Department of Civil Engineering, Institute of Technology, Nirma University, Ahmedabad, India

ABSTRACT: In this era of rapid industrialization and urbanization, road networks play a pivotal role in the movement of goods from one point to another. Hence, good roads are necessary for any type of infrastructure development and also for the development of a nation as a whole. It is often seen that, in the developing countries, the structural and functional performance of a pavement deteriorates significantly even before the designed service life of the pavement is reached. The most common cause for such pavement deterioration is the failure of the subgrade. Hence, a sound understanding of the properties of soil subgrade are essential for designing a pavement. Sometimes roads need to be built in regions where the subgrade soil strength is low. The properties of this soil can be improved by the inclusion of reinforcements in the subgrade soil. Different investigators have studied the effects of reinforcement on the physical properties of soil.

In this paper, the effect of the inclusion of jute fibers as reinforcement on the various properties, i.e. Optimum Moisture Content (OMC), Maximum Dry Density (MDD), and California Bearing Ratio (CBR) Values, are studied. The study also includes the effects of percentage reinforcement on these properties. The effects of reinforcement have been observed by designing a pavement subgrade for a known condition. The outcome of the design shows that the inclusion of reinforcements decreases the thickness of the subgrade layer.

1   INTRODUCTION

Ground improvement has been the primary application of many geotechnical construction techniques, permitting construction on weak soils by changing their characteristics. Soil mixing increases shear strength and reduces the compressibility and permeability of soft soils. Rigid inclusions reduce settlement and increase the bearing capacity of weak underlying stratum. At present, synthetic geotextiles are extensively used throughout the world for the protection of the banks and beds of waterways, strengthening of roads, stabilization of embankments, management of slopes, consolidation of soft soil, and other soil-related engineering applications. Geotextiles are a type of technical textile that are used in or on soil to improve its behavior and performance. Geotextiles were first developed in developed countries by making use of man-made polymers. Naturally available materials can also be used in the improvement of soil properties, for example bamboo fiber, coir fiber, and jute fiber.

Jute has been widely used in the textile industry for various applications. It is mostly used to make 'sackcloth' and carpet backing fabric. Jute is extracted from the inner bark of the genus Corchorus plant. Jute grows in India, Bangladesh, and China. It is available in various color varieties, from yellow to brown to dirty gray. The best quality fibers are smooth and feel soft. The size of an average single jute fiber is 2.5 mm in diameter and 12 mm in length, and weighs from 1.9 to 2.2 Tex. It is also known as geojute when it is used for ground improvement.

In this paper, the effects of the inclusion of randomly distributed geojutes on compaction parameters i.e. Optimum Moisture Content (OMC) and Maximum Dry Density (MDD), and strength parameters, i.e. CBR values, are considered for the study. The effect of geojute on the design of the subgrade is also incorporated in the study.

2 LITERATURE REVIEW

Chakrabarti and Bhandari (2004) reported on an experimental study on horizontally layered woven and non-woven geojute using a field CBR test and a plate load test on low strength highly compressive soil. The bearing capacity of the layer was found using a plate load test and the CBR value was found using field CBR at different layers. The model tests were done using a pit of size 1 m × 1 m × 0.6 m. The results were presented in the form of a Bearing Capacity Ratio (BCR), which is the ratio of the bearing capacity of a reinforced soil layer to that of an unreinforced layer. The BCR varied from 1.07 to 2.23 for different cases. The thickness of the base course of the flexible pavement was found using the results of the CBR test, and the thickness was found to be reduced by 32% in the case of woven geojute and 20% for non-woven geojute.

Sahu et al. (2004) carried out model testing to determine the aging effects on a geojute reinforced soil bed under cyclic load. Based on the test results, they concluded that the total and permanent settlement of the footings were found to reduce with the aging of the soil under various load cycles. An increase in cohesion and a reduction in moisture content were found to increase with the increasing aging period of the test bed. The overall performance of the test bed was found to improve with the aging of the soil, even after the complete biodegradation of the geojute.

Gosavi et al. (2004) reported that soil can be reinforced with low cost materials, such as the natural fibers of jute, coir, etc. CBR tests were performed by mixing geojute with silty sand. The CBR value was enhanced by about 50% from that of unreinforced soil. On a test on black cotton soil, they showed that the value of OMC increases and the value of MDD decreases with increased quantities of geojute fiber. Soaked CBR test results showed a considerable increase in the CBR value for black cotton soil when it was reinforced. The rate of increase in the CBR value with 2% addition of fibers is small, and the absolute value of CBR decreases further with the addition of more fibers.

3 EXPERIMENTAL STUDY

A series of experimental investigations were planned in order to study the influence of jute fiber reinforcements on various properties of soil. Siltys and from a nearby site was procured for the investigation. The particle size distribution curve is given in Figure 1. In this test, different fiber contents were considered by analyzing its effect strength. Five different fiber contents, i.e. 0% (unreinforced soil), 0.2%, 0.4%, 0.6%, and 0.8%, were considered in this study. A standard proctor test was done to analyze the effect of the fiber on the maximum dry density and optimum moisture content of the soil. A CBR test was then done to check the effect of the fiber content on the strength of the subgrade soil. Figure 2 shows jute fibers and a sol jute mixture.

3.1 *Standard proctor test*

A standard proctor test was performed in accordance with the specifications in IS 2720 (Part VII) – 1980. A steel mold with a volume of 1000 cm³ was used for the testing. The mold was placed on a solid base and the moist soil was compacted into the mold, with the extension attached in three layers of approximately equal mass, each layer being given 25 blows from a 2.6 kg rammer dropped from a height of 310 mm above the soil. The blows were distributed uniformly over the surface of each layer. The compacted soil was leveled off to the top of the mold. The same procedure was undertaken for different water in order to establish the optimum moisture content and maximum dry density.

3.1.1 *Effect of fiber content on OMC and MDD*
A standard proctor test gives the optimum moisture content and maximum dry density for a given soil. The variations of the optimum water content of different fiber content is as shown in Figure 3. The variation of the maximum dry density of different fiber content is as shown in Figure 4.

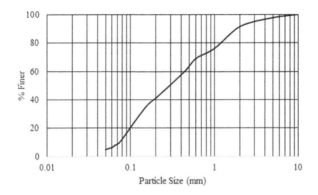

Figure 1.   Particle size distribution curve.

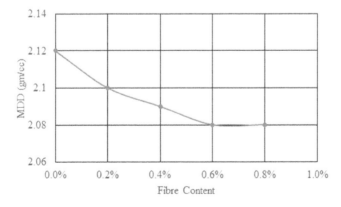

Figure 2.   (a) Sample of jute/fiber and (b) dry mix of soil and reinforcement.

Figure 3.   Variation of OMC with fiber content.

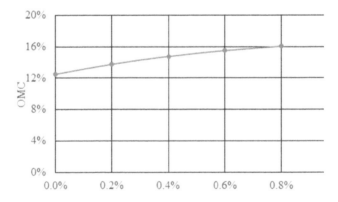

Figure 4.   Variation of MDD with fiber content.

107

3.2 CBR test

California Bearing Ratio (CBR) is widely used as an indicator to measure the strength of sub-grade soils. The CBR value is determined by using both laboratories and field approaches. Laboratory CBR was performed in this study.

3.2.1 Effect of fiber content on the CBR value

A series of CBR tests has been done with variations in the fiber content. A CBR test was done on a sample with a given fiber content at corresponding OMC and MDD. The utmost care was taken to get up to 95% of MDD for the soil prepared for the CBR test.

4 DESIGN OF PAVEMENT

4.1 Design of subgrade and assessment of thickness of pavement for highways

As per the recommendations of the Indian Road Congress (IRC 37:2012) for the design of flexible pavements, the subgrade is the top 0.5 m of the embankment. Any soil having a CBR of a minimum of 8% or above can be used for traffic consisting of 450 commercial vehicles per day. The soil needs to be compacted to a minimum 97% to avoid rutting due to densification over time. As seen from the graph shown in Figure 5, the unstabilized virgin soil was originally unfit to be used as a subgrade material, as it has a CBR value of around 7.2%. As the soil is rendered useless, a search for locally available soil with a CBR of more than 8% would have been necessary, which would have increased the cost of construction of the embankment due to lead and lift. Once the soil was stabilized with jute fibers, the CBR value increased. The addition of jute fibers in the range of 0.4%, 0.6%, and 0.8% increased the CBR values of the soil to 8%, 10%, and 11%, respectively. Hence, soil that is stabilized with jute fibers can be used for road embankments.

4.2 Pavement design

The pavement design is done in accordance with IRC 37:2012, using the design charts as mentioned in the code. The following table mentions the thickness of the pavement that is required for the stabilized soil and for virgin soil. Hence, it can be inferred that the stabilized soil requires less pavement thickness, thus making the pavement more economical and sustainable.

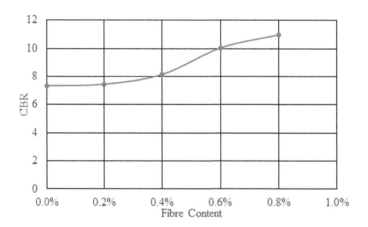

Figure 5. Variation of CBR value with fiber content.

Table 1. Assessment of pavement thickness for various percentages of jute fibers.

Sr. no	Traffic in MSA	Pavement thickness (GSB + SB + DBM + BC/SDBC) in mm		
			CBR value of soil stabilized with jute fibers	
		Virgin soil	0.4%	0.6%–0.8%
	CBR value	7%	8%	10%–11%
1	2	445	445	425
2	5	505	475	455
3	10	580	550	520
4	20	610	575	550
5	30	620	590	580
6	50	630	590	580
7	100	650	615	610
8	150	670	635	625

5 RESULTS AND DISCUSSION

Interpretations are one of the most important parts of the study. The result of the experimental study has been shown as variations of different properties obtained with fiber content. These properties were then used for the design of pavements in order to analyze the effect of fiber content on subgrade thickness.

5.1 *Effect of fiber content on compaction parameters*

The compaction parameters, i.e. OMC and MDD, show typical variations with added fiber content, as shown in Figures 3 and 4. The optimum moisture content increases with an increase in fiber content. This may be due to the fact that jute fiber would also do this. The moisture content increment reduces with an increase in fiber content. Maximum dry density reduces with an increase in fiber content. This may be due to the fact that the fiber content is taken as percentage weight, which will significantly affect the volume of the soil-fiber mix. Moreover, it was also observed that the reduction in MDD decreases with an increase in the fiber content. It was also observed that the effect of the fiber content is negligible on maximum dry density after certain fiber content.

5.2 *Effect of fiber content on strength parameters*

The variation of strength parameters under consideration, i.e. CBR value, with respect to fiber content, is as shown in Figure 5. It is evident that the CBR value increases with an increase in fiber content. This is due to the fact that the inclusion of fiber increases the angle of internal friction in the soil-fiber mix. It should also be noted that the jute, being water absorbing material, increases the apparent angle of internal friction.

5.3 *Effect of fiber content on pavement thickness*

It was observed that the virgin soil was unfit for use as subgrade material. The addition of geojute fibers increased the CBR value of the soil. Since the IRC method of pavement design is based on CBR values, it can be inferred that that the pavement thickness was reduced with the addition of geojute fibers, as the CBR value of the soil increased.

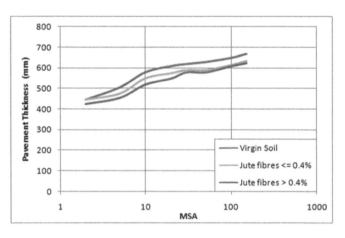

Figure 6.   Variation of pavement thickness with geojute fiber and MSA.

6   CONCLUSION

An experimental investigation to obtain the effect of randomly distributed geojute fiber on the design of pavement subgrade was attempted in this study. It was observed that the fiber content significantly affected the parameters under consideration, i.e. optimum moisture content, maximum dry density, and CBR values. In general, it was observed that MDD decreases with an increase in fiber content, while OMC and CBR values increase with an increase in fiber content.

The design of the subgrade and the thickness of the pavement was then studied using the data obtained from the experimental study. The effect of the fiber content on the thickness of the pavement subgrade was analyzed. It was observed that the pavement thickness reduces significantly with the addition of reinforcements in the subgrade soil.

REFERENCES

Chakrabarti, S. & Bhandari, G. (2004), *Influence of geojute reinforcement on thickness reduction of flexible pavement: A model study*. Jute and Jute Goetextiles, National Jute Board, Ministry of Textile, GoI.
Gosavi, M., Patil, K.A., Mittal, S. & Saran, S. (2004). Improvement of properties of black cotton soil subgrade through synthetic reinforcement. *Journal of the Institution of Engineers (India), 84*, 257–262.
IRC: 37, 2012. (2012, July). *Guidelines for the design of flexible pavements* (Third revision). Indian Road Congress.
IS 2720 part (VII). (1980). *Methods of test for soils: Determination of water content-dry density relation using light compaction*. Bureau of Indian Standards.
Sahu, R.B., Hajra, H.K. & Son, N. (2004). *A laboratory study on geojute reinforced soil bed under cyclic loading*. Indian Geotechnical Conference 2004, Kolkata.
Saran, S. (2006). *Reinforced soil and its engineering applications*. I.K International.

Technology Drivers: Engine for Growth – Mahajan, Modi & Patel (Eds)
© 2018 Taylor & Francis Group, London, ISBN 978-1-138-56042-0

Development of fragility curves for reinforced concrete buildings

D.K. Parekh & S.P. Purohit
Department of Civil Engineering, Institute of Technology, Nirma University, Ahmedabad, Gujarat, India

ABSTRACT: Earthquake scan cause damage to any type of man-made structure to varying degrees. Mitigation against seismic damages requires a scientific estimation of the hazards and their impact on structural systems. As structural systems are pushed beyond their linear limit of material and geometry, seismic hazard estimation becomes complex. Seismic hazard in the form of fragility curves can be developed using linear dynamic analysis and/or non-linear static analysis. In this study, a G+4 story Reinforced Concrete (RC) building with a moment resisting frame was considered in a high seismic zone. Fragility curves for a RC building for maximum story drift were developed using linear dynamic analysis with an emblem of earthquake ground motion. Fragility curves were also developed using non-linear static analysis based on FEMA 440 for maximum story drift. A comparative study between guidelines of ATC 40 to modified guidelines of FEMA 440 was also attempted. It was found that despite RC buildings being designed for seismic loading as per IS 1893 (Part1): 2002, they were deficient with an increase in PGA values for a set of ground motions.

1 INTRODUCTION

The behavior of RC buildings under earthquake conditions is still an open area of research, and decades by decades, improvements have been made in capturing the response of buildings. The seismic evaluation of RC buildings have been carried out by various methodologies, such as non-linear static analysis, linear time history analysis, and response spectrum. Fragility curves provide insight into the damaged state of a building and it is a well adopted form of seismic hazard evaluation. An emblem of earthquake ground motion is required to develop fragility curves. Hazus 2.1-MH Technical Manual along with ATC 40 or FEMA 440 can be used to develop fragility curves for a RC building. Limit state parameters like immediate occupancy, life safety and collapse prevention, along with maximum story displacement and/or story drift, may also be considered. Fragility curves show exceedance of a particular limit state under an emblem of seismic demand.

2 LITERATURE REVIEW

Gentruck et al. (2007) studied the fragility relationships for groups of buildings based on an inelastic response. In this study, they captured the response of wood frame structures in the USA, performed a pushover analysis, and developed fragility curves for a set of artificially generated ground motions that matched their soil origin. The study concluded that fragility curves, obtained with the help of conventional method, were more reliable than Hazus results and expert opinion.

Kirar and Maheshwari (2016) studied the correlation between Shear Wave Velocity (Vs) and SPT resistance by performing a survey of 10 sites in the Roorkey region of. Their study concluded that the regression done with uncorrected SPT values gave better correlation with Vs rather than corrected values of SPT, and an empirical relation was developed.

3 PROBLEM FORMULATION

3.1 *Reinforced concrete building configuration and input data*

A G+4 story RC building was considered with a plan of the building as shown in Figure 1. The geometric and material properties for the RC building studied are given in Table 1. The RC building was designed for gravity loads and seismic loads of Zone-5 with a special moment resisting frame as per IS 1893 (Part1): 2002. The PEER Strong Ground Motion Database was used to generate an earthquake ground motion database of varied nature.

To generate a database of ground motion for an Indian context, soil parameters, magnitude and types of fault were provided based on the work of Rastogi (2004) and Kirar and Maheshwari (2016). These sets of earthquake ground motions were an important input in order to perform linear dynamic analysis (linear time history analysis) and non-linear static analysis (pushover analysis) for a RC building.

3.2 *Linear dynamic analysis (linear time history analysis)*

Linear dynamic analysis was performed on a RC building to evaluate the maximum story drift. An emblem of about 30 earthquake ground motions are discussed in Section 3.1 and are considered typical of an earthquake hazard scenario in India. Typical earthquake ground motion is as shown in Figure 2. The RC building was modeled in the ETABS and SAP program and maximum story drift was determined. Figure 3 shows maximum story drift for the RC building that was subjected to a set of earthquake ground motions. It can be seen from Figure 3, that story drift increases to a maximum with an increase in height as well PGA values of earthquake ground motion.

3.3 *Non-linear static analysis (pushover analysis)*

Non-linear static analysis (pushover analysis) was also performed on the RC building as earthquake ground motion pushes the building beyond its elastic limit. Pushover analysis was

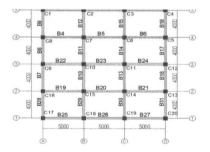

Figure 1.

Table 1. Geometric and material properties of a reinforced concrete building.

Structural details	Values	Unit
Plan dimension	25×16	M
Height of typical story	3	M
Concrete grade	25	MPa
Steel grade	415	MPa
Beam dimension	0.23×0.45	M
Column dimension	0.5×0.5	M
Slab thickness	135	mm
Soil type	Medium	–

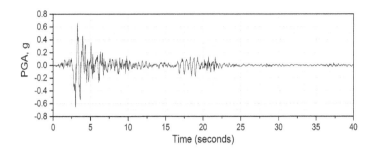

Figure 2. Typical earthquake ground motion.

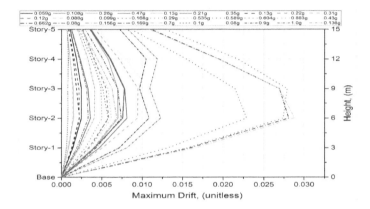

Figure 3. Maximum story drift for time histories with different PGAs.

Table 2. Spectral acceleration and displacement values at the performance point of the reinforced concrete building.

PGA (g)	FEMA 440 guidelines Sa (g)	Sd (m)	ATC 40 guidelines Sa (g)	Sd (m)	PGA (g)	FEMA 440 Sa (g)	Sd (m)	ATC 40 Sa (g)	Sd (m)
0.059	0.13	0.07	0.13	0.069	0.22	0.12	0.045	0.12	0.047
0.088	0.045	0.013	0.045	0.013	0.26	0.134	0.079	0.135	0.085
0.099	0.109	0.034	0.109	0.033	0.29	0.126	0.06	0.127	0.068
0.104	0.054	0.016	0.054	0.016	0.31	0.135	0.09	0.135	0.105
0.12	0.066	0.02	0.066	0.02	0.35	0.132	0.072	0.132	0.072
0.13	0.05	0.015	0.05	0.015	0.43	0.122	0.05	0.123	0.057
0.13	0.102	0.031	0.1	0.031	0.535	0.121	0.049	0.123	0.05
0.136	0.108	0.033	0.107	0.033	0.604	0.176	0.3094	–	–
0.149	0.068	0.02	0.068	0.02	0.7	0.121	0.251	–	–
0.168	0.118	0.042	0.118	0.042	0.883	0.127	0.189	–	–
0.169	0.13	0.069	0.13	0.069	0.9	–	–	–	–
0.21	0.13	0.068	0.13	0.069	1.0	–	–	–	–

carried out following both FEMA 440 and ATC 40 guidelines using the ETABS and SAP program. Table 2 shows the value of spectral acceleration and spectral displacement obtained at the performance point in ETABS and SAP following both the guidelines as discussed above.

The performance point for each pushover curve generated was calculated from the emblem of earthquake ground motion seismic demand.

4 DEVELOPMENT OF FRAGILITY CURVES

The fragility curve is a well adopted form of seismic hazard estimation worldwide. It gives the idea of probability of exceedance of a particular limit state corresponding to the response parameter taken into account for a given set of ground motions. In this study, fragility curves were developed for a RC building and analyzed using linear dynamic and non-linear static methods. The threshold values for a particular limit state considered were taken from Hazus 2.1-MH Technical Manual (Federal Emergency Management Agency, 2013) and FEMA 356 (Federal Emergency Management Agency, 2000). Probability of exceedance of a particular limit state was calculated using Equation 1 based on the report of Mid-America Center (Gentruck et al., 2007).

$$p_{(LSi/GMI)} = 1 - \Phi\left(\frac{\lambda_{CL}^{i} - \lambda_{D/GMI}}{\sqrt{\beta_{D/GMI}^{2} + \beta_{CL}^{2} + \beta_{M}^{2}}} \right) \tag{1}$$

where,
$\Phi[.]$ = standard normal cumulative distribution function.
$\lambda_{D/GMI}$ = natural logarithm of the demand parameter which can be determined by Equation 2.
$\beta_{D/GMI}$ = square root of the standard error which can be determined by Equation 3.

$$\lambda_{D/GMI} = \ln a_1 + a_2 \ln(GMI). \tag{2}$$

where,
a_1 and a_2 are the constants calculated through linear regression analysis.
$\ln(GMI)$ = natural logarithm of the ground motion indices taken as a demand.

$$\beta_{D/GMI} = \sqrt{\frac{\sum_{K=1}^{n}\left[\ln(GMI_k) - \lambda_{D/GMI}(GMI_k)\right]^2}{n-2}} \tag{3}$$

where,
N = number of data points taken into consideration.

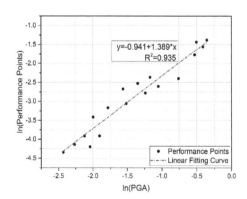

Figure 4. Linear regression analysis for the performance point (FEMA 440).

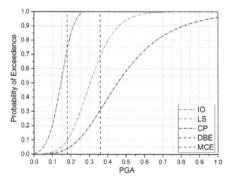

Figure 5. Fragility curve of the reinforced concrete building for a maximum story drift-linear dynamic analysis.

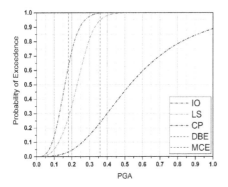

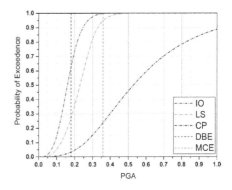

Figure 6. Fragility curve of there in forced concrete building for a maximum story drift-non-linear static analysis (FEMA 440).

Figure 7. Fragility curve of the reinforced concrete building for a maximum story drift-non-linear static analysis (ATC 40).

5 CONCLUSIONS

Following the major observations obtained during this study:

- Despite being designed for seismic loading as per IS: 1893 (Part1):2002, the RC building was found to be deficient with an increase in PGA values for a set of earthquake ground motions.
- Fragility curves designed for a RC building were analyzed using a non-linear static method based on FEMA 440 and ATC 40 guidelines. They were found to be identical in nature due to the normalization of the responses. This means that both the guidelines yield identical seismic hazards however, performance points derived using both guidelines were different and hence they have their own consequences on the behavior of the building.
- Comparison among fragility curves of the RC building using linear dynamic analysis and non-linear static analysis showed a wide difference. While fragility curves obtained for an RC building using linear dynamic analysis showed low probability of exceedance of IO and LS limit state upto 0.6 g PGA value of seismic demand, the high probability of exceedance of IO and LS state was found for the RC building analyzed with the non-linear static analysis method upto 0.4 g PGA value. However, probability of exceedance was found to be lower for the RC building under non-linear static analysis when compared to non-linear static analysis.

REFERENCES

Applied Technical Council. (1996). *Seismic evaluation and retrofit of concrete buildings* (ATC 40:1993), California: Author.
Bureau of Indian Standards. (2002). *Criteria for earthquake resistant design of structures* (IS 1893-1:2002), India: Author.
Federal Emergency Management Agency. (2000). *Prestandard and commentary for the seismic rehabilitation of buildings* (FEMA 356:2000), Washington D.C.: Author.
Federal Emergency Management Agency. (2005). *Improvement of nonlinear static seismic analysis procedures* (FEMA 440:2005), California: ATC.
Federal Emergency Management Agency. (2013). *Technical manual on multi-hazard loss estimation methodology* (Hazus 2.1-MH:2013), Washington D.C.: Author.
Gentruck, B., Elnashai, A.S. & Song, J. (2007). *A report on fragility relationships for populations of buildings based on inelastic response.* Illinois: Mid-America Center.
Kirar, B. & Maheshwari, B.K. (2016). Correlation between shear wave velocity (Vs) and SPT resistance (N) for Roorkee region. *Journal of Geosynthetics and Ground Engineering, 2,* 2–11.
Rastogi, B.K. (2004). Seismicity and earthquake hazard studies in Gujarat. *Journal of Earthquake Science and Engineering, 1,* 110–123.

Technology Drivers: Engine for Growth – Mahajan, Modi & Patel (Eds)
© *2018 Taylor & Francis Group, London, ISBN 978-1-138-56042-0*

Seismic performance evaluation of RC frame using incremental dynamic analysis

Jahanvi M. Suthar & Stuti M. Patel
Department of Civil Engineering, Institute of Technology, Nirma University, Gujarat, India

ABSTRACT: Performance-Based Earthquake Engineering is a modern approach. There are several static and dynamic non-linear analyses. Incremental Dynamic Analysis (IDA) is an emerging analysis method which offers seismic demand and capacity using multiple scaled suites of ground motion records. In this paper IDA is performed on G+4 storey bare frame. Seven ground motions are selected. IDA curves are generated and performance of frame is predicted.

1 INCREMENTAL DYNAMIC ANALYSIS

This chapter explains Performance-Based Seismic Design (PBSD) which is used to design of new buildings or for up gradation of existing buildings with a realistic understanding of the risk of life, occupancy and economic loss that may occur as a result of future earthquakes. There are two main parts of a performance objective, a damage state and a level of seismic hazard. To describe Seismic performance identified seismic hazard (earthquake ground motion) is used and design the maximum allowable damage state (performance level) is done.

This target performance objective is divided into Structural Performance Level and Non-structural Performance Level (ATC 40). Non-linear dynamic analysis includes Time History Analysis, Incremental Dynamic Analysis (IDA) and N2 method (Ghobarah, 2001). Here IDA is considered for predicting structure behaviour.

IDA was developed for seismic assessment considering the dynamic load. Earthquake ground motion are scaled from lower to higher intensity and a suite of ground motions are typically applied to the structure, to obtain statistics about the structures performance, characterized by displacement and eventually collapse, under a range of earthquake excitation. In IDA dynamic load is applied incrementally. IDA is applicable to evaluate the dynamic response and capacity of the frames not only in elastic region, but also beyond the linear regime up to global and local instability of the structures where either a soft storey mechanism takes place or connection fractures (Asgarian et al., 2010). In conceptualization of IDA, ground motions at multiple intensity levels are involved. Ground motion intensity is characterized by spectral acceleration (S_a) at the period of vibration of interest (T). IDA curves of structural response can be generated, as measured by an Engineering Demand Parameter (EDP, the maximum peak inter story drift ratio θ_{max}), versus the ground motion intensity level. Ground motion is measured by an Intensity Measure (IM, e.g., peak ground acceleration, PGA or 5% damped first mode spectral acceleration, $S_a(T_1,5\%)$). Then limit states (Immediate occupancy or Collapse prevention) can be defined on each IDA curve and the corresponding capacities can be calculated.

1.1 *Modeling, selection and scaling of ground motion*

G+4 storey RC bare frame is selected for IDA with 3 m storey height as shown in Figure 1.

Section properties for beams and columns are given in Table 1.

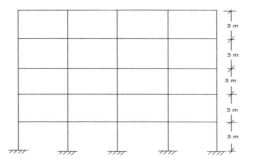

Figure 1. G+4 storey RC frame.

Table 1. Details of frame.

Floor	Beam (mm)	Column (mm)
GF	230×500	230×600
1st	230×500	230×600
2nd	230×450	230×500
3rd	230×450	230×500
4th	230×450	230×450

Table 2. Selected ground motions.

Sr. No.	Event	Year	Station	Magnitude (M)	Site class	PGA (g)
1	Imperial Valley-02	1940	EL Centro Array #9	6.9	D	0.34
2	North-west Calif-02	1941	Ferndale City	6.6	C	0.06
3	San Fernando	1971	Lake Hughes	6.61	C,D	0.21
4	Imperial Valley-06	1976	Aeropuerto Maxicali	6.5	B,C	0.33
5	Imperial Valley	1979	EL Centro Array #1	6.53	D	0.19
6	Imperial Valley	1979	Plaster City	6.53	C,D	0.05
7	Imperial Valley	1979	Superstition Hills	6.53	D	0.13

Ground motions for IDA are selected based on magnitude and distance from the fault of an event. Ground motions with magnitude between 6.4 to 6.9 and fault rupture distance between 10 km to 30 km of California region are selected. PEER (Pacific Earthquake Engineering Research) NGA database is used for scaling ground motions. (Table 2)

2 ANALYSIS

First of all the model (using SAP 2000) has been formed and the ground motion records have been selected. This entails appropriately scaling each record to cover the entire range of structural response, from elasticity, to yielding, and finally global dynamic instability. Every event is scaled with 0.15 g increment in spectral acceleration value. At every increment maximum inter storey drift ratio is calculated and later it is used for generation IDA curves.

3 POST PROCESSING

3.1 *Generation of IDA curves by interpolation*

Once the desired IM (intensity measured) and DM (Demand measure) values are extracted from the dynamic analysis, a set of points for each record are left. Without performing additional Dynamic Analysis just by interpolating these points, the entire IDA curve can be generated. Here IDA curves are generated by linear interaction. DM values at any IM level can be calculated.

It starts with a straight line in an elastic range and then shows the effect of early yielding and local damage by having some change in local tangent slope but generally it stays on elastic slope. At any given S_a level below 0.5 g graph remains about the same displacement as in an elastic system. Then after S_a value 0.5 g it starts softening, shows increment in tangent slope, reaching the flat line slightly above S_a value 1.2 g, where the structure responds with

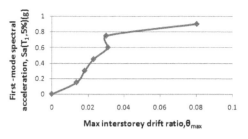

Figure 2 (a). IDA curve of ground motion-1.

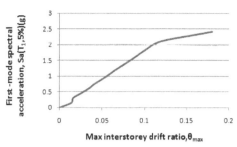

Figure 2 (b). IDA curve of ground motion-2.

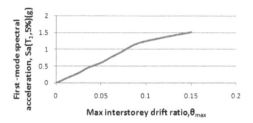

Figure 2 (c). IDA curve of ground motion-3.

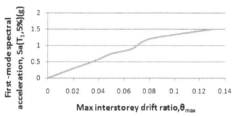

Figure 2 (d). IDA curve of ground motion-4.

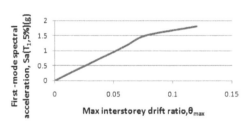

Figure 2 (e). IDA curve of ground motion-5.

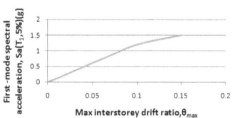

Figure 2 (f). IDA curve of ground moton-6.

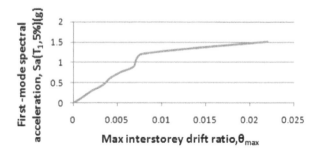

Figure 2 (g). IDA curve of ground moton-7.

practically infinite θ_{max} values and numerical non-convergence has been encountered during the analysis. This is when the structure has reached to global dynamic instability, when a small increment in IM level results in unlimited increase in DM-response. IDA curves are always not simple, as curve for record #1 is quite different than others.

3.2 Defining limit-states on IDA curves

In order to get performance calculation for PBD, defining limit-states is necessary on the IDA curves. Three limit states are chosen: Immediate Occupancy (IO), Collapse Prevention (CP) and Global dynamic Instability (GI). For this RC frame, IO limit-state is decided at θ_{max} = 1%.

FEMA-350 (Vamvatsikos & Cornell, 2005) is used to define CP point, which is not exceeded on the IDA curve until the final point where local tangent reaches 20% of the elastic slope or θ_{max} = 10%, whichever occurs first in IM term. CP limit-state is defined at a point where IDA curve is softening towards flat line but at low enough values of θ_{max} (less than 10%). At IO limit state, it is easy to calculate IM-values. For these, IM values that produce θ_{max} = 1% are calculated and if more than one value of IM is there at θ_{max} = 1%, lowest one will be selected. Considering Figure 3. (record #3), IO is violated for Sa(T1; 5%) ≥ 0:33 g or θ_{max} ≥ 1%.

As per CP limit-state definition, we have to find highest point where IDA tangent slope is equal to 20% of the elastic. These point usually lies on the softening segment that precedes the flat line. Another CP point is at θ_{max} = 10%. So whichever comes first, decides CP capacity. Simple shape of IDA curve of record #3 makes this easy. Here 20% -slope rule governs and generates one point. So, CP point is violated for $S_a(T_1; 5\%)$ ≥ 0:44 g or θ_{max} ≥ 2%.

Now considering record #1, it has complicated shape as shown in Figure 4. The IDA curve starts softening at about 0.2 g showing slope less than the elastic. After 0.2 g it hardens by local slope higher than the elastic. Also the response of frame at $S_a(T_1,5\%)$ = 0.6 g is higher than $S_a(T_1,5\%)$ = 0.75 g. IDA curve also starts softening after 0.8 g after it goes as a flat line. The lower CP point should be rejected, as it does not directly precedes flat line. These show that the frame is not as close to global collapse. So, other CP point definition should be required. So, CP point is exceeded for record #1 when $S_a(T_1; 5\%)$ ≥ 0:77 g. At these spectral acceleration value DM-value is 0.031 (Table 3).

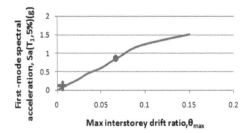

Figure 3.   Limit-states defined on record-3.

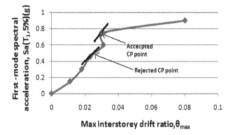

Figure 4.   Limit-states defined on record-1.

Table 3.   Limit-states data for all record.

	Sa (T₁, 5%)			θ_{max}		
Record No.	IO	CP	GI	IO	CP	GI
1	0.15	0.77	0.90	0.01	0.03	infinity
2	0.10	1.60	1.40	0.01	0.07	infinity
3	0.33	0.44	1.50	0.01	0.02	infinity
4	0.12	0.60	1.50	0.01	0.03	infinity
5	0.21	1.10	0.80	0.01	0.05	infinity
6	0.10	0.90	1.50	0.01	0.06	infinity
7	0.08	0.70	0.90	0.01	0.04	infinity

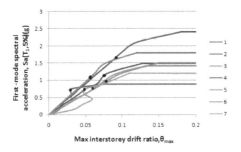

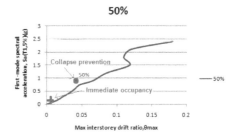

Figure 5. Summarizing IDA curves with limit-states.

Figure 6. Summary of IDA curves and corresponding limit-state capacities into 50% fractile.

3.3 *Summarizing IDA curves*

Figure 5 shows that IDA curves show different behavior, showing large record-to-record variability. The limit-state capacities can be summarized by some mean value and measure of dispersion like standard deviation. Figure 6 shows mean of all these seven IDA curves. As shown in Figure 6, Immediate occupancy occurs at $S_a(T_1,5\%) = 0.16$ g and $\theta_{max} = 1\%$, while Collapse prevention limit state occurs at $S_a(T_1,5\%) = 0.95$ g and $\theta_{max} = 3.9\%$.

4 CONCLUSION

The limit state capacities can be summarized by some mean value and measure of dispersion like standard deviation. 16%, 50% and 84% fractile values of IM and DM capacities for each limit states are selected. From this analysis, concluding points are:

1. In order to generate demand of $\theta_{max} = 4\%$, 50% records need to be scaled at level $S_a(T_1,5\%) = 0.95$ g means at $\theta_{max} = 0.04$ and $S_a(T_1,5\%) = 0.95$ g have forced frame to violate CP.
2. IO limit is shown at the intersection of each IDA curve with $\theta_{max} = 1\%$ and $S_a(T_1,5\%) = 0.16$ g line, CP point is represented by dots and GI occurs at flat line.

REFERENCES

Applied Technology Council (1996) ATC 40. *Seismic evaluation and retrofit of concrete buildings.* Redwood City, CA, ATC.

Asgarian, B., Sadrinezhad, A., and Alanjari, P. (2010) Seismic performance evaluation of steel moment frames through Incremental Dynamic Analysis. *Journal of Construction Steel Research. 178_190.*

Federal Emergency Management Agency (2000). FEMA 350. *Recommended seismic design criteria for new steel moment-frame buildings.* Washington D.C.

Federal Emergency Management Agency (2009) FEMA P695. *Quantification of building seismic performance factors,* FEMA P695. Washington, D.C.

Federal Emergency Management Agency (2012) FEMA P-58-1, *NEHRP Guidelines for the Seismic Performance assessment of Buildings.* Washington, D.C.

Ghobarah, A. (2001) Performance-based design in Earthquake Engineering: State of development. *Engineering Structures. 23(8), 878–884. Available from: doi: 10.1016/S0141-0296(01)00036-0.*

Naeim, F., Bhatia, H. (2001), Performance Based Seismic Engineering. *The Seismic Design Handbook.* Springer US. pp 757–792.

Priestley, M.J.N. (2000). Performance based seismic design. *Bulletin of the New Zealand Society for. Earthquake Engineering. 33(3), 325–346.*

Vamvatsiko, D., Jalayer, F., and Cornell, C.A. (2003). Application of Incremental Dynamic Analysis to an RC structures. *Proceedings of the FIB Symposium on Concrete Structures in Seismic regions, Athens.*

Vamvatsikos, D. and Cornell, C.A. (2005) Developing efficient scalar and vector intensity measures for IDA capacity estimation by incorporating elastic spectral shape information. *Earthquake Engineering and Structural Dynamics.* 1–22. Doi:10.1002/eqe.96.

Vamvatsikos, D. and Cornell, C.A., The Incremental Dynamic Analysis and its application to performance based Earthquake Engineering. *12th European Conference on Earthquake Engineering.* Elsevier Science Ltd. Barbican Centre, London, UK.

Vamvatsikos, D. and Fragiadakis, M. (2008). Seismic performance uncertainty of a 9-storey steel frame with non—deterministic beam-hinge properties. *The 14th World Conference on Earthquake Engineering, Beijing China.*

Technology Drivers: Engine for Growth – Mahajan, Modi & Patel (Eds)
© 2018 Taylor & Francis Group, London, ISBN 978-1-138-56042-0

Effect of temperature on the tensile strength of stainless steel wire mesh and fiber reinforced polymers

P.J. Shah, P.V. Patel & S.D. Raiyani
Department of Civil Engineering, Institute of Technology, Nirma University, Gujarat, India

ABSTRACT: The aim of the present experimental work is to investigate the mechanical property in terms of the tensile strength of Stainless Steel Wire Mesh (SSWM), Carbon Fiber Reinforced Polymers (CFRP) and Glass Fiber Reinforced Polymers (GFRP) when exposed to different elevated temperatures, ranging from 50°C to 300°C. Prior to testing all 21 coupons are subjected to the specified elevated temperature for 45 minutes and left for 24 hours to cool down up to ambient temperature. Stress-strain behavior, as well as failure modes are compared. The test results show that the decrease in the tensile strength of GFRP is more severe than those of CFRP and SSWM strips. The ultimate tensile strength of the SSWM, CFRP and GFRP at 300°C is decreased by about 22%, 20% and 29%, respectively, as compared to ambient temperature. The derived stress-strain relationship of SSWM, CFRP and GFRP at deferent temperature can be used as an input in numerical simulation of Reinforced Concrete (RC) member strengthened with SSWM, CFRP and GFRP.

1 INTRODUCTION

Structural repair and rehabilitation of concrete structures are necessary for all damaged or deteriorated structures to enhance or restore their load carrying capacity. Fiber-reinforced polymer (FRP) composite is widely adopted in rehabilitation and strengthening of structural members (Questha et al., 2016). Locally available Stainless Steel Wire Mesh (SSWM) material has a cost-effective solution for an external strengthening of concrete members (Kumar and Patel, 2016). The effect of temperature on the performance of FRP composites under axial tensile loads is more complex, where the performance of FRP composites is dependent on thermal softening of both the fibre and the polymer matrix. Chowdhury et al., (2007) experimentally show that performance of sheathed fibre reinforced polymer wrapped reinforced concrete columns exposed to fire and concluded that sheathed FRP strengthened column was able to resist elevated temperature for at least 90 min longer than the equivalent unsheathed FRP strengthened column. Hawileh et al., (2015) carried out an experimental program to investigate the mechanical properties in terms of tensile strength and elastic modulus of composite carbon (C), composite glass (G) sheets and their hybrid combinations (CG) when exposed to various elevated temperatures, ranging from 25°C to 300°C. The tensile behavior of CFRP sheets was investigated with two different epoxy resins at specified elevated temperatures up to 70°C by Wu et al. (2006). Cao et al. (2009) experimentally inspected the axial tensile behavior of CFRP and hybrid sheets at deferent temperatures ranging from ambient temperature to 200°C. The tensile strength was found to reduce by about 40% from that at an ambient temperature.

In general, information about the tensile strength of FRP laminates at higher temperature are lacking in the literature. Moreover, very less research has been performed on the use of SSWM for strengthening structural elements. This paper represents an experimental study carried out to investigate the effect of temperature on the tensile strength of SSWM, CFRP and GFRP composites. Researchers or engineers can use the outcome of this study as an input in numerical modelling to analyze the performance of FRP strengthened structural members subjected to elevated temperatures.

2 EXPERIMENTAL PROGRAM

A total of twenty-one coupons were prepared and tested to perceive the dilapidation of the tensile strength of SSWM, GFRP and CFRP under elevated temperature. Test specimens were divided into three different groups of SSWM, GFRP and CFRP strips. Seven coupon specimens for each group of strips were prepared for specified temperature exposure ranging from ambient temperature to 300°C. Preparation of coupon and further testing is carried out as per ASTM D 3039/D 3039M-08(2008). The strip is cut in size of 100 mm wide and 400 mm length. Schematic diagram of a specimen is shown in Figure 1(a). Wooden formwork is prepared to resemble the field application of FRP to the concrete surface as shown in Figure 1(b). Test specimens were fixed at the end by 100 mm wide and 100 mm long steel plate as shown in Figure 1(c).

Bonding material Sikadur 30 LP was used for bonding steel plate with strip specimens. Sikadur 30 LP was applied on two steel plates and above it, strip specimen was placed in such a way that no voids remained between FRP and steel plate. For proper bonding sufficient weight was put on steel plates. Each specimen was kept for curing up to 7 days at ambient temperature to get sufficient bond strength between steel plates and strip specimen. Then coupons were exposed to different temperatures of 50°C, 100°C, 150°C, 200°C, 250°C and 300°C, respectively in a standard muffle furnace for 45 min. The standard muffle furnace can achieve a maximum temperature of 1000°C. The muffle furnace temperature was controlled using a digital system as shown in Figure 2. Specimens after exposure of different elevated temperature are shown in Figure 3.

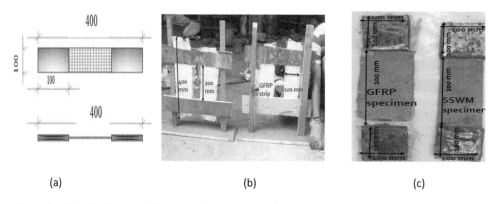

(a)   (b)   (c)

Figure 1.   Wooden formwork for preparing coupon specimens and cured coupon specimens.

Figure 2.   Muffle furnace.

Figure 3.   SSWM and GFRP specimens at elevated temperature.

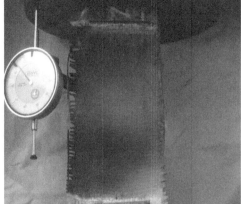

Figure 4.   Test set-up.

Table 1.   Tensile properties as reported by manufacture.

Material	Thickness (mm)	Tensile strength (MPa)	Elastic modulus (GPa)	Elongation at rupture (%)
SSWM	0.27	1100	194	2.5
CFRP	0.554	4000	360	1.5
GFRP	0.314	2000	310	4.5
Sikadur 30 LP (for SSWM)	–	18	10	1.2
Sikadur 330 (for CFRP and GFRP)	–	30	4.5	0.9

After exposure of specified temperature to the coupons, all specimens were retained to cool down for 24 h. Subsequently, they were tested at ambient temperature in the laboratory for tensile strength, as shown in Figure 4, using Universal Testing Machine (UTM). Elongation was measured using dial gauge. To provide the proper grip in UTM and a maintain clear gauge length of 200 mm for each specimen, a coupon was fixed at the end by 100 mm wide and 100 mm long steel plate.

The thickness, tensile strength, elastic modulus and elongation at failure as described by the manufacturers at ambient temperature for the Stainless Steel Wire Mesh (SSWM), Carbon Fiber Reinforced Polymer (CFRP) sheet and for the Glass Fiber Reinforced Polymer (GFRP) sheet are presented in Table 1. The two-part epoxy resins Sikadur 30 LP and Sikadur-330 are used to prepare SSWM and GFRP & CFRP coupon specimens respectively.

The mechanical properties of the epoxy are also given in Table 1 as described by the manufactures at ambient temperature.

3 RESULTS AND DISCUSSION

Three types of failure were perceived at different ranges of temperatures. The first type of failure (Type I) was experienced at a range of temperatures between 50°C – 150°C. The failure pattern of coupons is analogous to those tested at ambient temperature at different locations within the gauge length of specimens. The second type of failure (Type II) occurred at a range of temperature between 200°C –250°C. The epoxy (Sikadur 30 LP and Sikadur 330) matrix had become softer and the coupons failed due to partial impairment of epoxy adhesive followed by fibres splitting. The specimens with exposure at 300°C failed in Type III. In type III failure, the epoxy matrix was burned and led to a failure of the fibres. Figure 5 represents different types of failure. Figure 6 illustrates the stress-strain behavior of coupon samples of SSWM, CFRP, and GFRP at different elevated temperatures. From Figure 6 it can be observed that the ultimate tensile strength of SSWM, CFRP and GFRP coupons have been deteriorated when exposed to different raised temperatures. Figure 6 also illustrates the nature of the stress-strain curve of SSWM, CFRP and GFRP specimens at different temperatures. Stress-strain behavior of SSWM specimens has shown parabolic nature, while CFRP and GFRP specimens have shown almost linear nature.

However, Figure 7 shows a decline in the tensile strength of the tested coupons. In particular, at 100°C, the ultimate tensile strength of the SSMW, CFRP and GFRP coupons was decreased by 6%, 4%, and 8%, respectively as compared to ambient temperature. At 300°C,

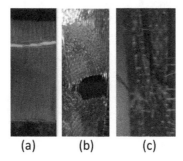

(a)   (b)   (c)

Figure 5. (a) Failure Type – 1 of SSWM specimen (b) Failure Type – 2 of GFRP specimen (c) Failure Type – 3 of CFRP specimen.

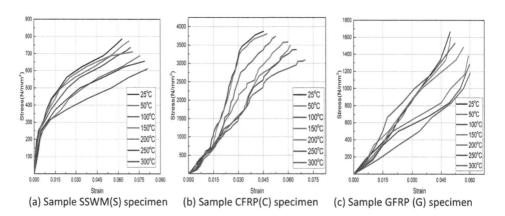

(a) Sample SSWM(S) specimen   (b) Sample CFRP(C) specimen   (c) Sample GFRP (G) specimen

Figure 6. Stress-strain behavior of the tested coupons.

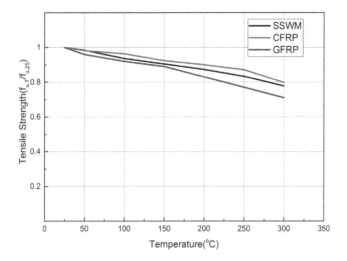

Figure 7.   Comparison of ultimate tensile strength ratio of coupon specimens.

the ultimate tensile strength of the SSWM, CFRP and GFRP coupons was decreased by 22%, 20% and 29%, respectively. Seemingly, the reduction in the ultimate tensile strength of the specimen GFRP is more severe than those of the SSWM and CFRP specimens.

4   CONCLUDING REMARKS

Experimental investigation on twenty-one coupons is presented to evaluate the tensile behavior and strength of SSWM, CFRP and GFRP specimens under various temperature conditions. Stress-strain behavior and different failure types of specimen are compared. Following concussions are drawn from the results of the experimental investigation:

– When specimens are subjected to a temperature range of 50°C–150°C, the failure pattern of coupons is analogous to those tested at ambient temperature at different locations within the gauge length of specimens. Specimens which are subjected to 200°C–250°C temperature range, have failed due to partial impairment of epoxy adhesive followed by fibres splitting. At 300°C, the epoxy matrix was burned and leads to a failure of the fibres.
– At higher temperature, CFRP specimens have better fire resistance as compared to SSWM and GFRP specimens.
– Temperature affects the slope of the stress-strain curves, which tends to decrease as the temperature increases. The descending curve of the CFRP and GFRP specimens becomes more linear as the temperature increases
– Stress-strain behavior of SSWM specimens has shown parabolic nature, while CFRP and GFRP specimens have shown almost linear nature. These behaviors can be used in numerical simulation as well as to prepare the analytical model.

REFERENCES

ASTM D3039/D 3039M (2008). *Standard test method for tensile properties of polymer-matrix composite materials.*
Cao S, Wu Z, Wang X. (2009). Tensile properties of CFRP and hybrid FRP composites at elevated temperatures. *Composites Materials.* 43(4), 315–330
Chowdhury EU, Bisby LA, Green MF, Kodur V. (2007). Investigation of insulated FRP wrapped reinforced concrete columns in fire. *Fire Safety Journal,* 42, 452–60.

Hawileh, R.A., Obeidah, A.A., Abdalla, J.A. and Tamimi A.A. (2015). Temperature effect on the mechanical properties of Carbon, glass and carbon-glass FRP laminates. *Construction and Building Materials,* 75, 342–348.

Kumar, V., and Patel, P.V. (2016). Strengthening of axially loaded circular concrete columns using stainless steel wire mesh (SSWM) – Experimental investigations. *Construction and Building Materials*, 124, 186–198.

Queshta M.I., Shafigh P. and Jummat M.Z. (2016). Research progress on the flexural behavior of externally bonded RC beams. *Archives of Civil and Mechanical Engineering*, 26(4), 982–1003

Wu Z, Iwashita K, Yagashiro S, Ishikawa T, Hamaguchi Y. (2006). Temperature dependency of tensile behavior of—CFRP sheets. *Composite Materials,* 32(3), 137–144.

Technology Drivers: Engine for Growth – Mahajan, Modi & Patel (Eds)
© *2018 Taylor & Francis Group, London, ISBN 978-1-138-56042-0*

An analytical approach on the evaluation of stress distribution beneath plain rigid wheels on Tri-1 lunar soil simulant

Pala Gireesh Kumar & S. Jayalekshmi
Department of Civil Engineering, National Institute of Technology, Tiruchirappalli, Tamil Nadu, India

ABSTRACT: Evaluation of stress distribution beneath plain rigid wheels on soil simulants is useful in analyzing the wheel-soil interaction of planetary rovers. Single wheels moving on a soil simulant are used in the prediction of wheel performance. In this paper, a cast iron small wheel of 160 mm in diameter, 32 mm width and a weight of 52.189 N, and a cast iron large wheel of 210 mm in diameter, 50 mm width and a weight of 67.444 N were studied. The lunar soil simulant on which the wheels moved upon was anorthosite based and called TRI-1. The normal and shear stresses beneath the wheels were determined. An analytical approach was used and the Reece model, Bekker model, Wong-Reece model and Iagnemma model were considered in estimating the normal stresses. The shear stresses were calculated using the Janosi and Hanamoto model. Determining the maximum angle was common for all three models (Reece, Bekker and Wong-Reece), whereas for the Iagnemma model, maximum angle was the average of the entry angle and exit angle. The geotechnical and mechanical properties of the TRI-1 soil simulant were obtained from the experiments in this study. The entry and exit angle for the two wheels were calculated for various models. The Wong-Reece equations were found to register the maximum shear stress (26.3394 kPa) for the small wheel and the Reece model was conservative for the large wheel (34.9298 kPa). The range of stresses corresponded to various soil conditions (loose and dense); the maximum normal stress for the small wheel was 32.1221 kPa (Wong and Reece models) and for the large wheel was 39.0156 kPa (Reece model), within the dense conditions.

1 INTRODUCTION

The surface terrain of planets like Mars and the Moon are covered mainly with fine grained, loose soil and sandstone rocks. A rover is required to operate perfectly in an unknown, unpredictable environment with lots of obstacles in order to explore any planetary surface. Designing and controlling a rover to explore these areas is a challenging task. The reason for this is that, the wheels of the rover can easily slip on loose soil, resulting in loss of traction between wheel and ground. This causes mission failures, such as that found in explorations Lunokhod 0 and Spirit 2010. This paper focused on the normal and shear stress distributions produced by a wheel at the wheel-soil interface. Hence, analysis of both normal and tangential/shear stress distributions were carried out for the prediction of wheel performance. A rigid wheel-soil interaction model was used to evaluate the normal and shear stress distribution at the wheel-soil interface on an anorthosite based lunar soil simulant,Tiruchirappalli-1 (TR-1) (Sreenivasulu, 2014), the properties of which are given in Section 2. This corresponds to Apollo16 site characteristics. Plate load tests were carried out on this simulant to yield the cohesion and friction moduli.

2 LITERATURE REVIEW

Ding Liang et al. (2010) found that the number of wheel lugs had little influence on wheel sinkage, the height of the wheel lugs had little influence on the flow of soil, and the difference of wheel sinkage, which was relatively small, was mainly caused by supporting and soil digging of lugs. From experimental results, the slip-sinkage principle of the lunar rover lugged wheels were analyzed and corresponding calculation equations were derived (Wong & Reece, 1967).

Sutoh et al. (2010 or 2012) used the linear traveling speed model. The wheel had lugs and guidelines for determining a suitable lug interval were described. Terramechanic stress models were used (Reece, 1965). This study aimed at optimizing the lug interval. When the number of lugs was increased from 3 to 12, the speed of the rover periodically changed, whereas for lugs more than 12, the speed remained constant.

Sutoh et al. (2010 or 2012) conducted experiments with a two-wheeled rover. Numerical simulations were also carried out. Increasing the wheel width from 50 mm to 150 mm, resulted in a decrease in the slip ratio to 0.3 (maximum change at slope angle equal to 17°). Hence, the wheel diameter was increased, but keeping the width constant. Contrary to general belief, as wheel diameter increases, the slip ratio decreases leading to better traveling performance. In the simulations, as the wheel width increased, the slip ratio decreased. Terramechanical stress models were also given (Bekker, 1969).

2.1 Stress disribution models

When a wheel travels over loose soil, normal and shear stress develops beneath the surface. These stresses are used in the calculation of forces. The motion performance of a rover is usually evaluated by its drawbar pull and driving torque, which is related to the normal and shear force distributions produced by the wheel at the wheel-soil interface. In this study, four models were considered for the analysis of a small and large wheel that traveled on TRI-1 soil simulant; a comparison of all the model results was made. The stress distribution model is shown in Figure 1. Four models which were considered are given in Table 1. Detailed analysis of stress distribution models and its comparison are explained in Section 2.

2.1.1 Normal stress distribution models

When a wheel travels over loose soil, normal stress develops beneath the surface. Maximum normal stress occurs at transition points between two zones; forward and rearward zones. Four models for normal stress distribution were considered and normal stress distribution models of Reece (1965) and Bekker (1969) are explained in Figure 1.

For the Iagnemma model

θ_m – Maximum Angle (the specific wheel angle where the normal stress is maximum):

$$\theta_m = (\theta_f + \theta_r)/2 \qquad (1)$$

θ_f – Entry Angle (the angle from the vertical to the point at which the wheel initially makes contact with the soil):

$$\theta_f = \cos^{-1}(1 - h/r) \qquad (2)$$

θ_r – Departure Angle (the angle from the vertical to where the wheel departs from the soil and the value is generally assumed to be zero):

$$\theta_r \cong 0 \text{ or } \theta_r = \cos^{-1}(1 - kh/r) \qquad (3)$$

θ_m – Maximum Angle (for the remaining models, it is given as

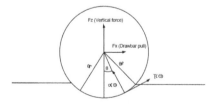

Figure 1. Stress distribution model of the wheel studied.

Table 1. Models for normal stress distribution.

Model	Normal stress	Remarks
Reece, 1965 (Model 1)	$\sigma(\theta) = \sigma_{max} (\cos\theta - \cos\theta_f)^n$	$\theta_m \leq \theta \leq \theta_f$
	$\sigma(\theta) = \sigma_{max} [\cos\{\theta_f - ((\theta - \theta_r)/(\theta_m - \theta_r)) (\theta_f - \theta_m)\} - \cos\theta_f]^n$	$\theta_r \leq \theta \leq \theta_m$
	$\sigma_{max} = (ck_c + \rho k_\phi b) (r/b)^n$	
Bekker, 1969 (Model 2)	$\sigma(\theta) = \sigma_{max} [(\cos\theta - \cos\theta_f)/(\cos\theta_m - \cos\theta_f)]^n$	$\theta_m < \theta < \theta_f$
	$\sigma(\theta) = \sigma_{max} [(\cos\{\theta_f - ((\theta - \theta_r)/(\theta_m - \theta_r)) *$	$\theta_r < \theta < \theta_m$
	$(\theta_f - \theta_m)\} - \cos\theta_f)/(\cos\theta_m - \cos\theta_f)]^n$	
	$\sigma_{max} = (ck_c + \rho k_\phi b) (r/b)^n (\cos\theta_m - \cos\theta_f)^n$	
Wong and Reece, 1967 (Model 3)	$\sigma(\theta) = ((k_c/b) + k_\phi) r^N (\cos\theta - \cos\theta_f)^N$	$\theta_m \leq \theta \leq \theta_f$
	$\sigma(\theta) = ((k_c/b) + \rho k_\phi) r^N [\cos\{\theta_f - ((\theta - \theta_r)/(\theta_m - \theta_r)) *$	$\theta_r \leq \theta \leq \theta_m$
	$(\theta_f - \theta_m)\} - \cos\theta_d]^N$	
Iagnamma et al., 2004 (Model 4)	$\sigma(\theta) = \sigma_1(\theta), \sigma_1(\theta) = ((k_c/b) + k_\phi) [r(\cos\theta - \cos\theta_f)]^n$	$\theta_m < \theta < \theta_f$
	$\sigma(\theta) = \sigma_2(\theta), \sigma_2(\theta) = ((k_c/b) + k_\phi) [r(\cos\{\theta_f - ((\theta - \theta_r)/(\theta_m - \theta_r)) *$	$\theta_r < \theta < \theta_m$
	$(\theta_f - \theta_m)\} - \cos\theta_f)]^n$	

Notes: h = wheel sinkage; n = sinkage exponent; b = wheel width; c = cohesion stress of the soil; r = wheel radius; ρ = soil bulk density; k_c, k_ϕ = pressure-sinkage moduli; W = wheel weight; h = defines how much the wheel initially compacts the soil when it contacts the soil surface; kh = defines how much the soil recovers in height followingits departure from the soil surface; and k = wheel sinkage ratio (which denotes the ratio between the front and rear sinkages of the wheel).

Table 2. Shear stress distribution model.

Janosi and Hanamoto, 1961	J (m)	Remarks
$\tau_x = (c + \sigma(\theta) \tan\phi)$	$j (\theta) = r[\theta_f - \theta - (1 - s)(\sin\theta_f - \sin\theta)]$	Without lugs
$[1 - e^{-j(\theta)/K}]$	$j (\theta) = r[\theta_f^1 - \theta - (1 - s) (\sin\theta_f^1 - \sin\theta)]$	
	$\theta_f^1 = \cos^{-1}((r - z)/(r + H))$	With lugs
	H = lug height; z = wheel sinkage	

Notes: θ = internal friction angle of the soil; K = shear deformation modulus (depends on the shape of the wheel surface); j = soil deformation; and s = wheel slip (given as ratio of wheel width to the wheel radius).

$$\theta_m = (a_0 + a_1 s) \theta_f \qquad (4)$$

where a_0 and a_1 are parameters and are dependent on the wheel-soil interaction).

$a_0 \cong 0.4$; $0 \leq a_1 \leq 0.3$ (assumed values (Wong, 1965)).

2.1.2 Shear stress distribution model

The shear stress distribution model is given by Janosi and Hanamoto (1961), and is used to find the shear stress developed beneath the wheel as shown in above Figure 1, which is applicable to the Reece, Bekker, Wong-Reece and Iagnemma models. The shear stress distribution model is as tabulated as in Table 2.

3 COMPARISON OF STRESS DISTRIBUTION MODELS

To carry out the analysis of various models, an investigation into the physical and mechanical properties of TRI-1 soil simulant and the wheel parameters (Section 2.1) were necessary. Geotechnical tests were carried out on the soil simulant to determine their properties

(Sreenivasulu, 2014). The physical and mechanical properties of TRI-1, which is an anorthosite based lunar soil simulant, are given as: Cohesion (c) 0.36 kpa; Internal Friction Angle (ϕ) 43.33°; cohesion modulus (k_c) 6.60 kN/m^{n+1}; friction modulus (k_ϕ) 139 kN/m^{n+2}; minimum density (ρ_{min})1.15 g/cc (denoted as roh in Figures 2 to 9); maximum density (ρ_{max}) 1.88 g/cc (denoted as roh 2 in Figures 2 to 9); soil exponent (n) 0.404; and shear deformation modulus (K) 1.02 ± 0.76 cm ($K_1 = 0.0026$ m, $K_2 = 0.0103$ m, and $K_3 = 0.018$ m).

Analytical work carried out for each model, for two different wheels on the TRI-1 simulant, and comparison of normal and shear stresses are illustrated in Figures 2 to 9 for both wheels (Section 2.2).

3.1 Wheel parameter details

A small wheel with a dimension of 160 mm × 32 mm and a large wheel with a dimension of 210 mm × 50 mm, with weights of 52.189 N and 67.444 N respectively, were considered for this study.

3.2 Figures

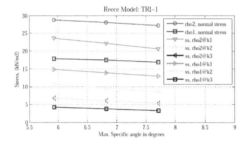

Figure 2.   Reece model/small wheel (Model 1).

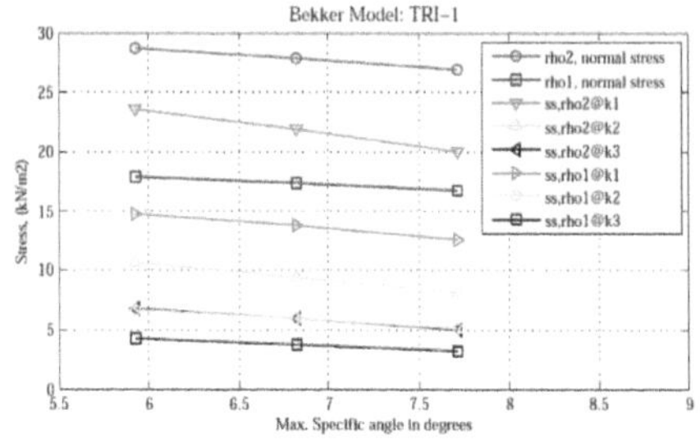

Figure 3.   Bekker model/small wheel (Model 2).

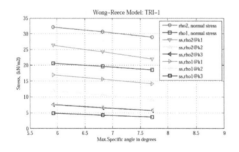

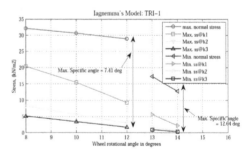

Figure 4.   Wong-Reece   model/small   wheel (Model 3).

Figure 5.   Iagnemma   model/small   wheel (Model 4).

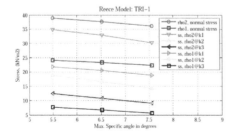

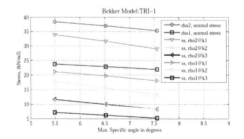

Figure 6.   Reece model/large wheel (Model 1).

Figure 7.   Bekker model/large wheel (Model 2).

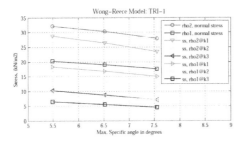

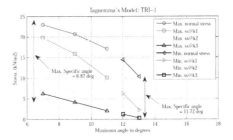

Figure 8. Wong-Reece model/large wheel (Model 3).

Figure 9. Iagnemma model/large wheel (Model 4).

4 RESULTS AND DISCUSSIONS

- From Figures 2 and 3, it was found that normal stress decreased with a rise in Maximum Specific Angle (θ_m). The minimum density induced minimum normal stress, whereas maximum density induced maximum normal stress. A similar phenomena was also found with the large wheel as shown in Figures 6 to 8.
- From Figures 2 and 3, it can be seen that shear stress decreases with an increase in the Shear Deformation Modulus (k). Shear stress at k_1 (θ_m = 5.93°) was found to be greater than the shear stress at k_3. It should also be noted that maximum normal stress and maximum shear stress were obtained at the minimum specific angle for the maximum density, rather than for an increase in θ_m.
- From Figure 4, it was found that the normal stress obtained was at its maximum (Wong-Reece model) when compared to other models. A similar phenomenon was observed when using other models.
- From Figure 5, it was found that the approach for finding normal and shear stress was similar, but finding the maximum specific angle differed. Maximum Angle (θ_m) was given as the average of the entry angle and exit angle. Thus, normal and shear stress distributions were found for the small wheel on the TRI-1 soil simulant.
- Compared to the small wheel, normal stress and shear Stress reached its maximum (Reece and Bekker models) for the large wheel on TRI-1 soil simulant.
- From Figure 7, it was found that a similar phenomenon of maximum normal stress at θ_m(5.49°) for dense soil was obtained. As well as for k_1, shear stress is maximum for the same case. In all other cases, shear stress decreased with an increase in θ_m.
- From Figure 9, it was found that normal and shear stress was less compared with the small wheel for Model 4 (as shown in Figure 5).
- Linear regression equations were developed for all the models (Reece, Bekker, Wong-Reece and Iagnemma) as well as R^2 values.
- It was found that the Wong-Reece model fitted well ($R^2 \cong 1$) compared to other models for both the plain small wheel and the plain large wheel.

5 CONCLUSIONS

Analytical work on the small wheel (160 mm × 32 mm) and large wheel (210 mm × 50 mm) were carried out using four models to determine normal and shear stress distributions beneath the wheel, when it interacted with soil. On comparing the results of both wheels for all models, it is concluded that the Wong-Reece model gave maximum normal and shear stress. Future scope exists in introducing lugs to the existing and in the comparison of both plain and lugged wheels. Work is currently progressing in this direction.

REFERENCES

Ding, L., Gao, H., Deng, Z. & Tao, J. (2010). Wheel slip-sinkage and its prediction model of lunar rover. *Journal of Central South University of Technology*, *17*, 129–135.

Iagnemma, K., Kang, S., Shibly, H. Dubowsky, S. (2004). Online terrain parameter estimation for wheeled mobile robots with application to planetary rovers. *IEEE Transactions on Robotics*, *20*(5).

Sreenivasulu, S. (2014). *Development and characterisation of TRI-1: An engineered lunar soil simulant and studies on wheel soil interaction* (Doctoral dissertation, Department of Civil Engineering, National Institute of Technology, Tiruchirappalli, India).

Sutoh, M., Nagatani, K. & Yoshida, K. (2012). Analysis of the travelling performance of planetary rovers with wheels equipped with lugs over loose soil. *Earth and space*, 1–10.

Sutoh, M., Yusa, J., Nagatani, K. & Yoshida, K. (2010). Travelling performance evaluation of planetary rovers on weak soil. *Journal of Field Robotics*.

Technology Drivers: Engine for Growth – Mahajan, Modi & Patel (Eds)
© 2018 Taylor & Francis Group, London, ISBN 978-1-138-56042-0

Experimental study of seismic response of different structural systems

Vineet Kothari

Department of Civil Engineering, Institute of Technology, Nirma University, Ahmedabad, Gujarat, India

ABSTRACT: During an earthquake a large amount of energy is released and structures are impacted by these forces. In order to reduce the response of a structure undergoing vibration it is necessary for the structure to absorb or dissipate the energy. There are two methods to improve seismic protection of a structure, either by increasing its stiffness (e.g. shear walls, bracings, MR-Frames dual systems) or by using modern methods to increase the performance of the structure (i.e. passive, active, semi-active, and hybrid control systems). This paper presents an experimental study of the dynamic responses of a shear wall model, viscoelastic damper model and bare frame model under different seismic excitations. Damping, stiffness, acceleration, velocity, and story displacement are calculated and compared for different indicative models. The results are presented and conclusions drawn on the basis of the study.

1 INTRODUCTION

Earthquakes are one of the major natural hazards for life on earth and have affected countless cities, towns and villages on almost every continent. Earthquakes cause damage to man-made structures. Conventional seismic design attempts to make buildings that do not collapse under strong earthquake shaking, but may sustain damage to non-structural elements and to some structural members in the buildings, which may cause a building to be non-functional after an earthquake. Basically, there are two methods to improve seismic protection of structures: first, traditional methods that increase the stiffness of structures (e.g. shear walls, bracings, MR-Frames dual systems); second, modern methods (i.e. passive, active, semi-active, and hybrid control systems).

It has been proved that viscoelastic dampers have been able to significantly increase the overall damping of a structure, hence improving the overall performance of dynamically sensitive structures (Samali & Kwok, 1995). Different experimental studies have also been conducted for full-scale model testing of different structural systems (Butterworth et al., 2004). It has also been proven that viscoelastic dampers provided better damping at higher temperatures (Chang et al., 1993). Some experimental studies have also been carried out on representative models with added bracing to increase seismic response control (Panchal & Purohit, 2012). The current paper focuses on the effectiveness of shear walls and viscoelastic dampers in structural response control.

1.1 *Structural system configuration*

For the experimental study, a three-story building model was fabricated. The frame of the model is square in plan as well as in elevation, and is constructed from steel bars and plates. This three-story model is illustrated in Figure 1 with different structural systems, where Figure 1b shows diagonally placed dampers, and Figure 1c shows steel plates acting as shear walls.

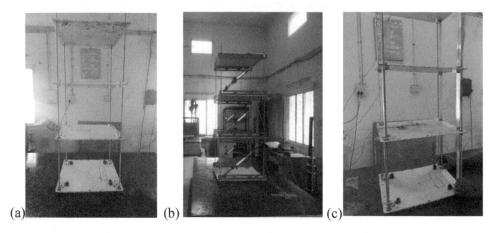

Figure 1. (a) Bare frame model; (b) viscoelastic damper model; (c) shear wall model.

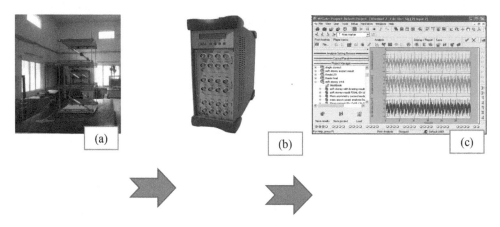

Figure 2. (a) Accelerometers attached to the model; (b) 16-channel vibration analyzer; (c) screen of NVGate software.

1.2 *Experimental setup*

Dynamic structural responses for three different structural models—bare frame model, shear wall model and viscoelastic damper model—were studied by incorporating free and forced vibrations. The responses were captured by OROS NVGate software. The data obtained were analyzed for frequency domain by calculating the natural frequency of the structure, and for time domain by calculating maximum acceleration, maximum velocity and maximum displacement for the structure. Elements of the experimental setup are shown in Figure 2. Uniaxial accelerometers are connected at three different stories, together with one accelerometer at the shake table, to calculate the response of the structure.

2 EXPERIMENTAL RESULTS AND DISCUSSION

All the experiments were first conducted on a bare frame model (Figure 1a). Then they were carried out on a viscoelastic damper model and a shear wall model (Figures 1b and 1c) in order to obtain the dynamic responses and properties of the different structures.

2.1 Calculation of stiffness

Stiffness was calculated for the different models with an assembly arranged as shown in Figure 3a. Load was increased on one side and displacement was calculated with the help of a dial gage. To calculate the stiffness of the third story, the first and second stories were locked with the help of a hook. Stiffness for the given model was calculated from a graph of slope of load vs deflection, as shown in Figure 3b. The calculated stiffnesses for the bare frame model, viscoelastic damper model and shear wall model were 17,691 N/m, 30,808 N/m and 43,491 N/m, respectively.

2.2 Calculation of damping ratio

Damping ratios for the three different models were determined using a free vibration test (logarithmic decrement method). In the case of a free vibration test, different accelerometers are attached at different story levels of the structure and an external displacement from its equilibrium position is applied at the top of the structure, as shown in Figure 4a. The structure is then released for damping-free vibration. The graph of acceleration versus time for the bare frame model is shown in Figure 4b, and based on this plot the damping ratio of the structure is calculated using Equation 1.The damping ratios calculated for the different models are shown in Table 1, with a comparison of all damping ratios shown in Figure 5.

The damping ratio can be determined from the following equation (Chopra, 2007):

$$\xi = \frac{1}{j^* 2\pi} \ln \frac{\ddot{u}_i}{\ddot{u}_{i+j}} \qquad (1)$$

where $\ddot{u}_i$ is peak acceleration for i^{th} cycle, and $\ddot{u}_{i+j}$ is peak acceleration for $i+j^{\text{th}}$ cycle where j is number of cycles.

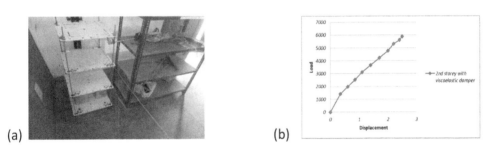

(a) (b)

Figure 3. (a) Assembly for stiffness calculation of third story; (b) calculation of stiffness from load vs displacement graph for viscoelastic damper model.

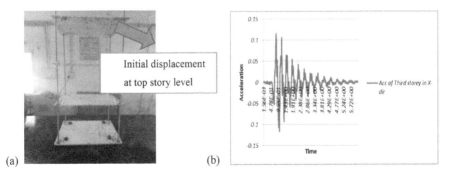

(a) (b)

Figure 4. (a) Bare frame model; (b) plot of acceleration vs time for bare frame model under free vibration.

Table 1. Damping ratio calculation for different models.

	Shear wall model	Bare frame model	Viscoelastic damper model
$\ddot{u}_i$	0.0508	0.0921	0.658
$\ddot{u}_{i+j}$	0.0256	0.078	0.264
ξ	0.064	0.153	0.145
$\xi(\%)$	5.34	1.53	14.51

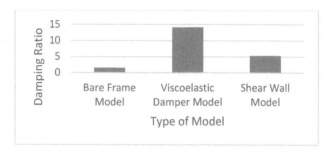

Figure 5. Comparison of damping ratios for different model types.

From Figure 5, it can be seen that the damping ratio increases dramatically from the bare frame model to the shear wall model, and from the shear wall model to the viscoelastic damper model. The highest damping ratio of 14.51% is achieved in the case of the viscoelastic damper model.

2.3 Calculation of natural frequency of the structure

To calculate the natural frequency of the structure, a Fast Fourier Transform (FFT) graph was plotted for the bare frame model using NVGate software, as shown in Figure 6. The fundamental natural frequency of the structure is calculated from this acceleration vs frequency graph. As shown in Table 2, the fundamental natural frequency is highest for the shear wall model, and lowest in the case of the bare frame model.

2.4 Comparison of acceleration velocity and displacement of different models at different forcing frequencies

In the case of the shake table test, two different models are given seismic excitation using forced vibration. The vibration properties of a structure are determined by varying the frequency of the shake table through a proportionate range. The amplitude of the maximum acceleration of the structure at each forcing frequency is measured. The original accelerations are integrated once to obtain velocity, and twice to obtain displacement of the structure. Different structural models are excited at forcing frequencies of 1.875 Hz and 2.5 Hz of the shake table and the response of the structures is captured using accelerometers attached to the shake table. Tables 3, 4 and 5 show the responses in terms of maximum acceleration and maximum displacement of structure for the bare frame model, viscoelastic damper model and shear wall model, respectively, for these two different forcing frequencies of the shake table (1.875 Hz and 2.5 Hz).

As shown in Figure 7, for 1.875 Hz frequency of shake table, percentage reductions in displacement for the viscoelastic damper model in comparison to the bare frame model for the three stories are 76.4%, 81% and 84.7%, respectively. For 2.5 Hz frequency of shake table, the percentage reductions in displacement for the viscoelastic damper model in comparison to the bare frame model for the three stories are 44%, 54% and 54.69%, respectively.

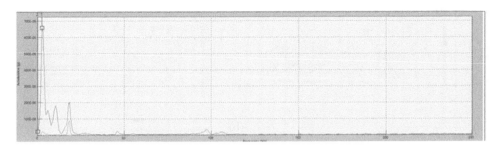

Figure 6.   FFT curve for bare frame model using NVGate software.

Table 2.   Natural frequency calculated for different structural systems using NVGate software.

	Bare frame model	Shear wall model	Viscoelastic damper model
Natural frequency (ω)	15.707 rad/sec	27.48 rad/sec	19.63 rad/sec

Table 3.   Response of bare frame model under different forcing frequencies of shake table.

Frequency (Hz)	1.875				2.5		
Story	Max acceleration (m/s²)	Max velocity (m/s)	Max displacement (cm)	Story	Max acceleration (m/s²)	Max velocity (m/s)	Max displacement (cm)
1st	1.027	0.274	0.81	1st	2.54	0.508	1.08
2nd	1.245	0.332	1.57	2nd	3.12	0.624	2.24
3rd	1.297	0.345	2.00	3rd	3.532	0.7064	3.064

Table 4.   Response of viscoelastic damper model under different forcing frequencies of shake table.

Frequency (Hz)	1.875				2.5		
Story	Max acceleration (m/s²)	Max velocity (m/s)	Max displacement (cm)	Story	Max acceleration (m/s²)	Max velocity (m/s)	Max displacement (cm)
1st	0.05373	0.01432	0.1911	1st	0.5826	0.11652	0.5652
2nd	0.0709	0.0189	0.2979	2nd	0.741	0.1432	0.882
3rd	0.1049	0.02797	0.3729	3rd	0.7662	0.15324	1.312

Table 5.   Response of shear wall model under different forcing frequencies of shake table.

Frequency (Hz)	1.875				2.5		
Story	Max acceleration (m/s²)	Max velocity (m/s)	Max displacement (cm)	Story	Max acceleration (m/s²)	Max velocity (m/s)	Max displacement (cm)
1st	0.06738	0.01786	0.1197	1st	0.1304	0.02608	0.1404
2nd	0.07689	0.0205	0.2534	2nd	0.1405	0.0281	0.2708
3rd	0.08608	0.0229	0.306	3rd	0.1675	0.0335	0.365

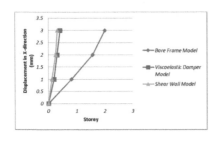 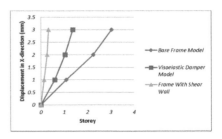

(a) (b)

Figure 7. Comparison of displacement responses for: (a) 1.875 Hz frequency of shake table; (b) 2.5 Hz frequency of shake table.

3 CONCLUSIONS

In this paper an experimental study of different structural systems is described. Different dynamic responses such as fundamental natural time period, damping ratio, acceleration, velocity and displacement are evaluated for a bare frame model, viscoelastic damper model and shear wall model. There is an increase in damping ratio for the viscoelastic damper model by comparison with the bare frame and shear wall models. Fundamental natural frequency and stiffness are highest in the case of the shear wall model. When compared with the bare frame model, the displacement calculated for three stories is lower in the case of the viscoelastic damper model, and lowest in the case of the shear wall model. Displacement reduction is higher at lower stories, showing that the viscoelastic damper model is more effective at lower stories.

REFERENCES

Butterworth, J., Lee, J.H. & Davidson, B. (2004). Experimental determination of modal damping from full scale testing. *13th World Conference on Earthquake Engineering, Vancouver, British Columbia, Canada, 1–6 August 2004* (Paper no. 310). Retrieved from http://www.iitk.ac.in/nicee/wcee/article/13_310.pdf.
Chang, K.C., Lai, M.L., Soong, T.T., Hao, D.S. & Yeh, Y.C. (1993). *Seismic behavior and design guidelines for steel frame structures with added viscoelastic dampers.* Buffalo, NY: National Center for Earthquake Engineering Research. Retrieved from http://mceer.buffalo.edu/pdf/report/93-0009.pdf
Chopra, A.K. (2007). *Dynamics of structures: Theory and application to earthquake engineering* (3rd ed.). Harlow, UK: Pearson Education.
Makris, N. & Constantinou, M.C. (1990). *Viscous dampers: Testing, modeling and application in vibration and seismic isolation.* Buffalo, NY: National Center for Earthquake Engineering Research. Retrieved from http://mceer.buffalo.edu/pdf/report/90-0028.pdf
Panchal, D. & Purohit, S. (2012). Dynamic response control of a building model using bracings. *Procedia Engineering, 51,* 266–273.
Samali, B. & Kwok, K.C.S. (1995). Use of viscoelastic dampers in reducing wind- and earthquake-induced motion of building structures. *Engineering Structures, 17*(9), 639–654.

Parametric study of multi-story buildings incorporating buckling-restrained braces

Kushal Parikh, Paresh V. Patel & Sharadkumar P. Purohit
Department of Civil Engineering, Institute of Technology, Nirma University, Gujarat, India

ABSTRACT: In recent years, the seismic design of buildings has undergone significant changes due to increasing demand for optimization of the structural systems of buildings to minimize the level of damage, economic loss, and structural repair costs following an earthquake. Buckling-Restrained Braces (BRBs) add more seismic energy dissipation capacity to conventional braced-frame systems. A typical BRB is made up of a yielding steel core, providing axial resistance which is confined by a concrete- or mortar-filled steel casing, providing flexural as well as buckling resistance. In this paper a parametric study of five- and ten-storied reinforced concrete buildings incorporating BRBs at various locations and in different configurations is presented. From the comparison of analysis results in terms of time period, story shear, story displacement and story drift, a more suitable configuration and location for BRBs are suggested for a multi-storied frame structure.

1 INTRODUCTION

A Buckling-Restrained Brace (BRB) is a structural element in a building that is designed to allow the building to withstand cyclical lateral loadings, typically induced by earthquake. It consists of a slender steel core in a concrete casing to support the core and prevent its buckling under axial compression. BRBs have improved energy dissipative behavior compared to Concentrically Braced Frames (CBFs). Bai and Ou (2016) developed a performance-based plastic design method for a dual system of reinforced concrete moment-resisting frames using BRBs. Sabelli and López (2004) discussed the methodology for design of buckling-restrained braced elements. Amiri et al. (2013) compared three steel frame structures of three, five and eight stories, retrofitted separately using tube-in-tube metal dampers and buckling-restrained braces, in order to compare their performance before and after the retrofitting. The current paper presents analysis of a multi-storied frame structure with BRBs to investigate the effectiveness of BRBs in controlling seismic response. A parametric study of five-storied and ten-storied reinforced concrete frame buildings incorporating BRBs at various locations with different configurations is carried out. Modal time period and various seismic parameters are compared for all buildings. Based on analysis of the results, the optimal location and configuration of BRBs are suggested.

2 BUILDING CONFIGURATION AND STRUCTURAL DATA

Regular framed structures with five and ten stories are considered in the study. The plan of the buildings is shown in Figure 1, and has dimensions of 23 m × 23 m. The size of each bay in both *x* and *y* dimensions is 4.6 m. Typical story height is 4.3 m, and the live load on a typical story is 3 kN/m². The floor finish on a typical story is 1.5 kN/m². Masonry walls of 115 mm thickness are assumed on all the beams. In accordance with IS 1893: Part 1 (BIS, 2002), the building is located in Zone III, supported on Medium soil and is assigned Importance Factor 1. The dimensions of structural elements and other structural data adopted for analysis and design are as shown in Table 1. Earthquake force on the buildings is estimated using an equivalent static method as per IS 1893: Part 1 (BIS, 2002). The buildings are modeled and analyzed

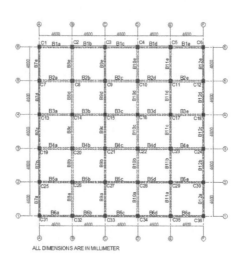

ALL DIMENSIONS ARE IN MILLIMETER

Figure 1. Plan view of the building.

Table 1. Structural data.

No. of stories	5	10
Building height (m)	21.5	43
Beam dimensions (mm)	230×465	230×465
Column dimensions (mm)	450×450	750×750
Slab thickness (mm)	165	165
Time period (in x direction) (secs)	0.403	0.807
Time period (in y direction) (secs)	0.403	0.807

using ETABS (Computers and Structures, Inc., Walnut Creek, CA, USA) engineering analysis software. Buckling-restrained braces are incorporated in the buildings at different locations in various configurations to characterize the behavior of the buildings.

3   CONFIGURATION AND LOCATION OF BRBs

In the present study, various configurations and locations of BRBs in five- and ten-storied structures are considered. The configurations of BRB considered, as shown in Figure 2, are Forward Diagonal (FD), Backward Diagonal (BD), Cross-Diagonal (CD) and Inverted-V Diagonal (IVD).

The BRBs are located on the periphery of the buildings. The various locations of BRBs considered in the study are a brace in the 3rd bay (Type 1), braces in the 1st, 3rd and 5th bays (Type 2), and braces in the 2nd and 4th bays (Type 3), as shown in Figure 3.

The design of Buckling-Restrained Braced Frames (BRBFs) is not governed by any building code, but recommendations of professional organizations are available. A joint AISC/SEAOC (American Institute of Steel Construction/Structural Engineers Association of California) group developed recommendations for design of a BRBF (Sabelli & López, 2004).

Axial forces in the bracings are extracted from the analysis results. Based on the axial forces and data from a manufacturer of BRBs (CoreBrace, 2016), various parameters are selected for modeling of BRBs in ETABS software. Table 2 shows the properties of BRBs assumed in the modeling of five- and ten-story buildings in ETABS.

142

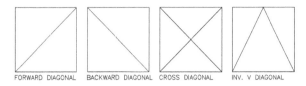

FORWARD DIAGONAL BACKWARD DIAGONAL CROSS DIAGONAL INV. V DIAGONAL

Figure 2.   Configurations of BRBs.

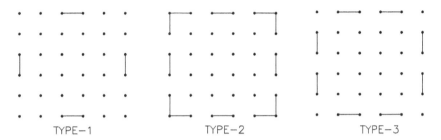

TYPE−1   TYPE−2   TYPE−3

Figure 3.   Locations of BRBs.

Table 2.   BRB properties.

Properties	5-Story	10-Story
Material of yielding core	Fe 262	Fe 262
Total BRB weight (kN)	3.106	3.915
Overall depth (mm)	203.2	254
Overall width (mm)	203.2	254
Area of yielding core (cm^2)	26	39
Length of yielding core (m)	3.97	3.97
Length of elastic segment (m)	2.33	2.33
Stiffness of elastic segment (kN/m)	804321	1013562

4   RESULTS AND DISCUSSION

Analysis results in terms of modal time period, story shear, story displacement, inter-story drift and internal forces in critical structural members are compared for a building without BRBs and buildings with BRBs in different configurations and locations.

4.1   *Modal time period*

A comparison of modal time period for buildings with BRBs in various locations and configurations is shown in Figure 4. It is clearly seen from Figure 4 that buildings with BRBs show a reduced time period compared to a bare frame (WB/Without BRBs). Further, the buildings with BRBs in a cross-diagonal configuration (Type 2) have the lowest modal time period.

4.2   *Story shear*

Story shear for buildings with various locations and configurations of BRB are shown and compared in Figure 5. Base shear and story shear are calculated as per IS 1893: Part 1 (BIS, 2002). Buildings with infill are considered for the calculation of time period of buildings with and without BRBs. Because the value of the time period of buildings according to the empirical formula given in IS 1893: Part 1 (BIS, 2002) remains the same, the values of story shear for the five- and ten-story buildings equates to the same for the bare frame buildings as for the frames with BRBs.

143

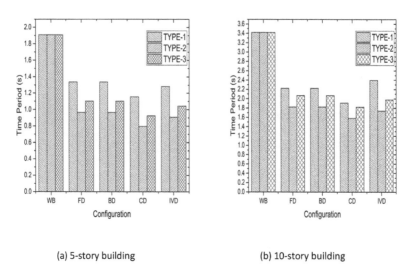

(a) 5-story building   (b) 10-story building

Figure 4. Comparison of modal time period.

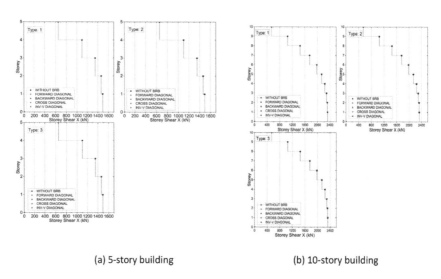

(a) 5-story building   (b) 10-story building

Figure 5. Comparison of story shear.

4.3 *Story displacement*

A comparison of story displacement for five- and ten-story buildings with and without BRBs is shown in Figure 6. The comparison shows that the maximum displacement for the five-story bare frame is 55.06 mm at the top story, and displacements for Type 1, Type 2 and Type 3 cross-diagonal configurations equate to 23.22 mm, 11.48 mm and 15.34 mm, respectively. The maximum displacement for the ten-story bare frame is 132.66 mm at the top story, and displacements for Type 1, Type 2 and Type 3 cross-diagonal configurations equate to 65.15 mm, 34.79 mm and 45.40 mm, respectively. Thus, inclusion of BRBs reduces the maximum top-story displacement.

4.4 *Story drift*

Story drift is defined as the ratio of displacement of two consecutive floors to the height between them. A comparison of story drift is shown in Figure 7 for buildings with various

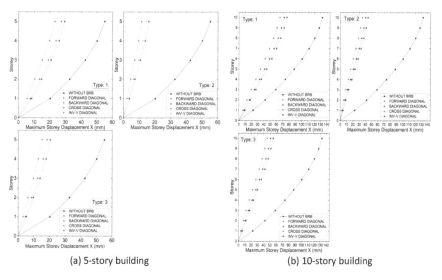

(a) 5-story building   (b) 10-story building

Figure 6.   Comparison of story displacement.

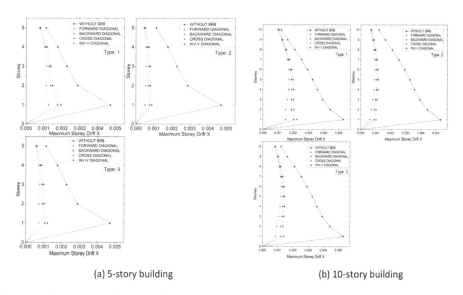

(a) 5-story building   (b) 10-story building

Figure 7.   Comparison of story drift.

locations and configurations of BRBs. From Figure 7 it is observed that the maximum drift for the five-story bare frame at story-1 is 0.0047, while story drifts for Type 1, Type 2 and Type 3 cross-diagonal configurations equate to 0.0013, 0.0005 and 0.0007, respectively. The maximum drift for the ten-story bare frame at story-1 is 0.0053, while maximum story drifts for Type 1, Type 2 and Type 3 cross-diagonal configurations equate to 0.0015, 0.0006 and 0.0009, respectively. Based on values of story displacements and drifts, Type 2 can be considered as the optimal location and cross-diagonal as the optimal configuration.

4.5   Axial forces and bending moments

Using load combinations as per IS 456 (BIS, 2000), axial forces are computed in columns and BRBs. The values showed that for a load case of 1.5 (Dead Load + Live Load) the maximum

145

axial force in column C33 is 997.78 kN, which is governing for design. In the load case of 1.5 (Dead Load ± Earthquake in x or y), the axial force in columns is reduced for buildings with BRBs. Thus, when the design of a building with BRBs is governed by lateral forces, the axial forces in columns are reduced by a significant amount. However, there was no significant change in axial forces of columns in the load case of 1.5 (DL + LL), even after placing BRBs. Thus it can be said that BRBs are effective only in response to lateral forces.

Bending Moment (BM) in Beam B6 where a BRB is placed is also compared for bare frames and frames with BRBs. The value of maximum BM in the case of a five-story bare frame with various load combinations is 187.92 kNm; on placing BRBs as per Type 2 with cross-diagonal configuration this is reduced to 56.63 kNm at supports. Similarly, the value of maximum BM from various load combinations in the case of a ten-story bare frame is 345.96 kNm, whereas on placing BRBs as per Type 2 with cross-diagonal configuration the figure reduces to 57.53 kNm at supports.

5 CONCLUDING REMARKS

Based on the analysis and design of five- and ten-story frame buildings with and without buckling-restrained braces of different configurations at different locations, the following conclusions are drawn:

– From the modal time period, story displacement & inter-story drift analyses it is found that Type 2 configuration of BRBs, that is, BRBs in 1st, 3rd and 5th bays, imparts higher stiffness to buildings under lateral loading.
– BRBs are effective when a building is subjected to lateral load due to earthquake. Under gravity loading they are not very effective.
– From the analysis and design results it is found that for Type 2 configurations, that is, BRBs in 1st, 3rd and 5th bays, a cross-diagonal system on the periphery is the optimal configuration and location for the five- and ten-story buildings considered in this study.

REFERENCES

Amiri, J.V., Mirzagoltabar, A.R. & Seifabadi, H.S. (2013). Effect of the height increasing on steel buildings retrofitted by buckling restrained bracing systems and TTD damper. *International Journal of Engineering, 26*, 1145–1154.
Bai, J. & Ou, J. (2016). Earthquake-resistant design of buckling-restrained braced RC moment frames using performance-based plastic design method. *Engineering Structures, 107*, 66–79.
BIS. (2000). *IS 456–2000: Plain and reinforced concrete code of practice.* New Delhi, India: Bureau of Indian Standards.
BIS. (2002). *IS 1893–2002 (Part 1): Criteria for earthquake design of structures: General provisions and buildings* (5th rev.). New Delhi, India: Bureau of Indian Standards.
CoreBrace. (2016). Resources (Design aids and test reports). West Jordan, UT: CoreBrace. Retrieved from http://www.corebrace.com/resources/.
Fahnestock, L.A., Sause, R. & Ricles, J.M. (2007). Seismic response and performance of buckling-restrained braced frames. *Journal of Structural Engineering, 133*(9), 1195–1204.
Kersting, R.A., Fahnestock, L.A. & López, W.A. (2015). *Seismic design of steel buckling-restrained braced frames: A guide for practicing engineers.* NEHRP Seismic Design Technical Brief No. 11. Gaithersburg, MD: National Institute of Standards and Technology. Retrieved from https://nvlpubs. nist.gov/nistpubs/gcr/2015/NIST.GCR.15-917-34.pdf.
Sabelli, R. & Aiken, I. (2004). U.S. building-code provisions for buckling-restrained braced frames: Basis and development. *13th World Conference on Earthquake Engineering, Vancouver, British Columbia, Canada, 1–6 August 2004* (Paper no. 1828). Retrieved from http://www.iitk.ac.in/nicee/wcee/article/13_1828.pdf.
Sabelli, R. & López, W. (2004). Design of buckling-restrained braced frames. *Modern Steel Construction, 44*(3), 67–73.

Technology Drivers: Engine for Growth – Mahajan, Modi & Patel (Eds)
© *2018 Taylor & Francis Group, London, ISBN 978-1-138-56042-0*

Parametric study of geopolymer concrete with fly ash and bottom ash activated with potassium activators

Sonal Thakkar & Praharshit Joshi
Department of Civil Engineering, Institute of Technology, Nirma University, Gujarat, India

ABSTRACT: Ordinary Portland cement is widely used cementitious material but is not eco-friendly. On the other hand, industrial wastes such as fly ash and bottom ash need to be more effectively utilized because they are produced in large volumes. In the present study, a concrete is produced by fully replacing Portland cement with fly ash and bottom ash. These source materials need to be activated by alkaline solutions. The alkaline liquids used in this study are solutions of potassium hydroxide and potassium silicate. An ambient curing technique was adopted, which allows its use for practical purposes. A parametric study was carried out to evaluate the effect of different parameters on the compressive strength of this geopolymer concrete. After achieving the compressive strength required, the concrete was also checked for split tensile strength and flexural strength and the results were slightly reduced when compared to the same grade of concrete formed from ordinary Portland cement.

1 INTRODUCTION

It is a well-known fact that the production of Ordinary Portland Cement (OPC) not only consumes significant amounts of natural resources and energy but also releases substantial quantities of carbon dioxide (CO_2) to the atmosphere. The amount of CO_2 released during the manufacture of OPC is the result of the calcination of limestone and combustion of fossil fuel and is in the order of one ton of CO_2 for every ton of OPC produced (Malhotra, 2000). Thus, it is essential to find alternatives to make environmentally friendly concrete. Bottom ash and fly ash are by-products of the combustion of pulverized coal in power plants. Fly ash is discharged in the precipitators and is obtained from the top of the power plant, while bottom ash is a coarser product and is generally discharged into ponds when washing of the residual ash is conducted. Presently, only 50% of fly ash is used while the rest is dumped as landfill; the use of bottom ash is very limited because it contains a large amount of unburnt coal particles and it is dumped as landfill (Xie & Ozbakkaloglu, 2015). Generally, bottom ash has a large particle size and a highly porous surface, resulting in higher water requirements and lower compressive strength. In geopolymer concrete, cement is totally replaced by any source material that is rich in silica (SiO_2) and alumina (Al_2O_3), and is made to react with binders which may be potassium- or sodium-based. These source materials and binders react to form chain-like polymers known as geopolymers and bind the materials (Davidovits, 1991). Geopolymerization generally requires heat for the binding reaction.

2 RESEARCH SIGNIFICANCE

Many studies have been carried out on geopolymer concrete using fly ash as a source material and it has been found to have excellent mechanical and durability properties, such as acid resistance, heat resistance, resistance to sulfate attack, and abrasion resistance (Hardjito et al., 2004; Kovalchuk et al., 2007). However, the use of bottom ash in alkali-activated concrete is restricted due to its particle size. Alkali-activated concrete gives very good compressive strength at an early age when temperature is applied. However, studies employing ambient temperature curing are lim-

Table 1. Chemical composition of materials.

Property	Unit	Fly ash	Bottom ash
Color	–	Light gray	Gray
SiO_2	%	61.44	60.48
Al_2O_3	%	31.80	32.16
CaO	%	1.20	1.04

ited in nature. Therefore, we make an attempt to study the parameters affecting mix design using ambient curing and potassium hydroxide (KOH) and potassium silicate (K_2SiO_3) as activators.

3 MATERIALS AND PREPARATION

3.1 *Materials*

Low-calcium, class-F fly ash and bottom ash, both obtained from a thermal power plant in Gandhinagar, India, were used in this study. Table 1 shows the chemical composition of the fly ash and bottom ash. Locally available 10 mm and 20 mm crushed aggregates have been used as coarse aggregates, while locally available river sand was used as a fine aggregate for concrete casting. Tests on both aggregates were conducted as per IS 2386 (BIS, 1963) and IS 383 (BIS, 1970), respectively, and corrections for moisture content and water absorption were carried out. The alkaline activators used were a combination of KOH and K_2SiO_3. The KOH was procured from a local market in flake form, and the K_2SiO_3 was in thick, sticky solution with a ratio of SiO_2/K_2O of 2.02 and specific gravity of 1.39.

3.2 *Preparation of alkaline solution and concrete*

The alkaline solution includes the particular molarity of KOH, which was diluted in water mixed with the K_2SiO_3. The amount of K_2SiO_3 was determined according to the KOH-to-K_2SiO_3 ratio. The solution for the geopolymer concrete was prepared one day before by dissolving the KOH flakes in the tap water available in the laboratory. The mass of KOH depends on its molarity in the mixture. Because there is no code for the design of geopolymer concrete mixture, a density method was used to determine the concrete ingredients. The method suggested by Hardjito et al. (2004) was used to design the mix, assuming density of concrete as 2400 kg/m³. Variation of mix parameters was conducted, as described in the following section, to evaluate the effect of each parameter on compressive strength. First, both coarse aggregate and fine aggregate were mixed in a pan mixture machine for four to five minutes. After this, source material was added in the form of fly ash and bottom ash and dry mixing was carried out for a further 3–4 minutes. The alkaline solution and extra water was added to the mix and further mixing was done to obtain a homogenous concrete. Subsequently, the concrete was poured into molds and compacted by means of a vibration table. The concrete was removed from the molds after a rest period of two days and then cured at ambient temperature.

4 PARAMETRIC STUDY

Concrete of M25 grade (having 25 N/mm² compressive strength after 28 days) using OPC was cast as a control; Table 2 describes the mixture design constituents on the basis of IS 10262 (BIS, 2009). For the parametric study, variation was made initially in the proportions of fly ash and bottom ash. It was found that an increase in bottom ash led to a decrease in compressive strength because it is more porous in nature. To utilize bottom ash more effectively in

Table 2. Mix proportions for OPC concrete.

Material	Unit	Content
Cement	kg/m³	334
Water	kg/m³	167
Coarse aggregate	kg/m³	1146
Fine aggregate	kg/m³	883

Table 3. Trial mixes of geopolymer concrete.

Mix	Fly ash (%)	Bottom ash (%)	Molarity	K_2SiO_3/KOH	Alkaline to cementitious material ratio
1	50	50	12	1	0.5
2	50	50	10	1	0.5
3	50	50	8	1	0.5
4	50	50	12	1.5	0.5
5	50	50	12	1	0.55
6	50	50	12	1.5	0.55

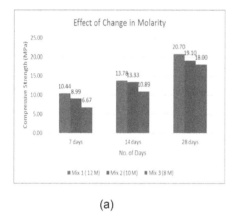

(a)

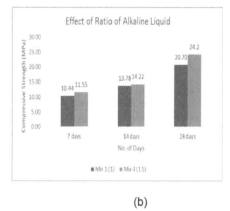

(b)

Figure 1. (a) Effect of different molarity ratios on compressive strength; (b) effect of change in alkaline liquid ratio on compressive strength.

construction, the optimum mixture of fly ash and bottom ash was determined. It was found that equal proportions of bottom ash and fly ash produced a mix design with M25 compressive strength. Further trial mixes were then generated by varying the molarity, the ratio of alkaline liquid, and the ratio of alkaline to cementitious material, as shown in Table 3.

5 RESULTS AND DISCUSSION

5.1 *Effect of different molarities of KOH solution*

The compressive strength results of Mixes 1, 2 and 3 are shown in Figure 1a, in which the molarity of KOH was varied from 12 M down to 8 M. It was observed that with a decrease in molarity the compressive strengths of the geopolymer concrete at 7, 14 and 28 days also decreased. This was due to less alkaline solution being available to react with the source material, leading to a decrease in strength.

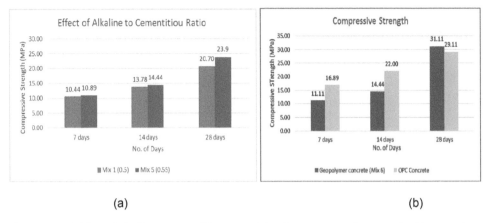

| (a) | (b) |

Figure 2. (a) Effect of alkaline to cementitious material ratio on compressive strength; (b) comparison of compressive strengths of OPC and geopolymer concretes.

5.2 *Effect of different ratios of alkaline liquid (K_2SiO_3/KOH)*

Figure 1b shows a comparison of compressive strength at different ratios of alkaline liquid at 7, 14 and 28 days. As the alkaline liquid ratio increased from 1 to 1.5, an increase in compressive strength was observed as higher amounts of K_2SiO_3 liquid were available for reaction with the source material, which increased the compressive strength.

5.3 *Effect of different ratios of alkaline to cementitious material*

Increase in ratio of alkaline liquid to cementitious material from 0.5 to 0.55 in Mix 1 and Mix 5, increases compressive strength at 7, 14 and 28 days as shown in Figure 2a. Increase in compressive strength was due to more amount of K_2SiO_3 which increases viscosity in the material helping to bind material better and increasing the strength.

6 MECHANICAL PROPERTIES

Variation was made in the dosage of super plasticizer and the amount of extra water added for workability purposes, and Mix 5 was arrived at, which achieved the target strength. Figure 2b shows compressive strength for the OPC and geopolymer concretes at various ages. At 28 days, the split tensile strength of the geopolymer concrete was 2.78 MPa, while that of the OPC concrete was 2.84 MPa; the flexure strength of the geopolymer concrete was 3.4 MPa and that of the OPC concrete was 3.8 MPa. Thus, both tensile and flexure strengths were higher for OPC concrete than geopolymer concrete, although the difference was small.

7 CONCLUSION

It can be concluded that a bottom ash and fly ash combination can be used to prepare medium-strength alkali-activated concrete. It was also found that an increase in molarity increases the compressive strength. Similarly, increases in the ratio of alkaline to cementitious material and the ratio of alkaline activator also increase the compressive strength. Through ambient curing, strength was achieved gradually and increased between 7 and 28 days. The split tensile strength and flexural strength of the geopolymer concrete were, respectively, 2.11% and 10.5% lower than those of OPC concrete. Thus, although a slightly lower strength was obtained in comparison to OPC concrete in terms of flexure and split tensile strength, compressive strength was approximately the same. This implies that medium-strength, ambient-cured geopolymer concrete with equal proportions of fly ash and bottom ash should be used because it satisfies strength criteria and is very sustainable from an environmental perspective. The ambient curing technique makes its application easier and means it can be readily applied in the field without any specific requirements.

REFERENCES

BIS. (1963). *IS 2386–1963: Methods of test for aggregates for concrete*. New Delhi, India: Bureau of Indian Standards.

BIS. (1970). IS 383–1970: *Specification for coarse and fine aggregates from natural sources for concrete*. New Delhi, India: Bureau of Indian Standards.

BIS. (2009). *IS 10262–2009: Guidelines for concrete mix design proportioning.* New Delhi, India: Bureau of Indian Standards.

Davidovits, J. (1991). Geopolymers: Inorganic polymeric new materials. *Journal of Thermal Analysis and Calorimetry*, *37*(8), 1633–1656.

Hardjito, D., Wallah, S.E., Sumajouw, D.M. & Rangan, B.V. (2004). On the development of fly ash-based geopolymer concrete. *ACI Materials Journal*, *101*(6), 467–472.

Kovalchuk, G., Fernández-Jiménez, A. & Palomo, A. (2007). Alkali-activated fly ash: Effect of thermal curing conditions on mechanical and microstructural development–Part II. *Fuel*, *86*(3), 315–322.

Malhotra, V.M. (2000). Introduction: Sustainable development and concrete technology. *Concrete International*, *24*(7), 1147–1165.

Xie, T. & Ozbakkaloglu, T. (2015). Behavior of low-calcium fly and bottom ash-based geopolymer concrete cured at ambient temperature. *Ceramics International*, *41*(4), 5945–5958.

Technology Drivers: Engine for Growth – Mahajan, Modi & Patel (Eds)
© 2018 Taylor & Francis Group, London, ISBN 978-1-138-56042-0

Study of shear lag effect in a hybrid structural system for high-rise buildings

Deep Modi, Paresh V. Patel & Digesh Joshi
Department of Civil Engineering, Institute of Technology, Nirma University, Gujarat, India

ABSTRACT: Advances in construction technology, materials, structural systems and analytical tools for analysis and design have facilitated the growth of high-rise buildings. The structural design of high-rise buildings is governed by lateral loads resulting from wind or earthquake. Lateral load resistance is provided by interior or exterior structural systems. Shear wall core and outrigger and belt truss are examples of interior structural systems, while tubular frame and diagrid are examples of exterior structural systems. Sometimes, the combined characteristics of two or more structural systems can be effectively used to satisfy the design criteria of high-rise buildings. This paper explores the shear lag effect in a hybrid structural system for a high-rise building: a G+60-story building with combined outrigger and diagrid structural systems is considered. Modeling and analysis of the building are carried out using ETABS software to evaluate the shear lag effect and compute the shear lag ratio for this hybrid high-rise structural system.

1 INTRODUCTION

With rapid urbanization and increasing scarcity of land, efficient structural systems are necessary for constructing high-rise buildings. For any high-rise building lateral loads such as wind and earthquake are governing, in addition to gravitational load. The higher the building, the more important is its lateral load-resisting system (Ali & Moon, 2007). The various lateral load-resisting systems are classified as interior or exterior structural systems. Outrigger and belt truss, Steel Plate Shear Wall (SPSW), and so on, are interior structural systems where lateral load resistance is provided by structural elements placed near the center of the building. Framed tube, diagrid, and so on, are exterior structural systems where lateral load is resisted via structural elements placed on the periphery of the building. A combination of interior and exterior structural systems, generally known as hybrid structural systems, have also evolved for high-rise buildings.

Researchers identified the shear lag effect in the 1930s and conducted studies on box beams. The shear lag effect can be assumed for any slender box element under lateral loading, such as airplane wing structures and box girders of bridges, as well as some of the structural systems of high-rise buildings such as core walls and framed-tube systems (Leonard, 2007). According to the assumptions of beam theory, plane sections remain plane after bending causing linear distribution of bending stress in the cross section of the beam. This assumption is only applicable in a box section if the shear stiffness of the cross section is very high or if there is no shear force in the box. However, in the presence of shear force in the box, shear flow develops across the flange and web panels. The presence of shear flow in panels causes longitudinal displacement of columns in such a way that the middle portion of the flange and web lag behind compared to the corner portion of box. This phenomenon causes non-linear longitudinal displacement of the flange and web, which results in non-linear stress distribution, as shown in Figure 1. Because of the shear lag effect, stress at the corner of the box increases, where web and flange meet, while the middle portion of the web and

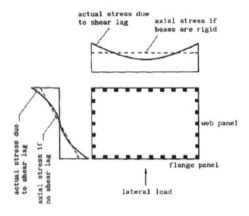

Figure 1. Axial stress in web and flange panel (Leonard, 2007).

flange panel has lower amounts of stress. This leads to increased lateral deflection of the structure, causing differential lateral deflection within the panels of the building (Leonard, 2007). The shear lag ratio for different structural systems such as braced-tube and framed-tube buildings ranges from 2.2 to 5.94 (Mazinani et al., 2014).

The shear lag effect reduces the efficiency of structural systems such as framed tube. The shear lag effect in hybrid structural systems is explored in this paper. A hybrid structural system can be derived by combining an outrigger and belt system with a diagrid structural system, and may consist of a SPSW core with an outrigger and belt truss system at the mid-height of the building and a diagrid system on the building's periphery. An outrigger and belt truss structural system has the advantage of reduced the bending moment in the shear wall core. Because a diagrid structural system resists lateral load through axial force in inclined columns at the periphery, it is more rigid. A hybrid of the outrigger and diagrid structural systems will have the advantages of both systems.

2 MODELING, ANALYSIS AND DESIGN OF HYBRID STRUCTURAL SYSTEM

A G+60-story building located in Ahmedabad, India, with a hybrid structural system is considered in the study. The building has plan dimensions of 36 m × 36 m, as shown in Figure 2a. The typical story height (h) is taken as 3.6 m. The building has a diagrid on its periphery and an outrigger and belt truss system at the 31st and 32nd floors, as shown in Figure 2c. It has a steel plate shear wall in the core region of the building. Dead load and live load are considered as per IS875 (Part 1) and IS875 (Part 2), respectively (BIS, 1987a, 1987b). Wind load is considered as per IS875 (Part3) (BIS, 2015), and earthquake load is considered as per IS1893 (Part1) (BIS, 2002). The building is modeled and analyzed in ETABS software (Computers and Structures, Inc., Walnut Creek, CA, USA). A portion of the model of the hybrid structural system in ETABS is shown in Figure 2b. Linear static analysis of the building is carried out and the time period of the building is found to be 6.82 seconds, and maximum lateral displacement and maximum inter-story drift of the building are, respectively, 431.23 mm and 2.41 mm, which are within permissible limits.

Various components of the structural system are designed for governing load cases as per IS 800 (BIS, 2007). A composite slab of steel deck and Reinforced Cement Concrete (RCC) of total thickness 150 mm are assumed. The sizes of the structural members assumed for the study are shown in Table 1 and cross sections of the members are shown in Figure 3. The sizes of the structural members are assumed to be uniform throughout the height of the building.

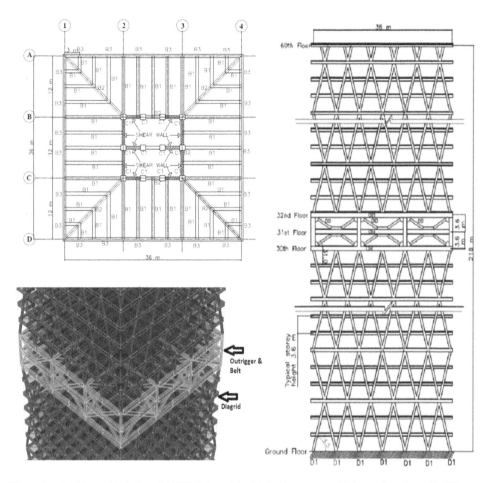

Figure 2.   (a) Plan of building; (b) ETABS model of hybrid structure; (c) front elevation of building.

Table 1.   Sizes of structural members of hybrid system.

Steel plate shear wall core	Plate thickness (mm)	10			
	Boundary elements, *C1* (*B × D × tk*) (mm)	1200 × 1200 × 50			
		Overall height, *D* (mm)	Flange width, *Wf* (mm)	Flange thickness, *Tf* (mm)	Web thickness, *tw* (mm)
Outrigger at mid-height	Diagonals, *OD*	600	500	50	50
	Beams, *OB*	650	500	50	50
Beams at floor level	*B1*	ISMB 600			
	B2	ISMB 600 with 60 mm top & bottom plates			
	B3	ISMB 500			
Diagonal column on periphery of building, *D1*	(*D × tk*) (mm)	275 × 30			

155

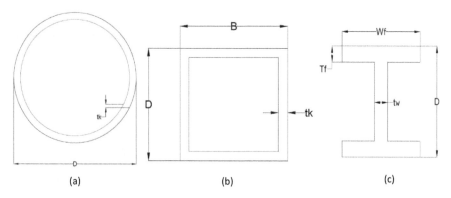

(a)   (b)   (c)

Figure 3. Cross sections of structural members: (a) diagrid member; (b) boundary element of steel plate shear wall; (c) outrigger and belt truss at mid-height.

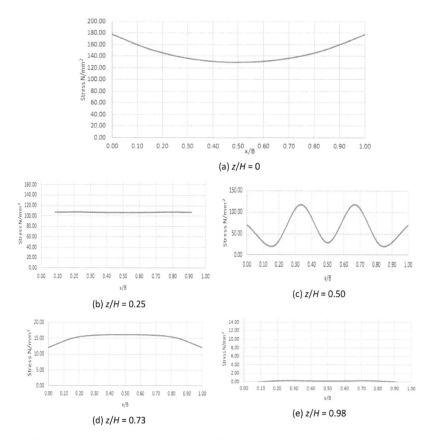

Figure 4. Axial stress vs (x/B) for flange panel at different heights in building.

3 SHEAR LAG RATIO

From the analysis results, axial forces in the peripheral columns are extracted to understand the shear lag effect. Stresses and forces in the flange and web of the building are evaluated assuming dynamic wind loading on the building in the x direction (DWLX) according to IS 875 (Part 3) (BIS, 2015).

156

Table 2. Shear lag ratio.

| z/H | Axial compressive stress (N/mm²) | | Shear lag ratio (f) |
	Corner column	Center column	
0	177.71	129.68	1.37
0.25	107.58	106.73	1.01
0.5	21.39	29.12	0.73
0.73	15.20	16.18	0.94
0.98	0.01	0.24	0.05

The axial compressive stress in flange columns at the base and at $0.25H$, $0.5H$, $0.73H$ and $0.98H$, where H is the height of the building, are calculated. Axial stress vs x/B (where x is horizontal distance from left side of building, and B is width of building) at various values of z/H are shown in Figure 4 (where z is height of interest). Because a diagrid does not have vertical elements, to obtain axial stresses in the longitudinal direction the vertical component of stress in the diagrid must be considered. At nodal points, two diagrid members intersect, hence the sum of the vertical components of axial stress of both diagrids at the point of interest is taken into consideration. From Figure 4, it can be seen that the effect of shear lag is maximal at the base, where the corner column has higher axial stress compared to the central column, as shown in Figure 4a. At one-quarter height of the building, all flange columns have almost equal axial stress, as shown in Figure 4b. At three-quarters height of the building, the corner columns are stressed less compared to the central column, representing a negative shear lag effect, as shown in Figure 4d. At mid-height of the building, because of the outrigger the variation in axial stress in the flange column is non-monotonic, as shown in Figure 4c.

Table 2 presents the maximum values of stress in the corner column and central column, as well as the shear lag ratio (f), which is the ratio of axial stress in corner column to central column.

4 SUMMARY AND CONCLUSIONS

In the present study the shear lag effect in a G+60-story building with 36 m × 36 m plan dimensions and 216 m height, having a hybrid structural system of diagrid and outrigger, is presented. Modeling and linear static analysis are carried out using ETABS software.

From the study the following conclusions are derived:

- There is existence of shear lag in the hybrid structural system. A positive shear lag effect is observed at the base of the building, while at three-quarters height of the building a negative shear lag effect is observed.
- The maximum shear lag ratio of 1.37 is observed at the base of building. At one-quarter height of the building the shear lag ratio approaches a value of 1, which indicates the absence of a shear lag effect. The presence of an outrigger and belt truss at mid-height causes non-monotonic change in the shear lag effect.
- The hybrid structural system experiences a lower shear lag effect compared to a framed-tube structural system, which means differences in axial stress between corner and central parts are lower, indicating a reduced amount of longitudinal displacement. In high-rise buildings, this will cause less damage to non-structural members such a sexternal cladding.

REFERENCES

Ali, M.M. & Moon, K.S. (2007). Structural developments in tall buildings: Current trends and future prospects. *Architectural Science Review*, 50(3), 205–223.

BIS. (1987a). *IS:875 (Part 1) –1987: Code of practice for design loads (other than earthquake) for buildings and structures:Dead loads.* New Delhi, India:Bureau of Indian Standards.

BIS. (1987b). *IS:875(Part 2) – 1987: Code of practice for design loads (other than earthquake) for buildings and structures:Imposed loads.* New Delhi, India: Bureau of Indian Standards.

BIS. (2002). *IS:1893 (Part 1) – 2002: Criteria for earthquake resistant design of structures.* New Delhi, India: Bureau of Indian Standards.

BIS. (2007). *IS:800–2007: Code of practice for general construction in steel.* New Delhi, India: Bureau of Indian Standards.

BIS. (2015). *IS:875 (Part 3) – 2015: Code of practice for design loads (other than earthquake) for buildings and structures:Wind loads.* New Delhi, India: Bureau of Indian Standards.

Leonard, J. (2007). *Investigation of shear lag effect in high-rise buildings with diagrid system* (Master's thesis, Dept. of Civil and Environmental Engineering, Massachusetts Institute of Technology, Cambridge, MA). Retrieved from http://hdl.handle.net/1721.1/39269.

Mazinani, I., Jumaat, M.Z., Ismail, Z. & Chao, O.Z. (2014) Comparison of shear lag in structural steel building with framed tube and braced tube. *Structural Engineering and Mechanics, 49*(3), 297–309.

Sarkisian, M.P. (2015). Fazlur Khan's legacy: Towers of the future. *Structure and Infrastructure Engineering, 12*(7), 802–821. doi:10.1010/15732479.1064969.

Technology Drivers: Engine for Growth – Mahajan, Modi & Patel (Eds)
© *2018 Taylor & Francis Group, London, ISBN 978-1-138-56042-0*

Development of design aid for non-rectangular reinforced concrete column–C shape

Arth Patel
Department of Civil Engineering, Institute of Technology, Nirma University, Ahmedabad, Gujarat, India

Rutvik Sheth
Department of Civil Engineering, Institute of Technology Dharmsinh Desai University, Nadiad, Gujarat, India

ABSTRACT: On day to day basis a structural designer comes across design of Non-Rectangular Reinforced Concrete (RC) column. Commonly the design of Non-Rectangular RC column except the circular section dealt with considering equivalent rectangular section due to limitation of codal stipulation IS 456:2000. A need has been realised to develop design aids for Non-Rectangular RC column which involve iterative & tedious design calculations. This paper aims to design Non-Rectangular RC short/slender column—C Shape, from first principle for axial load and biaxial bending moment. Apart from this computer module is developed in Visual studio 2012 for quick calculation, validation of Axial-Flexure (P-M) interaction diagram of developed computer program is carried out by comparing it with well-established computer program SAP 2000.

1 INTRODUCTION

Non-Rectangular RC columns such as C, L, T, and Cross (+) shaped sections are often used at outside and re-entrant building corners for architectural purposes (Subramanian, 2013). And as a structural member they are very important for the proper functioning of the building, and the failure at a critical location may result in the collapse of the entire building. So there should be reliable computer aid that can be directly used to design columns without trial and error process. And also the Non-Rectangular RC Columns can be classified as short and slender columns on the basis of their length, lateral dimensions, support condition, cross sectional shapes and nature of loading. Interaction curves for designing such kind columns subjected to axial load, axial load with uniaxial bending and axial load with biaxial bending have been developed by (Sinha, 1988) which requires interpolation (Subramanian, 2013). This motivate to develop a program module that can overcome this limitations and provide detailed and reliable results.

Strength of short column is generally governed entirely by the strength of the materials used and the cross sectional shape. But the slender columns are prone to buckling effect which depends on the columns cross section, lateral bracing system and the degree of resistance at the ends. The slender columns may be defined as a column that has significant reduction in its axial load capacity due to moment resulting from the lateral deflection of column. Slender concrete columns may fail by buckling in the elastic or inelastic stress state which is an instability failure or they may fail when the compressive strain in the concrete reaches its limits of 0.0035 which is the material failure (IS 456:2000).

2 THEORETICAL FORMULATION

The design of column section for given axial load, uniaxial and biaxial moment is commenced by assuming primary dimensions of cross section and suitable reinforcement pattern.

The principle of equilibrium, strain compatibility and material stress strain behavior are used to determine the axial load capacity of the column. If the section is inadequate to resist the given axial load and moment, the reinforcement percentage and size of cross section are revised. The formulation for C-Shaped RC column is shown in the paper.

The strains at different levels of reinforcement in steel and concrete are calculated from strain compatibility condition of plane section analysis. The stress values in concrete and steel at different levels is computed from stress strain curve according to the provisions of IS-456:2000.

In slender column there is significant reduction in axial load capacity because of the additional moment developed by deflection of member between its ends (Subramanian, 2013). This is incorporated in the computer program module as per the codal stipulation (SP 16:1978).

2.1 Governing equations

The ultimate load carrying capacity of the column Pu is calculated using strain and stress for various positions of neutral axis. The behaviour of members subjected to uniaxial and biaxial moment depends upon the amount of eccentricity or the magnitude of the bending moment in relation to the axial load which affects the position of neutral axis. Based on the nature of bending moment, the compressed edge can be either at flange portion or web portion. The formulation for highly compressed edge at flange portion is shown here and similarly it be done for highly compressed edge at web portion. Various situations which can arise based on the position of neutral axis, of which one of the situation neutral axis lies outside the section is mentioned as under.

2.1.1 When neutral axis lies outside the section

When the load is not axial but acting at a small eccentricity, the bending effect is very less compared to the effect of direct load. The strain distribution diagram becomes trapezoidal instead of rectangular. The stress distribution is rectangular from highly compressed edge to a depth where the strain is 0.002 and it is parabolic over the remaining region where strain is less than 0.002. When the neutral axis lies along the one edge of the section, the strain values linearly varies from 0.0035 at the highly compressed to zero at the opposite edge (Variyani & Radhaji, 2002). For purely axial compression, the strain is assumed to be uniformly equal to 0.002 across the section. The strain distribution lines for these two cases intersect each other at a depth of 3D/7 from the highly compressed edge. This point is assumed to act as a fulcrum for the strain distribution line when the strain at the highly compressed edge is 0.0035 minus 0.75 times the strain at least compressed edge as per IS: 456 clause 38.1(b). Here the load carrying and bending moment capacity of section is calculated by considering rectangular-parabolic stress block considering capacity of concrete and steel separately.

$$P_{ul} = 0.446 f_{ck} \times B_f \times D_f \tag{1}$$

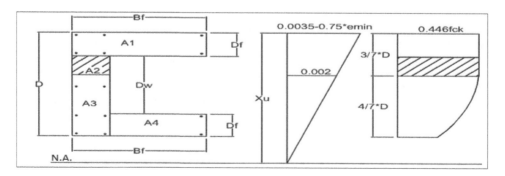

Figure 1. Location of N.A for the design of C-shape RC column along with stress and strain block for one of the possible case.

160

$$P_{u2} = 0.446 f_{ck} \times B_w \times \left(\frac{3}{7}D - D_f \right) \qquad (2)$$

$$P_{u3} = \int \left\{ B \times \left(-\frac{h}{g^2}x^2 + 2\frac{h}{g}x \right), kD - D, kD - D + D_f \right\} \qquad (3)$$

$$P_{u4} = \int \left\{ \left(B_f - B_w \right) \times \left(-\frac{h}{g^2}x^2 + 2\frac{h}{g}x \right), kD - D, kD - D + D_f \right\} \qquad (4)$$

$$P_u = \left\{ P_{u1} + P_{u2} + P_{u3} + P_{u4} \right\} + \sum_{i=1}^{n} \frac{P_i \times A_g}{100}\left(f_{si} - f_{ci} \right) \qquad (5)$$

$$A_g = 2 \times \left(B_f \times D_f \right) + \left(B_w \times D_w \right) \qquad (6)$$

$$X_1 = \frac{D_f}{2} \qquad (7)$$

$$X_2 = D_f + \frac{\frac{3}{7}D - D_f}{2} \qquad (8)$$

$$AX_3 = \int \left\{ B_w \times \left(-\frac{h}{g^2}x^2 + 2\frac{h}{g}x \right) \times x, kD - D, kD - \frac{3}{7}D \right\} \qquad (9)$$

$$X_3 = kD - \frac{AX_3}{P_{u3}} \qquad (10)$$

$$AX_4 = \int \left\{ \left(B_f - B_w \right) \times \left(-\frac{h}{g^2}x^2 + 2\frac{h}{g}x \right) \times x, kD - D, kD - D + D_f \right\} \qquad (11)$$

$$X_4 = kD - \frac{AX_4}{P_{u4}} \qquad (12)$$

$$M_u = P_{u1}\left(C_y - X_1 \right) + P_{u2}\left(C_y - X_2 \right) + P_{u3}\left(C_y - X_3 \right) + \sum_{i=1}^{n} \times \frac{P_i \times A_g}{100} \times \left(f_{si} - f_{ci} \right) \times Y_i \qquad (13)$$

Here, P_{u1}, P_{u2}, P_{u3}, P_{u4} = Axial load carrying capacity concrete sections A1, A2, A3, A4 respectively.
X_1, X_2, X_3, X_4 = Distance between the Center of Gravity (C.G.) of sections A1, A2, A3, A4 to top edge of whole section respectively.
P_i = Percentage of steel at Y_i Distance from C.G.
g = Geometrical property of parabolic stress block.

3 VALIDATION OF COMPUTER PROGRAM MODULE

The developed computer program module (Figure 4 and Figure 5) gives the interaction charts for given section of column and reinforcement data. It allows the input of vast range of load combinations in a single run. The program incorporates concrete grade of M20 and above, steel grade of Fe415, Fe500. It checks the adequacy of reinforcement provided based on codal provisions. It verifies the detailing criteria for cover, spacing of reinforcement as well as minimum and maximum percentage of reinforcement. Verification Problem taken for data: B = 1000 mm, B1 = 250 mm, D = 1000 mm, D1 = 250 mm, fck = 20 MPa, fy = 415 MPa, Pu = 4000 kN, Mux = 1250 kNm, Muy = 1250 kNm, and reinforcement detailing as shown in Figure 2. Validation of the interaction diagram of program module developed has been done with the SAP2000 shown in Figure 3.

161

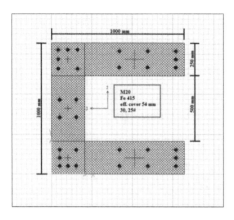

Figure 2.   Cross section of C shape column.

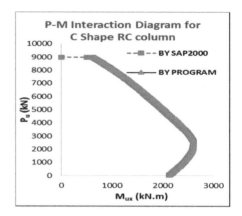

Figure 3.   Comparison of interaction diagram of C shape RC column.

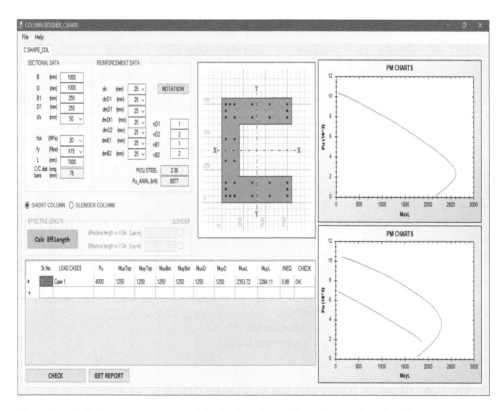

Figure 4.   Main form of program module developed which have inputs of sectional/geometrical data, reinforcement details, short/slender section and axial & biaxial moment

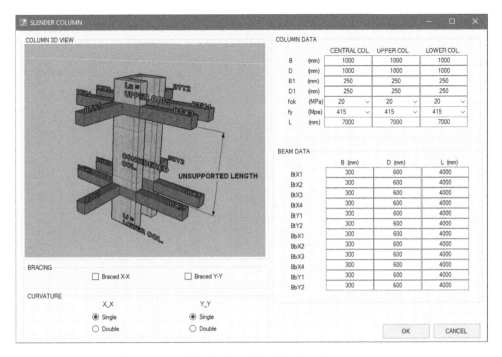

Figure 5. Sub-form of program module developed for slender column.

4 SUMMARY AND CONCLUSION

Formulation of Non-Rectangular short/slender RC column – C Shape has been carried out considering the various positions of neutral axis and apart from this a computer program module has been developed to design for axial load and biaxial bending moment.Program module developed is validated with well-established computer program SAP 2000.

This program module developed can be well utilized as it is very quick and reliable to use. Such a work can also be extended for design of other Non-Rectangular Shaped sections i.e. L, T & Cross (+).

REFERENCES

Bureau of Indian Standards (1978) SP 16:1978. *Design Aid for Reinforced concrete to IS 456: 1978.* New Delhi, BIS.
Bureau of Indian Standards (2000) IS 456:2000. *Plain and Reinforced Concrete Code of Practice.* New Delhi, BIS.
Sinha, S.N. (1988). *Handbook of Reinforced Concrete Design.* Tata McGraw Hills.
Subramanian, N. (2013). *Design of Reinforced Concrete Structures.* Oxford University Press.
Variyani, U.H., & Radhaji, A. (2002). *Design Aids for Limit State Design of Reinforced Concrete Members.* Khanna Publishers.

Mechanical engineering track

Technology Drivers: Engine for Growth – Mahajan, Modi & Patel (Eds)
© 2018 Taylor & Francis Group, London, ISBN 978-1-138-56042-0

Dynamic modeling of a two-link flexible manipulator using the Lagrangian finite elements method

Natraj Mishra & S.P. Singh
Department of Mechanical Engineering, UPES, Dehradun, Uttarakhand, India

ABSTRACT: In this paper, the mathematical modeling of a two-link flexible manipulator is given using the Lagrangian finite elements method. Plane frame elements are used for modeling the links of the flexible manipulator. The equation contains the scope for consideration of viscous damping, centrifugal, Coriolis and gravitational forces, and coupling between rigid and flexible motions. Simulation results are obtained by directly solving the matrix equations. The joint responses and tip responses of the links are obtained for stepped input torques. The paper considers the effect on natural frequencies of the flexible manipulator due to change in its configuration. The direct solving of the matrix equation of motion requires great computational effort. To overcome this drawback, it is suggested that the assumed modes method may be used along with the finite elements method.

1 INTRODUCTION

The presence of vibration in robotic arms decreases their accuracy. The controlling of vibrations is a great challenge. The vibrations occur for various reasons, such as the presence of inertia and stiffness in the links, or sudden change in motion of the links. Various authors have done research in the areas of dynamic modeling and the control of flexible manipulators. From the available literature, it has been found that there are two main approaches for modeling the dynamics of flexible manipulators: the Lagrangian assumed modes method and the Lagrangian Finite Elements Method (FEM). The present work focuses on the use of the Lagrangian finite elements method. Since the 1980s, most authors have used the Lagrangian finite elements method for dynamic modeling. Among a variety of researchers, some of those who have used the finite elements method for modeling flexible manipulators are Sunada and Dubowsky (1981), Dado and Soni (1986), Naganathan and Soni (1986), Usoro et al. (1986), Bayo (1987), Simo and Vu-Quoc (1987), Tzou and Wan (1990), Chedmail et al. (1991), Alberts et al. (1992), Gaultier and Cleghorn (1992), Hu and Ulsoy (1994), Stylianou and Tabarrok (1994a, 1994b), Theodore and Ghosal (2003), and Fotouhi (2007).

2 MATHEMATICAL MODELING

Mishra et al. (2015) have provided a mathematical model of two-link flexible manipulators using the Lagrangian assumed modes method. They describe an expression for joint torques, which is given by Equation 1:

$$\tau_j(t) = M_{j1}\ddot{\theta}_1 + M_{j2}\ddot{\theta}_2 + M_{j3}\ddot{w}_1 + M_{j4}\ddot{w}_2 + M_{j5}\ddot{w}_1^*$$
$$+ H_j + G_j + C_{j1}\dot{\theta}_1 + C_{j2}\dot{\theta}_2 + C_{j5}\dot{w}_1^* + K_{j5}w_1^* + K_j^{\#} \tag{1}$$

where subscript '*j*' denotes the joint number (*j* = 1,2 for the present case). The terms containing M_j, H_j, G_j, C_j and K_j correspond to inertia, centrifugal/Coriolis, gravity, viscous damping and restoring torques/forces. Equation 1 was derived using Lagrangian dynamics. In FEM, the system is divided into finite elements having nodes, and complete focus is now upon the nodes. By increasing the number of finite elements, the number of nodes increases and hence the accuracy of the solution also increases.

Figure 1 shows the dynamic modeling of the two-link flexible manipulator using two finite elements. From the figure, it can be seen that there are three reference frames: one global frame of reference (X-Y) and two local frames of reference (X_1–Y_1 and X_2–Y_2). The rigid rotations of the neutral axes (X_1 and X_2) of the links are represented by θ_1 and θ_2. The elastic motion of any point on Link 1 and Link 2 are represented by w_1 and w_2 respectively. The motion w_1^* represents the elastic motion of the end point of Link 1. The second link is attached at that point. The flexible manipulator is modeled using two plane frame elements. A plane frame element has two nodes with each node having three degrees of freedom: one axial and two bending (deflection and slope). The equation of motion for damped vibrations for an 'n' degrees of freedom system is as follows.

$$[M_{ff}]_{n \times n}\{\ddot{q}_f\}_{n \times 1} + [C_{ff}]_{n \times n}\{\dot{q}_f\}_{n \times 1} + [K_{ff}]_{n \times n}\{q_f\}_{n \times 1} = \{Q_f\}_{n \times 1} \tag{2}$$

In Equation 2b, M_{ff} is the global mass matrix, C_{ff} is the global damping matrix, K_{ff} is the global stiffness matrix, Q_f is the global force vector, and q_f is the vector of global degrees of freedom for elastic motion. If in a local coordinate system, the local stiffness matrix (order 6×6) is represented by k^{re}, the local mass matrix (order 6×6) is represented by m^{er} and the local force vector (column matrix 6×1) is represented by f' then in the global coordinate system, the global stiffness matrix will be: $k^e = L^T k^{re} L$, the global mass matrix will be: $m^e = L^T m^{er} L$ and the global force vector will be: $Q_f = L^T f'$, where L is the transformation matrix. The transformation matrix, L, depends upon the direction cosines of the links under consideration. For the manipulator described in the present work, the direction cosines are related to the joint rotations. The complete formulation for the frame element may be seen in Chandrupatla and Belegundu (2002). The final equation of motion, having both rigid and elastic motions of the links, can be expressed by Equation 3:

$$\begin{aligned}
&\begin{bmatrix} M_{rr} & M_{rf} \\ M_{fr} & M_{ff} \end{bmatrix}_{(n+N) \times (n+N)} \begin{Bmatrix} \ddot{q}_r \\ \ddot{q}_f \end{Bmatrix}_{(n+N) \times 1} + \begin{bmatrix} K_{rr} & K_{rf} \\ K_{fr} & K_{ff} \end{bmatrix}_{(n+N) \times (n+N)} \begin{Bmatrix} q_r \\ q_f \end{Bmatrix}_{(n+N) \times (n+N)} \\
&+ \begin{bmatrix} H \\ 0 \end{bmatrix}_{(n+N) \times 1} + \begin{bmatrix} G \\ 0 \end{bmatrix}_{(n+N) \times 1} = \begin{Bmatrix} Q_r \\ Q_f \end{Bmatrix}_{(n+N) \times 1}
\end{aligned} \tag{3}$$

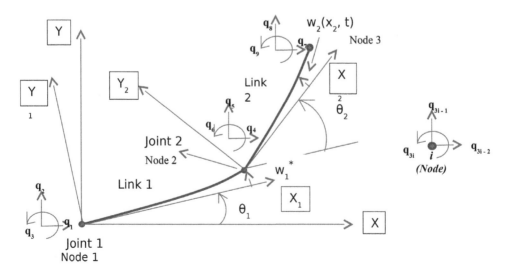

Figure 1.   Dynamic modeling of a two-link flexible manipulator using two finite elements.

168

In Equation 3, subscripts r and f stand for rigid and flexible, respectively. N represents the number of rigid degrees of freedom present in the system and n represents the elastic degrees of freedom obtained from finite element formulation. For the present case, because there are two flexible links, we have $N = 2$. Hence, M_{rr} consists of two diagonal elements: M_{11} and M_{22}. M_{rf} and M_{fr} represent the coupling between rigid and elastic motions. One plane frame finite element per link is considered for the present analysis. Thus, there are three nodes and nine degrees of freedom. The right-hand side of Equation 3 is represented by the following expressions:

$$Q_r = \left\{ \begin{array}{c} \tau_1 \\ \tau_2 \end{array} \right\} \quad \text{and} \quad Q_f = \{0\,0\,0\,0\,0\,0\}^{\gamma} \tag{3a}$$

where τ_1 and τ_2 are Joint 1 and Joint 2 torques, respectively.

3 RESULTS AND DISCUSSION

In this section, the natural frequencies of the flexible manipulator in various configurations, and the joint responses and tip responses of the links are provided. The simulation results are obtained by solving Equation 3 using the *ode45* ordinary differential equation solver (Math-Works Inc., Natick, MA). In the present work, during solving, the terms involving coupling between rigid and elastic motions (M_{rf} and K_{rf}) are not considered. The effect of gravity and damping are also not considered.

Table 1. The variation of the first two natural frequencies of the two-link flexible manipulator with change in configuration of the links.

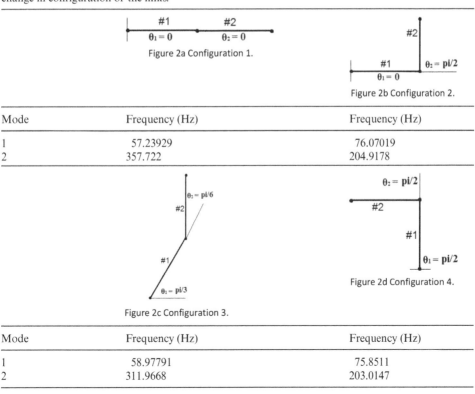

Figure 2a Configuration 1.

Figure 2b Configuration 2.

Mode	Frequency (Hz)	Frequency (Hz)
1	57.23929	76.07019
2	357.722	204.9178

Figure 2c Configuration 3.

Figure 2d Configuration 4.

Mode	Frequency (Hz)	Frequency (Hz)
1	58.97791	75.8511
2	311.9668	203.0147

Table 2. Physical parameters of both the links of the two-link flexible manipulator.

Physical parameter	Value
Density	7,850 kg/m^3
Length	1 m
Young's modulus	2.1×10^{11} N/m^2
Area of cross section	4.387×10^{-6} m^2
Area moment of inertia	$27.0550427 \times 10^{-6}$ m^2
Joint 1 torque	-0.1 Nm (step)
Joint 2 torque	-0.01 Nm (step)

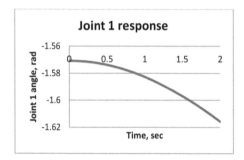

Figure 3. Variation of Joint 1 angle with time.

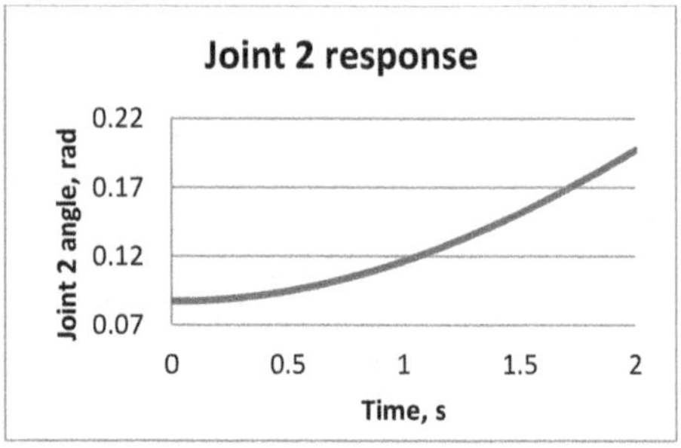

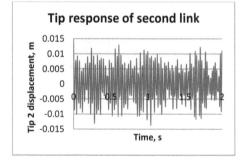

Figure 4. Variation of Joint 2 angle with time.

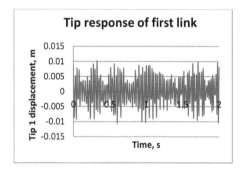

Figure 5. Tip response of the first link.

Figure 6. Tip response of the second link.

3.1 *Natural frequencies of the flexible manipulator*

After mathematical modeling, the Eigen values of the flexible manipulator are found for different configurations. Table 1 gives the first two natural frequencies of the two-link flexible manipulator for four different configurations.

From Table 1, it can be seen that configuration 2 (Figure 2b) exhibits the maximum value of the first mode frequency, while configuration 1 (Figure 2a) exhibits the minimum value of the first mode frequency for the two-link flexible manipulator. On the other hand, the value of the second mode frequency is the highest in configuration 1 and the lowest in configuration 4. It is also observed that significant changes in natural frequencies occur only when the second link moves. Thus, the natural frequencies of a flexible manipulator change with change in configuration of the links. This effect has also been shown by Mishra et al. (2015) and Ghosal (2006).

3.2 *Joint and tip responses*

The response of the flexible manipulator system is obtained by directly solving the matrix equation described by Equation 3. For solving, the coupling terms $\mathbf{M}_{rf}$ and $\mathbf{K}_{rf}$ are set to zero. The joint response of the flexible manipulator is found by subjecting both the joints to stepped torques. Table 2 gives the physical parameters of the flexible manipulator.

The joint responses are shown in Figures 3 and 4, while the tip responses are shown in Figures 5 and 6.

Figure 3 and Figure 4 show the joint responses of the first and second joints of the flexible manipulator, respectively. Figure 5 and Figure 6 show the tip responses of both the links of the flexible manipulator. From Figures 5 and 6, it can be inferred that both the links exhibit vibrations. The tip displacement of the second link is more than that of the first link.

4 CONCLUSION

The mathematical modeling of the two-link flexible manipulator has been presented. For this, the Lagrangian finite elements method has been used. Plane frame elements are used for modeling the flexibility of the links. Thus, the links are modeled as frames (Saha, 2014) and not beams. The natural frequencies of the manipulator change as its configuration changes with time. The finite elements method automatically takes care of this. In the present work, the matrix equations are solved directly. This increases the computation time because there are as many second-order nonlinear differential algebraic equations as the number of degrees of freedom in the system. To overcome this drawback, the assumed modes method may be used together with FEM. In this method, the first few modes of the system are found, using the mass and stiffness matrices obtained by using the finite elements method. Then by modal analysis (Grover & Nigam, 2003), the response of the system can be found. There are as many second-order differential equations as the number of assumed modes. This greatly reduces the computational effort. Meghdari and Ghassempouri (1994) have described the advantages and disadvantages of both the Lagrangian assumed modes method and the Lagrangian finite elements method.

REFERENCES

Alberts, T.E., Xia, H. & Chen, Y. (1992). Dynamic analysis to evaluate viscoelastic passive damping augmentation for the space shuttle remote manipulator system. *Journal of Dynamic Systems, Measurement and Control, 114*(3), 468–475.

Bayo, E. (1987). A finite-element approach to control the end-point motion of a single-link flexible robot. *Journal of Robotic Systems, 4*(1), 63–75.

Chandrupatla, T.R. & Belegundu, A.D. (2002). *Introduction to finite elements in engineering* (3rd ed.). Upper Saddle River, NJ: Pearson.

Chedmail, P., Aoustin, Y. & Chevallereau, C. (1991). Modelling and control of flexible robots. *International Journal for Numerical Methods in Engineering, 32*(8), 1595–1619.

Dado, M. & Soni, A.H. (1986). A generalized approach for forward and inverse dynamics of elastic manipulators. In *Proceedings of the IEEE International Conference on Robotics and Automation* (pp. 359–364).

Fotouhi, R. (2007). Dynamic analysis of very flexible beams. *Journal of Sound and Vibration, 305*(3), 521–533.

Gaultier, P.E. & Cleghorn, W.L. (1992). A spatially translating and rotating beam finite element for modeling flexible manipulators. *Mech. Mach. Theory, 27*(4), 415–433.

Ghosal, A. (2006). *Robotics: Fundamental concepts and analysis*. Noida, India: Oxford University Press India.

Grover, G.K. & Nigam, S.P. (2003). *Mechanical vibrations* (7th ed.) (pp. 249–254). Roorkee, India: Nem Chand & Bros.

Hu, F.L. & Ulsoy, A.G. (1994). Dynamic modeling of constrained flexible robot arms for controller design. *Journal of Dynamic Systems, Measurement and Control, 116*(1), 56–65.

Meghdari, A. & Ghassempouri, M. (1994). Dynamics of flexible manipulators, *Journal of Engineering, Islamic Republic of Iran, 7*(1), 19–32.

Mishra, N., Singh, S.P. & Nakra, B.C. (2015). Dynamic modelling of two link flexible manipulator using Lagrangian assumed modes method. *Global Journal of Multidisciplinary Studies, 4*(12), 93–105.

Naganathan, G. & Soni, A.H. (1986). Non-linear flexibility studies for spatial manipulators. In *Proceedings of the IEEE International Conference on Robotics and Automation* (pp. 373–378).

Saha, S.K. (2014). *Introduction to robotics* (2nd ed.). New Delhi, India: McGraw Hill Education.

Simo, J.C. & Vu-Quoc, L. (1987). The role of non-linear theories in transient dynamic analysis of flexible structures. *Journal of Sound and Vibration, 119*(3), 487–508.

Stylianou, M. & Tabarrok, B. (1994a). Finite element analysis of an axially moving beam, part I: Time integration. *Journal of Sound and Vibration, 178*(4), 433–453.

Stylianou, M. & Tabarrok, B. (1994b). Finite element analysis of an axially moving beam, part II: Stability analysis. *Journal of Sound and Vibration, 178*(4), 455–481.

Sunada, W. & Dubowsky, S. (1981). The application of finite element methods to the dynamic analysis of flexible spatial and co-planar linkage systems. *Journal of Mechanical Design, 103*(3), 643–651.

Theodore, R.J. & Ghosal, A. (2003). Robust control of multilink flexible manipulators. *Mechanism and Machine Theory, 38*(4), 367–377.

Tzou, H.S. & Wan, G.C. (1990). Distributed structural dynamics control of flexible manipulators—I. Structural dynamics and distributed viscoelastic actuator. *Computers & Structures, 35*(6), 669–677.

Usoro, P.B., Nadira, R. & Mahil, S.S. (1986). A finite element/Lagrange approach to modeling lightweight flexible manipulators. *Journal of Dynamic Systems, Measurement and Control, 108*(3), 198–205.

Technology Drivers: Engine for Growth – Mahajan, Modi & Patel (Eds)
© 2018 Taylor & Francis Group, London, ISBN 978-1-138-56042-0

Trajectory planning for a four-legged robot

Mihir M. Chauhan & Mohit Sharma
Department of Mechanical Engineering, Institute of Technology, Nirma University,
Ahmedabad, Gujarat, India

ABSTRACT: This paper presents trajectory planning of a leg's end-point position for a four-legged robot. A kinematic model of the leg is developed using the Denavit–Hartenberg (DH) method. The dynamic equation of motion is developed using the Euler–Lagrange formulation. Based on a mammal's gait pattern, the leg's trajectory is planned mathematically. Computer-Aided Design (CAD) model simulation is used to visualize the planned trajectory. The planned trajectory is tested for walking of a small four-legged robot working model. The working model of the robot is also tested for stable walking on different ground terrains and with maximum payloads.

1 INTRODUCTION

In the early 20th century, the concept of a humanoid machine first came into consideration, but today it is possible to imagine human-sized robots with the competence for cognition and movement. Robots with wheeled and tracked locomotion have found application in many places but are still limited due to their requirement for continuous path contact. Due to this limitation, many difficult terrains cannot be traversed by robots using wheel and track locomotion systems. Compared to wheeled and tracked robots, legged robots have the potential to walk in a much wider assortment of terrains, just like the biological legged animals that are in contact with nearly all of the earth's land surfaces (Manko, 1992). This has motivated a lot of research on legged robots in recent years. Despite the great number of achievements, legged robots still fall far behind the capabilities of their bionics. Looking at this need, an attempt is made here to develop a four-legged walking robot. Various literature is studied to identify the walking patterns of four-legged animals (Yi, 2010). A mathematical model of a robot leg has been developed. Forward and inverse kinematic models of a leg have been developed using the Denavit–Hartenberg (DH) method. Based on the gait cycle of mammals (Son et al., 2010), the leg's joint positions are identified. Using the forward kinematic model, the positions of the end point of the leg are obtained and they are mapped to a cubic curve (Zhuang et al., 2014). Using a cubic trajectory, the intermediate positions of the end point are interpolated, and the corresponding joint angles are derived using inverse kinematic solutions. These joint angles are used in a dynamic model and the joint torque is evaluated. Based on the maximum torque requirement, a servomotor is selected in the fabricated model. A Computer-Aided Design (CAD) model simulation (Kumar & Pathak, 2013) has been carried out to generate the end-point trajectory, which is compared with that of the planned trajectory. The end-point trajectory is validated through experimental testing on a working model.

2 MATHEMATICAL MODELING

Mathematical modeling is an analytical description of the spatial geometry of motion with respect to a fixed reference frame/base frame. Figure 1a shows a schematic diagram of the four-legged robot, with joint frames identified according to the DH method. The DH parameters are identified on the basis of the axis representation, as shown in Figure 1b. Here,

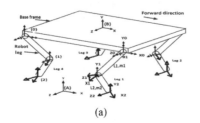

(a)

Link i	A_i	α_i	D_i	θ_i
0-1	L_1	0	0	θ_1
1-2	L_2	0	0	θ_2

(b)

Figure 1. (a) Schematic diagram of robot; (b) DH parameters.

L_1 and L_2 are the length of the upper limb and lower limb, respectively, and θ_1 and θ_2 are the angles for the hip joint and knee joint, respectively. Based on the DH parameters, the transformation matrices that define the joint frames with respect to the previous frame are obtained. The final transformation matrix $^0T_2$ is as follows:

$$^0T_2 = \begin{bmatrix} C_{12} & -S_{12} & 0 & L_2C_{12}+L_1C_1 \\ S_{12} & C_{12} & 0 & L_2S_{12}+L_1S_1 \\ 0 & 0 & 1 & 0 \\ 0 & 0 & 0 & 1 \end{bmatrix} \text{ where } C_{12} = \cos(\theta_1+\theta_2); S_{12} = \sin(\theta_1+\theta_2). \quad (1)$$

Equation 1 shows the description of the end point of a leg with respect to frame {0} (i.e. hip joint). For a given position and orientation of the leg, with respect to base/reference frame, it is required to find a set of joint variables (joint angles) that would bring the leg into the specified position and orientation.

$$\theta_1 = A\tan2\left(\frac{r_{24}}{r_{14}}\right) - \phi; \quad \theta_2 = A\tan2(S_2,C_2) \quad (2)$$

where $\phi = \tan^{-1}\left(\frac{L_2S_2}{L_1+L_2C_2}\right)$, $C_2 = \left[\frac{(r^2_{14}+r^2_{24})-(L_1^2+L_2^2)}{2L_1L_2}\right]$, $S_2 = \sqrt{1-\left[\frac{(r^2_{14}+r^2_{24})-(L_1^2+L_2^2)}{2L_1L_2}\right]^2}$, and r_{ij} shows the elements of the end-point position matrix.

During the work cycle, the four-legged robot should accelerate, move at constant speed, and decelerate. The time-varying position and orientation of the four-legged robot depends on its dynamic behavior. In this paper, the mathematical model for the dynamic behavior of the quadruped walking robot is developed using Euler–Lagrange formulation. The Equation of Motion (EOM) of the single leg, using Euler–Langrange formulation, is as follows:

$$\begin{bmatrix} \tau_1 \\ \tau_2 \end{bmatrix} = \begin{bmatrix} M_{11} & M_{12} \\ M_{21} & M_{22} \end{bmatrix}\begin{bmatrix} \ddot{\theta}_1 \\ \ddot{\theta}_2 \end{bmatrix} + \begin{bmatrix} H_1 \\ H_2 \end{bmatrix} + \begin{bmatrix} G_1 \\ G_2 \end{bmatrix}, \text{kg-cm} \quad (3)$$

where m_1, m_2 = mass of the upper limb and lower limb; h_1, h_2 = height of the upper limb and lower limb; w_1, w_2 = width of the upper limb and lower limb; τ_1, τ_2 = joint torques of the hip joint and knee joint; M_{ii} = effective inertia at joint i where the driving torque τ_i ascts; M_{ij} = coupling inertia between joint i and joint j; H_1, H_2 = the Coriolis and centrifugal coefficient matrix at hip joint and knee joint; G_1, G_2 = the gravity matrix at hip joint and knee joint;

$$M_{11} = \frac{1}{12}m_1\left[25L_1^2 + w_1^2\right] + \frac{1}{12}m_2\left\{7L_2^2 + w_2^2 + L_1^2 + 6L_2h_2 + 30L_1L_2C_2 + 6L_1h_2(2C_2 - 3S_2)\right\}$$

$$M_{22} = \frac{1}{12}m_2\left[25L_2^2 + w_2^2\right]; M_{12} = M_{21} = \frac{1}{12}m_2\left[25L_2^2 + w_2^2 + 6L_1L_2C_2 - 6L_1h_2S_2\right];$$

174

$$\begin{bmatrix} H_1 \\ H_2 \end{bmatrix} = \begin{bmatrix} \dfrac{-m_2 L_1}{2} \left(h_2 C_2 - L_2 S_2 \right) \left[\dot{\theta}_1 \dot{\theta}_2 - \dot{\theta}_2^2 \right] \\ \dfrac{m_2 L_1}{2} \left[L_2 S_2 + h_2 C_2 \right] \dot{\theta}_1^2 \end{bmatrix} ; \quad \begin{bmatrix} G_1 \\ G_2 \end{bmatrix} = \begin{bmatrix} \dfrac{-m_1 g_1}{2} \left(3 C_{12} L_2 + 3 C_1 L_1 + 2 S_1 - h_1 S_1 - h_2 S_{12} \right) \\ -\dfrac{m_2 g_2}{2} \left(3 C_{12} L_2 - S_{12} h_2 \right) \end{bmatrix}$$

3 ROBOT SPECIFICATION AND GAIT PLANNING

Based on observation and study of the literature, the working model specifications are decided (Hirose, 1984). A ground clearance of 150 mm is selected. Given a ratio of lower limb length to upper limb length of 1.5, the link lengths are fixed at 60 mm and 90 mm for the upper and lower limbs, respectively. A trot-type gait pattern as adopted by mammals for normal walking is used, as shown in Figure 2a. The gait cycle of a mammal (cat) is observed and its intermediate leg positions are identified, as shown in Figure 2b, with the dimensions specified.

4 CAD MODELING AND SIMULATION

Before the work proceeds from the design stage to the fabrication stage, it is necessary to develop a CAD model. A CAD model of a quadruped robot has been developed on a SOLIDWORKS-2015 software platform. Development of a CAD model helps in fixing the dimensions of the quadruped robot and calculating its mass properties. Figure 3 shows different views of the final assembled CAD model of the four-legged robot.

To improve stability during locomotion, leg segment motion analysis is carried out on the basis of angle of motion of the mammal (i.e. cat). During locomotion, the left front

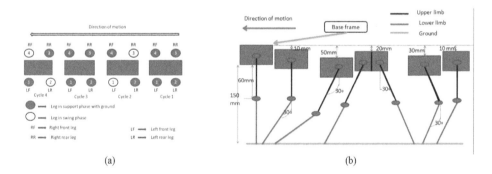

(a) (b)

Figure 2. Gait planning of a four-legged robot.

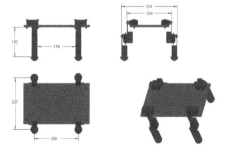

Figure 3. CAD model of the four-legged robot.

leg and the right rear leg are in support position when the right front leg and the left rear leg are in swing position. When the robot is trotting, the highest height of leg rise above the earth (h) is 0.05 m. The hip joint's swing amplitude is 30° and the knee joint's swing amplitude is 30°.

5 EXPERIMENT AND TESTING

A working model of a four-legged robot has been fabricated on the basis of the CAD model. The motors are selected as per the maximum torque requirement at each joint, obtained from the dynamic model. A factor of 1.5 is used to adjust the payload, joint friction and other variables. Figure 4 shows the fabricated model with all of its electronic elements.

Aluminum box sections are used to fabricate the legs. A very lightweight hardboard is used as the body to house the various electronic components. A total of eight 11 kg-cm stall torque, 5-volt DC servomotors are used, which are controlled through an Arduino Mega 2560 controller. An ultrasonic sensor is attached at the front to detect obstacles. The robot is programmed in a manner to follow various kinds of terrain, by maintaining stability and stopping walking as soon as any obstacle is detected within a distance of 200 mm. Figure 5 shows the walking of the robot on different ground surfaces, with a payload of 300 g, and on a 20° slope.

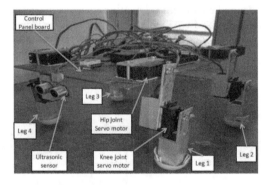

Figure 4. Working model of the four-legged robot.

Figure 5. Locomotion of the working model over different terrains.

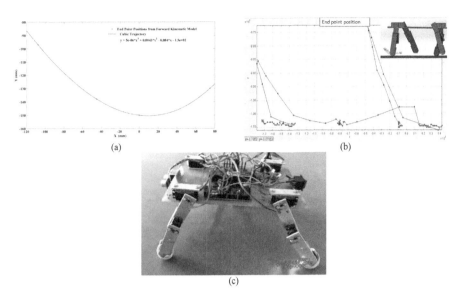

Figure 6. End-point trajectory: (a) analytical; (b) simulation; (c) working model.

6 RESULTS AND DISCUSSION

As shown in Figure 2b, the identified joint angles are input to the forward kinematic models (Equation 1) for the six positions of the legs, and the corresponding six positions of the end point of the legs are obtained as shown in Figure 6a. A cubic spline trajectory is mapped through these points and based on cubic equations the intermediate positions of the end point are interpolated. These interpolated positions are input to the inverse solutions (Equation 2) to obtain the sequential joint angles.

The sequential joint angles obtained from the inverse solutions are set as segment input to the corresponding servomotor in CAD model simulation through SOLIDWORKS. Figure 6b shows the end-point trajectory obtained through SOLIDWORKS motion simulation. It is observed that the trajectory is similar to the cubic spline as predicted in the mathematical model. The working model of the four-legged robot is programmed for the same sequential joint angle as obtained from the inverse solution and it is tested to walk on a flat smooth surface. Using video tracking software, the point on the leg is selected and its trajectory is plotted for one complete cycle of the gait, as shown in Figure 6c. It is observed that the end-point trajectory is the same as that planned and expected.

7 CONCLUSION

A four-legged robot has been developed and demonstrated successfully. It can walk on different ground surfaces. It can satisfactorily walk on a slope of 20°. It is also observed that the end point of the leg follows the same trajectory as that planned mathematically and obtained from CAD model simulation. The maximum payload for the robot is 300 g. The weight of the robot is 1.25 kg.

REFERENCES

Hirose, S. (1984). A study of design and control of a quadruped walking vehicle. *Robotics Research*, *3*(2), 113–133.

Kumar, G. & Pathak, P.M. (2013). Dynamic modeling & simulation of a four legged jumping robot with compliant legs. *Robotics and Autonomous Systems, 61*(3), 221–22.

Manko, D.J. (1992). *A general model of legged locomotion on natural terrain*. Dordrecht, The Netherlands: Kluwer Academic Publishers.

Son, D., Jeon, D., Nam, W.C., Chang, D., Seo, T. & Kim, J. (2010). Gait planning based on kinematics for a quadruped gecko model with redundancy. *Robotics and Autonomous Systems, 58*(5), 648–656.

Yi, S. (2010). Reliable gait planning and control for miniaturized quadruped robot pet. *Mechatronics, 20*(4), 485–495.

Zhuang, Y.F., Liu, D.Q. & Wang, J.G. (2014). Dynamic modeling and analyzing of a walking robot. *The Journal of China Universities of Posts and Telecommunications, 21*(1), 122–128.

Technology Drivers: Engine for Growth – Mahajan, Modi & Patel (Eds)
© *2018 Taylor & Francis Group, London, ISBN 978-1-138-56042-0*

Modeling and experimental investigation of reciprocating vibro separator

J.V. Desai, V.B. Lalwani & D.H. Pandya
Department of Mechanical Engineering, LDRP Institute of Technology and Research, Kadi Sarva Vishwavidyalaya, Gujarat, India

ABSTRACT: In this paper, a model of a reciprocating vibro separator is prepared using Creo Parametric design software and is further analyzed with ANSYS software. The motion behavior of a reciprocating vibro separator model and experimental setup is observed. Tools such as Fast Fourier Transform (FFT) and time domain data graphs are used for the observation. From this work, the authors conclude that the experimental setup matches the computational model. The motor angle of the vibro separator varies from 25 degrees to 35 degrees for a constant motor speed of 1,000 rpm.

1 INTRODUCTION

We live in a century in which the use of technology is getting more and more important, so researchers are trying to put their knowledge for improvement into all day-to-day processes. Take, for example, the simple process of separating wheat, where in olden days we had to put in more effort, but now gradation plants have been developed. Currently, we are facing the problem of multiple uses of such plants. Some of the work on this particular process is being done by researchers such as Zhao et al. (2016), who have noticed that the average velocities of simulations with both spherical and non-spherical particle models in each case shows similar trends with the tests. For most cases, the velocities of spherical particles are more highly overestimated than those of non-spherical particles because of the simplification of particle shapes. Du et al. (2014) have used a separator which has an unbalanced rotor mass at its center. By analyzing the screening process of three different vibration screens they observed that the variable linear vibration screen has better power distribution and screen surface movement. The flexible screen surface can increase the amplitude of the screen surface and reduce the material-blocking phenomenon. The experimental results for the two styles of surface vibration screens shows the huge advantage of the flexible screen surface over the fixed screen surface in screen efficiency and in avoiding material crush; it also provides a powerful proof to verify the correctness of the simulation work (Li & Ma, 2014). The research results that were reported by Li and Ma showed that, in terms of a nonlinear vibration system supported by a soft nonlinear characteristics spring, the amplitude value of the nonlinear system can be automatically compensated when the vibrating mass of the vibrating system is fluctuating in small-scope, which makes the approximate amplitude remain constant. Zhao et al. (2010) indicate that the amplitude and the vibration direction angle have a great effect on the average particle velocity and the average throw height considered over the normal range of linear screen parameters. The vibration frequency and the inclination angle of the screen plate have a small influence. To obtain the ideal sieving effect for materials that are difficult to sieve, the frequency and amplitude of vibration, the inclination angle of the screen plate, and the vibration direction angle should be 13 Hz, 6.6 mm, 6 degrees and 40 degrees, respectively. Soldinger (2002) has used the Monte Carlo simulation to check the effect of angle between the base line and the separator box bottom layer line; she then concluded that for the β value of 5 degrees, the rpm speed ranges from 1,000 to 800. Zhao et al. (2011) tried to establish the optimum angle between the base line and the separator box bottom layer line β; they concluded that the increment of screen deck has the same effect on a banana screening process as does the inclination of the discharge end, and that when the values of inclination of discharge and increment of screen deck inclination are between 10 and 5 degrees, the banana screening process shows a good screening performance in a simulation.

Liu et al. (2013) studied linear, circular and elliptical motion of the screen and reported that the travel velocity of the particles during linear screening was the fastest. This results in a thin material layer but gives the lowest overall screening efficiency. The circular mode gives the lowest particle velocity along the screen but the highest screening efficiency. In this case, the material layer is thick but the interaction between particles and the penetration effect are enhanced.

Golovanevskiy et al. (2011) argued that the description of specific vibration-induced phenomena presented in their paper provides the basis for development of materials-handling technologies and process equipment to affect bulk material flow behavior with vibration. To ensure the highest vibration-aided bulk granular material separation efficiency, they suggested that the amplitude and frequency of vibration should be selected on the basis of providing a vibration overloading factor w, where values of w are ≈ 3 or slightly higher. Further research needs to focus on the development of an overall model describing the behavior of granular material under vibration. For a particle on a screen moving from the front to the back of the screen, the regimes of the different particle behaviors, such as stable periodic motion, period-doubling bifurcation motion, bifurcation motion, and chaotic motion, were determined by Wang et al. (2017). Chaotic motion was found to be beneficial in separating particles effectively from other threshed agricultural materials, and at the same time avoids particles accumulating on the screen, enhancing the probability that the particle penetrates the screen holes. A detailed investigation of this aspect (passage rate) was not performed and remained an open problem. Li et al. (2002) concluded that for a screening system involving granular materials, it has been demonstrated that the critical feeding rate or bed depth for the most effective screening operation can be determined by conducting a Discrete Element Modeling (DEM) simulation. Further work will focus on the implementation of advanced experimental techniques to measure the process and to validate the model. He and Liu (2009) studied a vibro separator and found that its motion follows an elliptical trace. A theoretical kinematic analysis of the vibrating screen was done to study how varying different parameters affected the motion of the screen. Kinematic parameters of the vibrating screen for motion traces that are linear, circular or elliptical were obtained. Their work also concludes that the position of the exciter axle center relative to the center of gravity of the vibrating screen is extremely important for screening efficiency. Thus, by adjusting the relative position of the axle center we can design a vibrating screen with higher processing capacity without increasing power consumption. Zhao et al. (2009) report that the dynamic response analysis shows that adding stiffening angle and longitudinal stiffeners to both side plates decreases the transverse deformation of the side plate and ameliorates the twist deformation of the screen frame. The maximum transverse displacement of the vibrating screen is 0.13 mm.

The present work is to investigate the dynamic motion behavior of a reciprocating vibro separator by using ANSYS software, in addition to using the DEM simulation method.

2 COMPUTATIONAL MODEL AND ANALYSIS

The reciprocating vibro separator has three types of vibration mode, of which the elliptical vibration mode gives the best result for screening. To observe this elliptical motion a computational

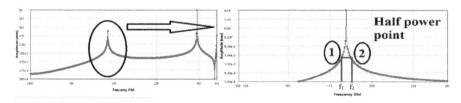

Figure 1. Frequency response graph.

model of the reciprocating vibro separator is created using Creo Parametric 3.0 3D modeling software and analyzed using ANSYS 14.5 software.

3 EXPERIMENTAL SETUP

An experimental setup was prepared as per the computational work. Here the two vibro motors run at 1,000 rpm. The variation in motor speed is ±20 rpm. Each vibro motor has 0.5 hp power. The experimental setup is run between 980 and 1,020 rpm motor speed with a 30 degree motor angle. The experimental work was performed at Gajanand Industries in Unjha (see Figure 2).

4 RESULTS AND DISCUSSION

In the present work, the material properties of the rubber foundation and the motor angle (α) are considered as the input parameters. After adjusting the input parameters, the motion of the reciprocating vibro separator is observed. Different tools are available to observe the motion of the vibro separator, from which the time domain data and Fast Fourier Transform (FFT) are taken into consideration for motion analysis. One of the input parameters, the foundation rubber elasticity, is changed. For computational work, the elasticity of the foundation rubber is assumed to be 10 MPa, 25 MPa, or 40 MPa. The computational model is analyzed on the basis of a motor angle (α) of 30 degrees and at 1,000 rpm speed for all three elasticities.

Figures 3a and 3b are acceleration graphs that indicate that the acceleration value of the vibro separator is higher in the vertical direction than in the horizontal direction. Figures 3c and 3d indicate that in the vertical direction there is no influence from either subharmonic or superharmonic responses.

Figure 2.   Experimental setup at Gajanand Industries (left), and a vibro analyzer (right).

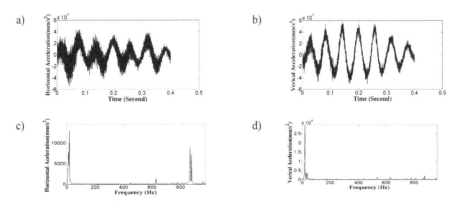

Figure 3.   Computational dynamic motion analysis at 30 degrees and 1,000 rpm with 40 MPa elasticity.

Figures 4a and 4b indicate that the acceleration value of the vibro separator is approximately double in the horizontal direction compared to the vertical direction. Figures 4c and 4d clearly indicate the same behavior as observed in Figures 3c and 3d. In addition, all four figures indicate that if we increase the elasticity, then the system becomes stable.

Figures 5a and 5b show that the value of acceleration in the horizontal direction is nearly the same as in the vertical direction. Figures 5c and 5d are the FFT plots that indicate that the motion is periodic with some vibration.

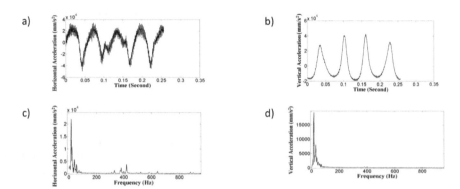

Figure 4. Computational dynamic motion analysis at 30 degrees and 1,000 rpm with 25 MPa elasticity.

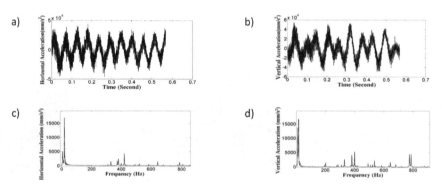

Figure 5. Computational dynamic motion analysis at 30 degrees and 1,000 rpm with 10 MPa elasticity.

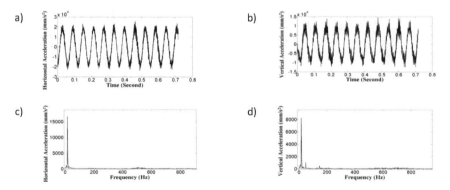

Figure 6. Experimental dynamic motion analysis at 30 degrees and speed varied between 980 and 1,020 rpm.

Figures 6a and 6b indicate that the experimental time responses have concluded stable and periodic response to system. Figures 6c and 6d are the FFT plots that indicate that the motion is periodic.

5 CONCLUSIONS

In the present work the dynamic motion is observed for the horizontal and vertical directions using a dynamic analyzer:

1. Experimental Figures 6a and 6b clearly indicate that the acceleration value should be more in the horizontal direction. The acceleration value must be lower in the vertical direction than in the horizontal direction.
2. The stability of the system varies as we change the elasticity. For instance, the model with 40 MPa elasticity is more stable than the model with 10 MPa elasticity, as shown in Figures 3a, 3b, 4a, 4b, 5a and 5b.

ACKNOWLEDGMENT

This work is financially supported by the Gujarat Council of Science and Technology (GUJCOST) (Grant no GUJCOST/MRP/2016-17/527).

REFERENCES

Cleary, P.W., Sinnott, M.D. & Morrison, R.D. (2009). Separation performance of double deck banana screens – Part 1: Flow and separation for different accelerations. *Minerals Engineering, 22*(14), 1218–1229. doi:10.1016/j.mineng.2009.07.002.

Du, C., Gao, K., Li, J. & Jiang, H. (2014). Dynamics behavior research on variable linear vibration screen with flexible screen face. *Advances in Mechanical Engineering, 2014*, 957140.

Golovanevskiy, V.A., Arsentyev, V.A., Blekhman, I.I., Vasilkov, V.B., Azbel, Y.I. & Yakimova, K.S. (2011). Vibration-Induced phenomena in bulk granular materials. *International Journal of Mineral Processing, 100*(3–4), 79–85. doi:10.1016/j.minpro.2011.05.001.

He, X.M. & Liu, C.S. (2009). Dynamics and screening characteristics of a vibrating screen with variable elliptical trace. *Mining Science and Technology (China), 19*(4) 508–513.

Li, J., Webb, C., Pandiella, S.S. & Campbell, G.M. (2002). A numerical simulation of separation of crop seeds by screening—Effect of particle bed depth. *Food and Bioproducts Processing, 80*(2), 109–117.

Li, X. & Ma, M. (2014). Dynamics analysis of the double motors synchronously exciting nonlinear vibration machine based on acceleration sensor signal. *Sensors & Transducers, 176*(8), 290–295.

Liu, C., Wang, H., Zhao, Y., Zhao, L. & Dong, H. (2013). DEM simulation of particle flow on a single deck banana screen. *International Journal of Mining Science and Technology, 23*(2), 273–277. doi:10.1016/j.ijmst.2013.04.007.

Soldinger, M. (2002). Transport velocity of a crushed rock material bed on a screen. *Minerals Engineering, 15*(1–2), 7–17.

Wang, L., Ding, Z., Meng, S., Zhao, H. & Song, H. (2017). Kinematics and dynamics of a particle on a non-simple harmonic vibrating screen. *Particuology, 32*, 167–177. doi:10.1016/j.partic.2016.11.002.

Zhao, L., Liu, C. & Yan, J. (2010). A virtual experiment showing single particle motion on a linearly vibrating screen-deck. *Mining Science and Technology (China), 20*(2), 276–280. doi:10.1016/S1674-5264(09)60197-6.

Zhao, L., Zhao, Y., Bao, C., Hou, Q. & Yu, A. (2016). Laboratory-scale validation of a DEM model of screening processes with circular vibration. *Powder Technology, 303*, 269–277. doi:10.1016/j.powtec.2016.09.034.

Zhao, L., Zhao, Y., Liu, C., Li, J. & Dong, H. (2011). Simulation of the screening process on a circularly vibrating screen using 3D-DEM. *Mining Science and Technology (China), 21*(5), 677–680. doi:10.1016/j.mstc.2011.03.010.

Technology Drivers: Engine for Growth – Mahajan, Modi & Patel (Eds)
© 2018 Taylor & Francis Group, London, ISBN 978-1-138-56042-0

Automation and standardization of processes used in packing and logistics

Parth Kasarekar
Department of Mechanical Engineering, Hasmukh Goswami College of Engineering, Vahelal, Ahmedabad, India

Nilesh Ghetiya
Department of Mechanical Engineering, Institute of Technology, Nirma University, Ahmedabad, India

Jaykishan Baldania
KHS Machinery Pvt. Ltd., Ahmedabad, India

ABSTRACT: Research and development in the packing and logistics sector is very important. In departments of logistics, various types of automation of systems are required to avoid time wastage; such examples are the generation of models and drawing of wooden boxes, stuffing-plan generation and calculation of cubic footage. The main objective of this automation and standardization process is to automate all the processes, such as automatic model generation, automated stuffing plans, automatic costing calculation, and, eventually, the optimization of box design. It will also reduce the manufacturing lead time. Much of the inefficiency arises from nonstandard working systems such that effective utilization of time and money cannot be achieved. In the present study, an attempt is made to reduce nonproductive time during various packing and logistics processes through automation. This can be achieved by using various elements of standardized software such as Creo Parametric and VB.Net.

1 INTRODUCTION

The packing and logistics department is a most important department. It handles 'post-production' work such as packing and transportation of materials. There are various processes such as generation of the container model and boxes, standardization, stuffing-plan generation, and transportation-mode selection. These are very important parts of the packing and logistics department activities. A vast amount of time is wasted in trial and error and various nonproductive processes due to nonstandardization of these activities. The main goal of this work is to reduce such nonproductive time and create standardization of the associated processes. This will reduce the cost by effective utilization of resources. This can be achieved by the customization of software such as Creo Parametric and Microsoft Excel with the help of programming languages such as Visual Studio. This also saves time in inputting data. In stuffing-plan generation, the main objective is to utilize the maximum space in a container for transportation with optimum safety, so that the number of containers can be reduced by the effective placement and orientation of boxes in the container. This can be carried out through the use of Visual Basic programming for automation and standardization.

2 LITERATURE REVIEW

The purpose of automation and standardization in this area is to save time, particularly in the generation and production of wooden containers for various components of machines. All industries require higher productivity and shorter working cycle times. This can be achieved in part by saving time in the design and development of the product. Jadeja and Bhuptani (2014) showed that the use of design software such as Creo Parametric, customized by a programming language (e.g. Java), could save most time in design. Scaffidi et al. (2008) described the use of web macro tools, such as Microsoft Office and Visual Basic macros, to automate lengthy procedures which have to be performed repeatedly. This leads to time-saving and improved process economy. Zaware and Mirza (2015) claimed that a total of 83% of the design lead time and modeling time for various types of heat transfer fins was saved. Software plays a significant role in supply chain management planning in the current industrialization and globalization age. Helo and Szekely (2005) described the use of web and software tools for automation of warehouse and transport management so that accurate planning can take place. Automation in design and development requires some research activity, especially in the customization of various pieces of modeling software (e.g. Creo Parametric). Franke and Piller (2003) described the problems in applying toolkits to the integration of processes in mass customization. The main goal of our work is to reduce nonproductive time through the standardization of various packaging and logistic processes. This will reduce cost through more effective utilization of resources.

3 METHODOLOGY

3.1 *Requirement for automated bill of materials and drawings generation*

In industry, the great variety of machines contain many different components. These components are first packed into wooden boxes of differing sizes, and then placed in containers. If a machine is to be transported to a customer, there must be availability of the appropriate containers at that time. For this, a Bill Of Materials (BOM) and information drawings need to be given to the manufacturer for these boxes. However, the calculations for the BOM and generation of drawings in design software is often very time-consuming and tedious. By programming appropriate automation in Visual Studio, the BOM calculations and drawings generation can take place just by clicking on the name of the particular machine. This reduces the production and purchase lead times in logistics, and increases the productivity of the system.

3.2 *Selection of programming language*

There are many programming languages available for the programming of various tasks. Among these, Java and Visual Basic provide interfaces for tools such as Creo Parametric and Microsoft Excel. However, the Java platform is very hard to learn and is not very user-friendly, and the. NET platform is too sophisticated for our required task. Hence, VB.Net has been selected for the present work.

3.3 *Selection of software for interfacing with VB.Net*

Microsoft Excel is chosen because it organizes data well in tabular form. An example of an Excel master file is shown in Figure 1. Creo Parametric has better modeling capability and is also to some extent better at calculating tabular forms of data. It is also used for generation of drawings from a particular 3D model and for BOM generation.

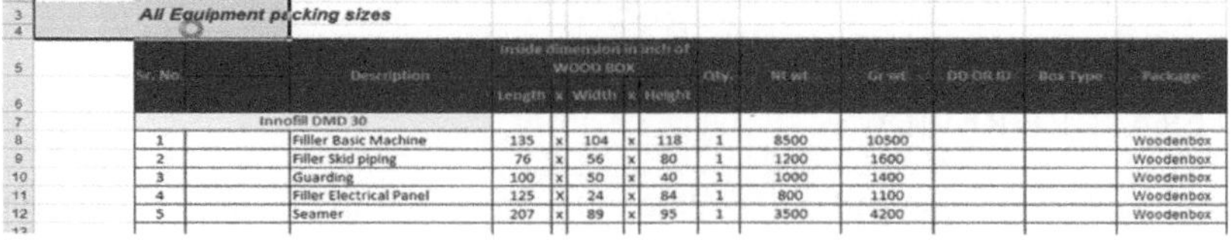

CAN Line
All Equipment packing sizes

Sr. No.	Description	Length	x	Width	x	Height	Qty.	Nt wt	Gr wt	DD OR ID	Box Type	Package
	Innofill DMD 30											
1	Filler Basic Machine	135	x	104	x	118	1	8500	10500			Woodenbox
2	Filler Skid piping	76	x	56	x	80	1	1200	1600			Woodenbox
3	Guarding	100	x	50	x	40	1	1000	1400			Woodenbox
4	Filler Electrical Panel	125	X	24	x	84	1	800	1100			Woodenbox
5	Seamer	207	x	89	x	95	1	3500	4200			Woodenbox

(Length x Width x Height: Inside dimension in inch of WOOD BOX)

Figure 1. Excel master file.

3.4 Programming algorithm

The algorithm for generating a program is shown in Figure 2.

3.5 Software execution

In executing the software, the processes are run in parallel (e.g. front end, back end). The front end is in front of the user, so that the user can input data. All the data of the machines are taken from the master file, as shown in Figure 1.

3.6 Front end

The front end is developed in VB.Net, and the main console is shown in Figure 3. It contains the following features, each of which has various options:

– Master login;
– Customer;
– Technology;
– Machine;
– Save.

All components for the selected machine and their various specifications can be generated, as shown in Figure 4. Examples of relevant characteristics include length, width and height.

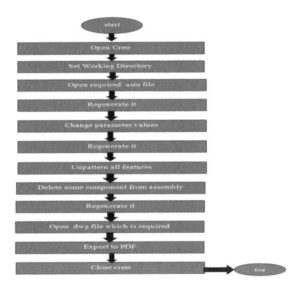

Figure 2. Programming algorithm.

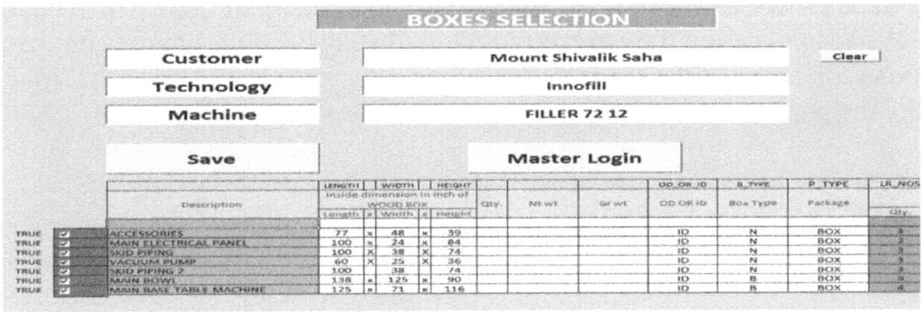

Figure 3. Front end console.

187

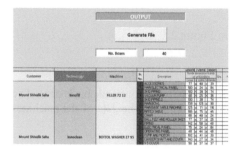

Figure 4.   Output of front end console.

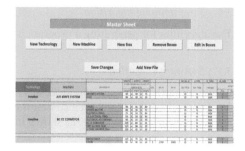

Figure 5.   Alteration of front end console.

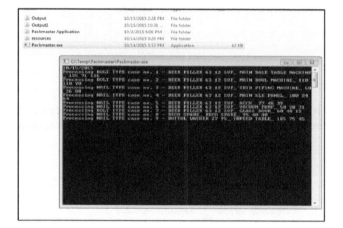

Figure 6.   Back end.

If a particular component is not required, then it can be ignored just by unchecking the associated checkbox, as shown in Figure 5. Then, by clicking the 'SAVE' button, all components and their specifications will be saved. Subsequently, the 'OUTPUT' button can be pressed and the associated front end is generated. Then by pressing the 'GENERATE FILE' button, all the descriptions of the components are generated in a file, as shown in Figure 5.

3.7   Back end

In the back end the following processes take place:

– compilation of Visual Basic Application Programming Interface (API) for Creo Parametric;
– connection with Creo Parametric and other supporting software;
– looping through the list of boxes to produce output.

During these processes, other applications may run in Excel and give rise to bugs, preventing the back end application from executing. Thus, during the programming process, the start and end times are noted and a program for killing other processes during that time period is created. So, the program execution in the back end is as shown in Figure 6.

4   RESULTS AND DISCUSSION

The current software ('Packmaster') saves considerable time and also increases productivity in generating the BOM and drawings for many components of particular machines. After the software program has completed, the output shown in Figure 7 is obtained; these folders

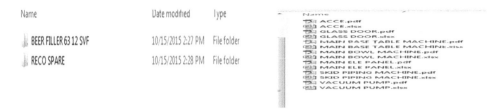

Name	Date modified	Type
BEER FILLER 63 12 SVF	10/15/2015 2:27 PM	File folder
RECO SPARE	10/15/2015 2:28 PM	File folder

Name

ACCE.pdf
ACCE.xlsx
GLASS DOOR.pdf
GLASS DOOR.xlsx
MAIN BASE TABLE MACHINE.pdf
MAIN BASE TABLE MACHINE.xlsx
MAIN BOWL MACHINE.pdf
MAIN BOWL MACHINE.xlsx
MAIN ELE PANEL.pdf
MAIN ELE PANEL.xlsx
SKID PIPING MACHINE.pdf
SKID PIPING MACHINE.xlsx
VACUUM PUMP.pdf
VACUUM PUMP.xlsx

Figure 7.   Output folder of Packmaster.   Figure 8.   Output folders.

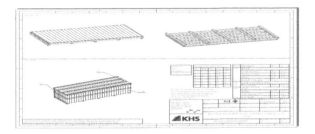

Figure 9.   Isometric view.

Pos	Sr.	Descriptic	Width	Thickness	Length	Qty.	Vol. (cubi	Vol. (cubic feet)
		1 BOTTLE WASHER 35 95_ CHAIN_ 80 40 24				1	31724	18.359
	10	2 BASE				1	9240	5.347
		3 BOARD_B.	8	1.5	40	9		
		4 BOARD2_I	8	1.5	40	1		
		5 LONG RUN	4	3	82	3		
		6 JOIST	3	2	40	2		
		7 CROSS RU	4	3	42	2		
	20	8 END WALL				2	1560	0.903
		9 BOARD_E!	8	1	25.5	4		
		10 BOARD2_I	8	1	25.5	1		
		11 BATTEN_F	4	1.5	45	2		
	30	12 SIDE WALL				2	3321	1.922
		13 BOARD_S!	8	1	28.5	9		
		14 BOARD2_!	10	1	28.5	1		
		15 BATTEN_F	4	1.5	82	2		
	40	16 TOP				1	5610	3.247
		17 BOARD_T!	8	1	45	9		
		18 BOARD2_'	13	1	45	1		
		19 BATTEN_F	4	1.5	85	2		
		20 TOP_BATT	3	1.5	85	2		
	90	21 Loose members				1	7112	4.116
		22 Loose mei	3	2	40	6		
		23 Loose mei	4	1	40	32		
		24 Loose mei	3	2	23	4		

Figure 10.   Bill of materials.

refer to the machine name and contain the BOM and drawings of all components associated with that machine. For example, in the two folders shown in Figure 7, namely 'BEER FILLER 63 12 SVF' and 'RECO SPARE', all the drawings and the BOM of all the components are stored. All the drawings are stored in PDF files, and the BOMs are stored in Excel files, as shown in Figure 8.

The drawings include an isometric view, top view and base view, as shown in Figure 9. The title block of the drawing shows the various components used, again as shown in Figure 9. The BOM in Excel format is shown in Figure 10. In this way, all the components can be documented in the form of the BOM and drawings in one go.

5   CONCLUSION

Generalized software has been developed for logistics processes, which is useful to the organization in automatically generating details such as an automatic bill of materials and drawings for wooden boxes for all the components of a particular machine. In addition, just by clicking on the machine name, a stuffing plan and wooden boxes for the components are generated automatically for various containers. Creo Parametric and Microsoft Excel have been linked with Visual Basic for this purpose. Finally, optimum linkage between the stuffing plan and the BOM and drawings generation is achieved, which greatly reduces nonproductive time. As a result, data entry efforts will be greatly reduced.

REFERENCES

Franke, N. & Piller, F.T. (2003). Key research issues in user interaction with configuration toolkits in a mass customization system. *The International Journal of Technology Management*, *26*(5–6), 1–30.

Helo, P. & Szekely, B. (2005). Logistics information systems: An analysis of software solutions for supply chain co-ordination. *Journal of Industrial Management & Data Systems*, *105*(1), 5–18.

Jadeja, I.J. & Bhuptani, K.M. (2014). Developing a GUI-based design software in VB environment to integrate with CREO for design and modeling using case study of coupling. *International Journal of Engineering Sciences & Research Technology*, *3*(4), 4089–4095.

Scaffidi, C., Cypher, A., Elbaum, S., Koesnandar, A. & Myers, B. (2008). Using scenario-based requirements to direct research on web macro tools. *Journal of Visual Languages & Computing*, *19*(4), 485–498.

Zaware, A.B. & Mirza, M.M. (2015). Customization of UG NX Software for 3D modelling of fins. Presented at *National Conference for Engineering Post Graduates RIT, 15 June 2015* (pp. 67–72).

Technology Drivers: Engine for Growth – Mahajan, Modi & Patel (Eds)
© 2018 Taylor & Francis Group, London, ISBN 978-1-138-56042-0

Design of involute cam and follower mechanism for Frisbee-throwing machine

V.J. Vekaria, P.H. Prajapati, M.S. Pandya, M.M. Chauhan, A.I. Mecwan & D.K. Khothari
Institute of Technology, Nirma University, Ahmedabad, Gujarat, India

ABSTRACT: This paper deals with the design of a special involute cam and follower mechanism for a Frisbee-throwing machine. In this particular application, the problem statement was to provide a specified angular displacement to the follower. The main problem is to provide a free path to the follower when the stored energy in the spring releases the Frisbee; at that time no portion of the cam profile should be in contact with the follower. Thus, for this application a special involute cam profile has been developed, which will disengage with the follower at the rise time. There will be no contact between the cam and follower during the recess time. This special design is described in the paper.

1 INTRODUCTION

This machine has been prepared at the Institute of Technology, Nirma University, Ahmedabad, Gujarat, India. Various types of throwing machines are commercially available at present, but this machine is unique because a totally different mechanism for the development of the throwing machine has been used. Until now, most throwers have used motorized rollers (see Figure 1) to provide angular velocity to the Frisbee, which is pushed into the path of rotating rollers (Lima & Zabka, 2010). However, this mechanism was tricky and not always reliable for accurate throwing of the Frisbee. The complexity of high-speed rollers has been avoided in our mechanism, and instead an aluminum box section mounted on a torsional spring has been used to throw the Frisbee.

Many other Frisbee throwers are available commercially for skeet (clay pigeon) throwing (see Figure 2). Almost all the throwers use a tension spring for throwing the Frisbee with the desired angular speed, and motors are used to load the tension spring. In other types of Frisbee thrower, the Frisbee is gripped from the upper surface with the help of a suction-cup gripper; the gripping arm is rotated at high speed, and then at a certain point the Frisbee is

Figure 1. Motorized-roller Frisbee thrower (straight guide) (Team254, 2013).

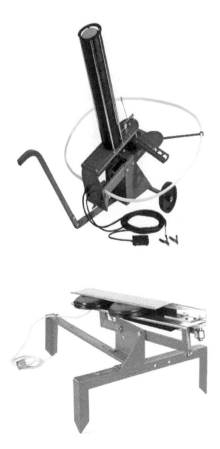

Figure 2.   Commercial skeet throwers (Champion, 2017; Allen, 2017).

released and it achieves the desired trajectory. This method is very difficult to calibrate for certain fixed trajectories and is time-consuming.

For our Frisbee thrower, a power-window motor was customized to drive the cam. The backside cap of the motor was opened, which was lying opposite the output shaft of the motor, and the output shaft was removed. The cam was then coupled directly to the worm wheel of the motor with nut-bolts and thrust bearings. To couple the worm wheel and cam, a special coupling was designed and manufactured using nylon material on a lathe. The other side of the follower was press-fitted into the aluminum box section and the assembly was completed using two 3 mm nut-bolts.

During the rise time, the cam loads the aluminum box section mounted on the torsional spring onto an automatic latch, which is run by a 10-rpm geared DC motor. When the latch unlatches, the aluminum section with the follower tries to regain its position because of the energy stored in the torsional spring. At this time, due to the special 'No Contact' arrangement during the recess period, the follower is helped to retrace its path. Then the process is repeated.

1.1   *Motivation and background*

The traditional Frisbee-throwing machines either use a continuously rotating wheel mounted on a high-torque motor, or potential energy stored in a spring with the help of a high-torque motor. Both of these methods require a motor with very high torque and the power consumption is extremely high during peak load conditions. The motor has to carry the weight of the rotating wheel. In addition, during the Frisbee-throwing procedure the rotating wheel directly transfers the impact jerk to the motor. These factors affect the working life of the motor, and the overall power consumption, reliability and accuracy of the machine. The use of an involute cam and follower system provides a smoother and more gradual action in the machine, reducing the torque requirement due to mechanical leverage, and hence the power consumption decreases. There is no contact between the cam and follower during the Frisbee-throwing procedure; hence the motor does not receive any type of jerk or impact through the cam and follower. This provides a longer working life for the motor, leads to a reduction in the overall power consumption, and provides smoother functionality of the machine.

2 INVOLUTE CAM DESIGN

The involute cam profile was designed with the help of SolidWorks software, and by using the following step-by-step procedure as illustrated in Figures 3, 4 and 5:

1. The involute profile was sketched with the help of the circle tangent method using proper dimensions.
2. The profile was extruded and another sketch to cut the portion of the cam which restricts the free motion of the follower was sketched over the previous profile to cut that portion out.
3. To reduce the weight of the cam, slots were sketched over it and cut out.
4. Final cam design achieved.

Similarly, the follower was designed to trace the profile of the cam and obtain the smooth and required function of the mechanism. The size of the cam was fixed by the involute base

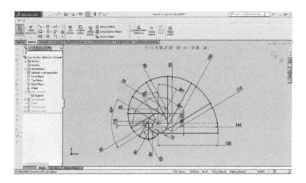

Figure 3.   Involute cam profile.

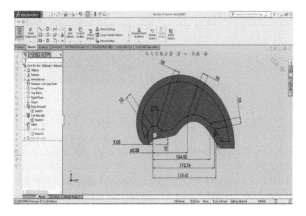

Figure 4.   Final dimensions of cam.

193

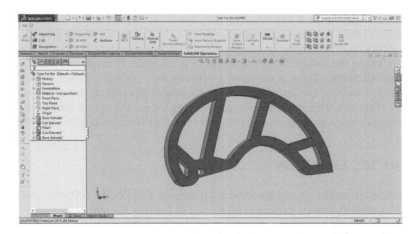

Figure 5.   Final design of cam.

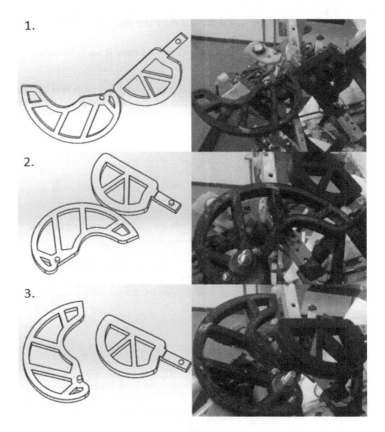

Figure 6.   Various contact positions of cam and follower.

circle diameter, but the cam shaft was, nevertheless, mounted a little offset from the center of the involute base circle. Based on this eccentricity, the center distance between the motor shaft and the rotating spring-mounted aluminum box section, and the rotation of the follower that was required (in degrees) decided the size and shape of the follower.

As shown in position 1 of Figure 6, the follower just starts experiencing angular displacement as the cam touches it through the rotary motion of the cam in a counterclockwise direction. At this point the cam has already rotated through 40 degrees. Position 2 shows the

194

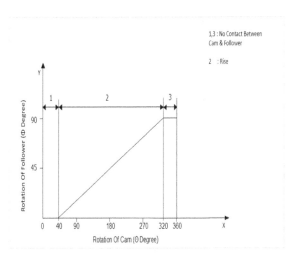

Figure 7. Displacement of follower with respect to the cam.

Table 1.

Spring constant (k)	Material (%)
1.3	15
1.7	15
2.16	16
2.7	18
3.4	20

intermediate situation, and in position 3 of Figure 6 the cam loses contact with the follower and the follower-mounted aluminum box section gets latched as a result of which the follower remains at 90 degrees, whereas the cam completes its revolution and the motor is stopped by the proximity sensors that detect the cam. The displacement graph of the cam and follower in Figure 7, approximated from the data obtained by practical experiment, shows the corresponding three regions. In practice, the graph should be slightly curved and there should not be any sharp corners.

3 DEVELOPMENT OF CAM

Following the design process, the cam and follower were manufactured using a 3D printing machine. The filamentous material used for 3D printing was Acrylonitrile Butadiene Styrene (ABS plastic). A latticed honeycomb-shaped structure was chosen to provide good strength, giving a material density of 20%. This could be increased if a higher-strength cam was required to work in high-torque conditions. Table 1 shows the experimental data obtained for the spring constant and the corresponding material density required.

4 SUMMARY AND CONCLUSIONS

Because contact between the cam and follower is not maintained during the 0–40 degree time interval, and during that interval the cam does not come across the follower's free path, there is no impact between any parts of the machine, which ultimately reduces jerk and improves the accuracy of the trajectory. The desired angle of rotation of the follower can be obtained by changing the center distance between the cam and follower. Based on the different strengths (spring constant k) of the torsional spring and the degree of rotation required, the material density of the 3D material used in manufacturing the cam can be determined using Table 1. This particular type of cam could have a wide scope in robotic applications.

REFERENCES

Allen. (2017). Allen Claymaster Target Thrower. Retrieved from http://www.ebay.com/itm/Clay-Disk-Thrower-Target-Practice-Trap-Skeet-Shooting-Hunt-Pigeon-Allen-Claymast/222649874687

Champion. (2017). Champion WheelyBird Auto-Feed Trap. Retrieved from http://www.ebay.com/itm/Auto-Automatic-Skeet-Thrower-Clay-Pigeon-Trap-Shotgun-Shooting-Practice-Target-/282152083782

Lima, M.F. & Zabka, P. (2010). Design and analysis of conjugate cam mechanisms for a special weaving machine application. Proceedings of International Conference on Innovations, Recent Trends and Challenges in Mechatronics, Mechanical Engineering and New High-Tech Products Development, Bucharest, 23–24 September 2010, 2, 24–32. Retrieved from http://www.incdmtm.ro/mecahitech2010/articole/pag.24–32.pdf

Team254. (2013). 2013 FRC Build Season Blog – Day #15: Improved Shooter. Retrieved from https://www.team254.com/day-15/

Technology Drivers: Engine for Growth – Mahajan, Modi & Patel (Eds)
© *2018 Taylor & Francis Group, London, ISBN 978-1-138-56042-0*

Tracking variations in the natural frequencies of a fixture-workpiece system during pocket milling

M.H. Yadav & S.S. Mohite
Department of Mechanical Engineering, Government College of Engineering, Shivaji University, Karad, Maharashtra, India

ABSTRACT: Contact deformations and structural deformations are inherent in a fixture-workpiece system, resulting in dimensional and form inaccuracies in machined components. During metal removal, the mass and compliance of the workpiece vary in different proportions depending on the volume and geometry of metal removal. This leads to a gradual increase in workpiece deformations and a shift in the natural frequencies (ω_n) from that of the initial un-machined workpiece. This causes a change in dynamic magnification factor (transmissibility) and vibration amplitudes. Thus, as machining progresses, the undesirable workpiece deformations and inaccuracies creep in, particularly in solid workpieces involving substantial metal removal and thin-walled workpieces. Accurate identification of changing dynamic behavior in the workpiece and control of these vibrations can maintain machining quality and productivity. In this work, we have developed a fully numerical approach to track and quantify the frequency shift and counter its effect by tweaking the clamping and cutting forces of the fixture-workpiece assembly. To validate the numerically obtained values of natural frequencies, experiments are performed on an electrodynamic shaker at different stages of pocket milling and for various clamping pressures. The natural frequencies obtained numerically show good agreement with experimental values, with a maximum error of ±10%. Subsequently, the effect of metal removal on workpiece vibrations is presented for different pocket geometries and clamping pressures. Finally, to control increasing vibrations due to decreasing workpiece stiffness, a compensation scheme is presented based on the manipulation of clamping pressures.

1 INTRODUCTION

The dynamic response of a fixture-workpiece system changes significantly due to decreases in the mass and stiffness of the workpiece occurring together during pocket milling. The decrease in the mass pushes the natural frequencies forward whereas the decrease in stiffness of the workpiece pulls them backward. The effect of decrease in stiffness is far more significant than that of decrease in mass, leading to a net backward shift in natural frequency, particularly for thin-walled workpieces and workpieces involving a large percentage of metal removal. This causes an increase in the transmissibility (dynamic magnification factor), a manifold rise in vibration amplitude and, ultimately, manufacturing inaccuracies. It is, however, a challenging task to quantify workpiece compliance and mitigate its effect on workpiece deformation, because of such issues as the arbitrary shapes of workpieces, nonlinear contact behavior, and friction at the interface. For a problem as complex as this the numerical approach of modeling becomes indispensable, particularly with the advent and access to the computational tools of recent times.

In the past, a finite element model of the fixture-workpiece system was developed by many researchers to investigate the influence of compliance of the fixture body on workpiece static deformation (Siebenaler & Melkote, 2006). A combination of analytical model for contact and numerical model for structure was developed to study the effects of material removal on the dynamic properties and performance of the machining fixture-workpiece system (Deng, 2006; Mali et al., 2015). In addition, various techniques are reported in the literature to minimize workpiece vibrations during machining. An intelligent fixturing system for adjusting clamping forces adaptively is used for controlling deformations during thin-walled machining (Župerl et al., 2011). Damping involving a thin flexible layer mounted with distributed discrete masses attached with a viscoelastic layer during thin-wall milling are found to reduce vibrations to a great extent (Kolluru et al., 2013). The electromagnetic induction principle can also be used to minimize vibrations of thin-walled workpieces (Yang et al., 2015). In spite of these efforts, there is a need for a robust predictive tool that accurately computes the contact and structural deformations, and the modal properties, of workpieces and fixture assemblages, including the metal-removal effect under dynamic machining conditions. In addition, a corrective mechanism, such as manipulating cutting/clamping forces, needs to be developed to counter the workpiece deformations in a calibrated manner. The development and incorporation of an integrated framework to predict and correct dynamic workpiece deformations would make an intelligent Computer-Aided Fixture Design (CAFD) system (Yadav & Mohite, 2011).

In this study, a rigid fixture is designed and its compliance is investigated. Figure 1a shows, schematically, a prismatic workpiece of Al 6061-T6 with six location points (L1–L6) and two clamping points (C1 & C2). Spherical-headed locators and clamping elements maintain frictional contact with the flat faces of the prismatic workpiece and exert clamping forces. The fixture-workpiece system is modeled as a flexible workpiece resting on flexible fixture elements (locators/clamps). At the interface of these contacts, three contact springs, viz. one in normal direction to contact and two in lateral directions for frictional contact, are modeled in three orthogonal directions. The overall stiffness in any given direction is contributed to by three sources: fixture element (k_{ijf}), contact (k_{ijc}) and workpiece (k_{ijw}). The composite stiffness at fixture-workpiece contact is a series combination of these, as shown in Figures 1b and 1c.

The contact stiffness is obtained analytically by the well-established Hertzian contact theory, based on simplifying assumptions. The contact stiffness, which is a function of clamping pressure, is computed analytically and found to vary from 3.3 to 5.25×10^8 N/m for 800–3,200 N force. Similarly, the stiffness of the fixture element is 6.6×10^9 N/m, which remains constant. The structural stiffness is computed approximately by assuming the workpiece wall as a plate with one edge free and three edges fixed. It is found to be of the order of 10^7 N/m. However, excepting simple geometries, computations of workpiece stiffness cannot be done analytically due to their arbitrary shapes. Continuous change in shape of the workpiece makes this task even more challenging. A numerical approach, therefore, becomes essential to find the change in workpiece stiffness as well as to model the highly nonlinear contact problem. The next section briefly explains the numerical modeling procedure.

2 NUMERICAL MODEL

A solid aluminum 6061-T6 workpiece measuring 100 mm × 100 mm × 70 mm is located in a 3–2–1 fixture and clamped by two hydraulic clamps. Pocket milling is performed, starting with the 80 mm × 80 mm lateral dimensions of the pocket and increasing pocket depth gradually, creating 20%, 50%, and 60% metal removal. This represents the case of substantial metal removal. Next, the pocket size is enlarged laterally, for 64%, 70% and 76% metal removal, to create a case of a thin-walled workpiece, as shown in Table 1. Modeling frictional contacts, static and modal analyses are carried out in ANSYS® Workbench 18.1 to investigate the

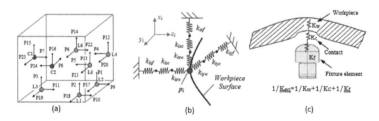

Figure 1. (a) Contact forces in normal and lateral directions at workpiece-fixture element contacts (p_i); (b) composite stiffness for i^{th} contact; (c) workpiece (K_w), contact (K_c) and fixture element stiffness (K_f).

Table 1. Pocket sizes for substantial metal removal and thin-walled machining cases.

Type	Substantial metal removal (z direction)			Thin-walled machining (x-y direction)		
% Metal removal	20%	50%	60%	64%	70%	76%
Size (mm)	$80 \times 80 \times 22$	$80 \times 80 \times 55$	$80 \times 80 \times 66$	$87 \times 87 \times 59$	$91 \times 91 \times 59$	$95 \times 95 \times 59$

Table 2. Settings for contact modeling, meshing and analysis for numerical model in ANSYS®.

Contact model	Option	Meshing	Option	Analysis setting	Option
Contact type	Frictional	Method	Hex dominant	Auto time step	Off
Frictional coeff.	0.375	Face size	2.5 mm	Solver	Direct
Behavior	Symmetric	Size fun.	Curvature	Weak springs	Off
Formulation	Aug. Lagrange	Relevance	Fine	Large deflection	Off
Pin ball rad.	1 mm	Quality	0.8	Fixed support	Loc. rest
Detection	Node normal from contact	Elements	SOLID 186, CONTA 174, TARG 170	Clamp pressure for C1 and C2	10–40 bar

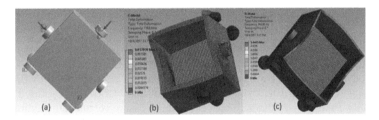

Figure 2. (a) Solid model showing fixed supports at six locators and clamping pressures applied on two clamps; (b) modal shape of workpiece with 64% metal removal; (c) modal shape of workpiece with 76% metal removal.

workpiece deformations and natural frequencies for various percentages of metal removal and clamping pressures of 10, 25 and 40 kg/cm². The flat faces of the six locators L1 to L6, which rest on the fixture plate, are constrained in terms of movement (u_x, u_y, $u_z = 0$) and the spherical faces maintain a frictional contact with the workpiece. Clamping forces are applied at the back of two clamps (see Figure 2a).

Table 2 shows the settings used for static and modal analysis in ANSYS® for frictional contact modeling, meshing and analysis. Illustrative modal shapes are shown in Figures 2b and 2c. In the next section, the experimental procedure used for validation is explained in brief.

3 EXPERIMENTAL MODAL ANALYSIS

To validate the numerical model, an experimental modal analysis of the workpiece-fixture assembly is performed on a horizontal slip table of an electrodynamic vibration shaker, and a Resonance, Search, Track and Dwell (RSTD) test is conducted. The experiments are conducted within the frequency range 500–3,000 Hz with a sweep rate of 1 octave/min and acceleration 0.5 g. The accelerometer used is a model 353B02 with sensitivity of 9.81 mv/g. The aluminum workpieces with 0 to 76% metal removal are held in a 3–2–1 fixture with three pressures (10, 25 and 40 kg/cm²) to obtain the experimental natural frequencies (see Figures 3a and 3b). The results of the numerical and experimental investigations are presented in the following section.

199

Figure 3. (a) Experimental test setup of fixture-workpiece mounted on a horizontal slip table of an electrodynamic shaker; (b) rigid fixture based on a 3-2-1 principle and two hydraulic clamps.

Table 3. Comparison of simulated (f_{xs}) and experimental frequencies (f_{xe}) of x-mode.

Type	Substantial metal removal				Thin-walled machining		
% Pocket	0	20	50	60	64	70	76
Simulated freq. f_{xs}(Hz)	1113	1185	1389	1414	1369	1227	913
Experimental freq. f_{xe}(Hz)	1052	1084	1521	1459	1216	1151	950
% Error	−5.48	−8.52	9.50	3.18	−9.42	−6.19	4.05

4   RESULTS AND DISCUSSION

The results are presented in three subsections. In the first subsection, the results of modal analysis of the proposed numerical model are validated with the experimental results obtained on an electrodynamic shaker by a sine sweep RSTD test. This is followed by the plot of the effect of metal removal on the dynamic response of the workpiece, captured in terms of transmissibility. Finally, with the help of a clamping pressure versus deformation graph, it is explained how the deformations can be checked by tweaking the clamping force.

4.1   Validation of numerical model by experiment

Table 3 shows a comparison of simulated and experimentally obtained frequencies of x-mode, presented for various percentages of metal removal at 800 N clamping force.

For the substantial material removal case, the natural frequency is found to increase by 27%, showing that the loss of stiffness is at a slower rate (see Figure 5a, K_1, K_2 and K_3) than that of the mass. On the contrary, for the thin-walled case, the frequency is found to decrease by about 35%, showing that the loss of stiffness is faster (see Figure 5b, K_1', K_2' and K_3') than that of the mass. The variation between the simulated and experimentally obtained frequencies is found to be around ±10%, which is quite good. This variation is attributed to differences in assumptions made in simulations and actual experimental conditions such as the presence of damping, the release of prestress during vibration testing, approximations in applying boundary conditions in the numerical model, and variations in clamping pressure and temperature of the workpiece. Thus the match is quite good, and variation can be reduced by a stringent control of experimental conditions and more accurate numerical modeling. Thus, the objective of this investigation is to provide a fully numerical model for modal analysis of a fixture-workpiece assembly. The effect of frequency shift on vibration amplitude using transmissibility is presented in the next section.

4.2   Variation of transmissibility with percentage metal removal

The end mill exerts dynamic cutting forces consisting of spindle speed and its harmonics (i.e. multiples due to number of cutter flutes). The vibration response of the workpiece under its changing dynamic characteristics during metal removal is captured well by the term transmissibility (magnification factor), given by:

$$\frac{X}{X_{st}} = \frac{1}{\sqrt{(1-r^2)^2 + (2\xi r)^2}} \tag{1}$$

where $r = \frac{\omega}{\omega_n}$ is the frequency ratio, consisting of ω, the spindle speed (in rad/sec), and ω_n, the natural frequency. Further, ξ is the damping factor consisting of the damping coefficient and $C_c = 2\sqrt{KM}$, the critical damping coefficient. Note that the natural frequency (ω_n) and the damping coefficient (C_c) of the workpiece are the functions of the stiffness K and the mass M of the workpiece. The metal-removal process is characterized by variation in the frequency ratio 'r' and the damping factor 'ξ' of the workpiece as the stiffness K and the mass M continuously decrease during the machining process. Transmissibility is the measure of vibration intensity and depends upon 'r' and 'ξ', as given in Equation 1. Figure 4 shows the simulated results of percentage metal removal versus transmissibility (dynamic magnification factor) for different spindle speeds.

The effect of the rise and fall in the natural frequency, respectively, for the case of substantial metal removal and thin-walled workpiece, as discussed in the previous subsection, leads to changing vibration levels. This is plotted in Figure 4 in terms of transmissibility for different percentages of metal removal. There is a drop in the transmissibility for the substantial metal-removal case, whereas for the thin-walled case the transmissibility rises, meaning higher vibration levels. There is a significant increase in dynamic magnification factor (transmissibility) for higher cutter speeds, viz. 3,000, 6,000 and 9,000 rpm, which are 2, 5 and 25%, respectively. A higher percentage increase in the dynamic magnification factor leads to higher vibration levels, causing errors in geometric accuracies and surface finish of the workpiece. This problem can be mitigated by changing the cutting parameters and clamping pressure on the workpiece in a predetermined manner. This is explained in the next subsection.

4.3 Controlling deformation by clamping force manipulation

Figure 5a shows the variation of clamping forces versus workpiece deformation at clamp C2 for the case of substantial metal removal (i.e. 0–60%). Although there is substantial metal removal, the drop in the stiffness is slow, as is evident from the steep lines and gradual decreasing slopes (i.e. stiffness). With reference to the stiffness of the solid workpiece, the stiffness drops by half for 0–60% metal removal from the solid workpiece. Figure 5b shows the variation of clamping forces versus deformation for the thin-walled workpiece (i.e. 64–76% metal removal). In this case, the stiffness drops by approximately one order of magnitude for a net 12% metal removal (64–76%) in the thin-walled case.

The family of characteristic curves having diminishing slopes with increasing metal-removal percentages leads to increased deformation of the workpiece. This leads to an undesirable workpiece deformation and manufacturing inaccuracies. The characteristic can be

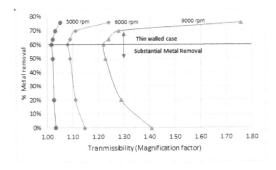

Figure 4. Case variations showing decrease in transmissibility for substantial metal-removal case (0–60%) and increase in transmissibility for thin-walled case (64–76%) for three spindle speeds.

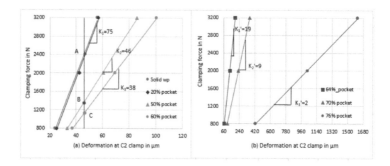

Figure 5. Variation of simulated clamping forces vs. deformation at C2 for: (a) substantial metal removal: 0–60% pocketed workpiece; (b) thin-walled case: 64–76% pocketed workpiece.

used in the countering of this effect by reducing clamping forces (as well as adjusting cutting forces by tuning cutting parameters). This can be done by interpolating along a vertical line AC as shown in Figure 5a, and the same can be implemented for the thin-walled machining case. The conclusions of this work are summarized in the next section.

5 CONCLUSIONS

The changing mass and compliance of a workpiece during machining cause considerable deformations and shift in natural frequencies of the workpiece, leading to manufacturing inaccuracies. In this work, a fully numerical approach is proposed to track the contact and structural deformations, as well as the dynamic transmissibility of the workpiece, as metal is being removed during pocket milling. The results of numerical simulations are validated with the experimental results on a vibration shaker using a sine sweep and resonance, search, track and dwell test. A corrective strategy is presented in the form of force-deflection characteristic curves for different percentages of metal removal to compensate for the changing workpiece deformations by tweaking clamping and cutting forces. The proposed methodology can be integrated with a computer-aided fixture design system to predict the in-process dynamic behavior of a fixture-workpiece system and compensate for same by changing the clamping pressure, thereby maintaining quality and productivity in manufacturing.

REFERENCES

Deng, H. (2006). *Analysis and synthesis of fixturing dynamic stability in machining accounting for material removal effect* (Doctoral thesis, Georgia Institute of Technology, Atlanta, GA). Retrieved from http://smartech.gatech.edu/bitstream/handle/1853/14038/deng_haiyan_200612_phd.pdf.
Kolluru, K., Axinte, D. & Becker, A. (2013). A solution for minimising vibrations in milling of thin walled casings by applying dampers to workpiece surface. *CIRP Annals, 62*(1), 415–418.
Mali, S., Yadav, M.H. & Mohite, S.S. (2015). Theoretical and experimental studies of effect of metal removal on dynamic behavior of fixture-workpiece system. In *International Conference on Emerging Trends in Mechanical Engineering, Kochi, India, January 2015* (pp. 285–292).
Siebenaler, S.P. & Melkote, S.N. (2006). Prediction of workpiece deformation in a fixture system using the finite element method. *International Journal of Machine Tools and Manufacture, 46*(1), 51–58.
Yadav, M. & Mohite, S.S. (2011). Development of a decision support system for fixture design. In K. Shah, V.R. Lakshmi Gorty & A. Phirke (Eds), *Technology systems and management* (pp. 182–189). Berlin, Germany: Springer.
Yang, Y., Xu, D. & Liu, Q. (2015). Vibration suppression of thin-walled workpiece machining based on electromagnetic induction. *Materials and Manufacturing Processes, 30*(7), 829–835.
Župerl, U., Čuš, F. & Vukelić, Đ. (2011). Variable clamping force control for an intelligent fixturing. *Journal of Production Engineering, 14*(1), 19–22.

Technology Drivers: Engine for Growth – Mahajan, Modi & Patel (Eds)
© *2018 Taylor & Francis Group, London, ISBN 978-1-138-56042-0*

Design of a cost-effective fly ash brick manufacturing machine

Sagar Ponkia, Shubham Mehta, Hiren Prajapati & Reena Trivedi
Department of Mechanical Engineering, Institute of Technology, Nirma University, Gujarat, India

ABSTRACT: Fly ash is generated as a by-product after the combustion of coal, and it is an environmental pollutant. An effective way to utilize fly ash is in the construction sector, after a rise in demand in housing corresponding to the increasing population. This increasing demand can be fulfilled by using automation in brick manufacturing. However, the existing automatic fly ash brick manufacturing machine cost is around 18–20 lakh rupees. In Gujarat, micro-small-medium brick porters are asking for machines costing below 10 lakh rupees. Hence an attempt is made to develop a low-cost fly ash brick manufacturing machine. This paper deals with the design and finite element analysis of the supporting structure, punch, de-mold punch, mold box, punch plate, and main frame support rectangular bar. The design and calculation of the conveying belt, feeder, and pan mixer are described. Here the machine is designed for three cavities per stroke to reduce cost.

Keywords: mold; frame; punch; de-mold; pan mixer; conveyor; feeder; hopper

1 INTRODUCTION

The disposal of such a huge quantity of ash is a serious issue. In this machine we used a feeder, punch, de-mold punch, pan mixer, mold box, conveyor, hydraulic cylinder, pump, and a motor with complete automation. Nataatmadja (2008) has compared various combinations of fly ash-based bricks by varying the proportion of fly ash, sand, sodium silicate, lime and water. Nataatmadja has suggested the fly ash-based cured brick as being a good competitor to clay-based fired bricks. The brick composition recommended by a Gujarat-based fly ash brick manufacturer is fly ash (35%), sand (45%), lime (12%) and cement (8%). Water is added (5–6%) to the homogeneous mixture of the material. As per the IS 1077-1992 standard (BIS, 1992), the size of the brick is selected as being 230 mm × 110 mm × 75 mm.

2 DESIGN AND ANALYSIS OF MACHINE COMPONENTS

In this section various components of a fly ash brick-making machine are discussed. The various machine parts are as follows.

2.1 *Punch*

The punch is used for compacting and converting the brick material into the form of a brick. Here, power is provided by a hydraulic cylinder. The brick material is in wet powder form. EN8 material is used to make the punch. Standard quad mesh is used and the maximum element size is 5 mm. The top plate of the punch (*mnop* in Figure 1) is mounted with the hydraulic cylinder. 200 kN force is reacted with the three bottom plates (*abcd*, *efgh*, and *ijkl*) of the punch. The maximum von Mises stress of 179.49 MPa is generated in the upper side between two plates and the selected material has a permissible stress limit of 235 MPa. Thus, the proposed design is safe. The maximum deformation is 0.082 mm at the bottom plate of the punch at both the sides.

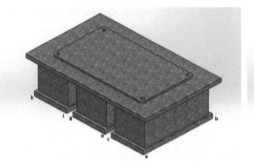

Figure 1.   Punch.

Figure 2.   De-mold punch.

2.2   *De-mold punch*

The de-mold punch is used here as a platform where brick material is conveyed through the feeder. When the punch presses the brick material to form a brick, the total compacting force passes to the de-mold punch through the brick material. Hence the material of the de-molding punch should be stronger than that of the molding punch. This functional requirement of the de-molding punch can be fulfilled using EN8 steel. This punch has to bear the compacting pressure as well as the brick weight. The length of the bar is 165 mm. The mesh used is quad mesh and the maximum element size is 5 mm. The bottom plate of the de-mold punch (*mnop*) has a fixed geometry, and a force of 200 kN is applied to the upper side of the de-mold punch (*abcd*, *efgh* and *ijkl*).

The maximum von Mises stress generated is 125.96 MPa, between the bar and the top plate. The selected material's permissible stress limit is 235 MPa (High Peak Steels, 2016). Thus, the proposed design is safe. The deformation that occurs is 0.091 mm. Thus, the plate is safe.

2.3   *Mold box*

The dimensions of the mold box cavity are $230 \times 110 \times 125$ mm. Quad mesh is used and the maximum element size is 5 mm. A nut and bolt are used to fix both the workpiece and the bottom supports. Force is applied in the mold box cavity from all twelve sides of the mold box. The maximum von Mises stress of 129.64 MPa is generated in the bottom edge of the mold box and the selected material has a permissible stress limit of 235 MPa. Thus, the proposed design is safe. The maximum deformation is 0.11 mm in the bottom end side of the mold box cavity.

2.4   *Frame support rectangular bar*

The column support is the main support of the whole machine. As per the loading condition, the cross section dimension is selected as per standard dimensions. The maximum element size is 5 mm and is fixed at the bottom surface of this stand. Force is applied at the cross section of the bar (1, 2, 3 and 4 in Figure 4). The maximum von Mises stress generated is 102.96 MPa between the bars, and the selected material has a permissible stress limit of 250 MPa. Thus, the proposed design is safe and the factor of safety is 2.42. The deformation that occurs is 0.1 mm. Thus, the bar is safe.

2.5   *Frame*

The main machine body is known as the main frame. In this frame all the operations of the machine are performed. In the main frame the mold, punch plate, de-mold and hydraulic cylinder are fitted. The EN8 mild steel material is selected because of its high strength and

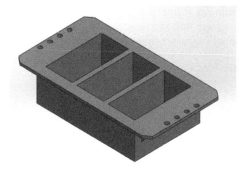

Figure 3.   Mold box.

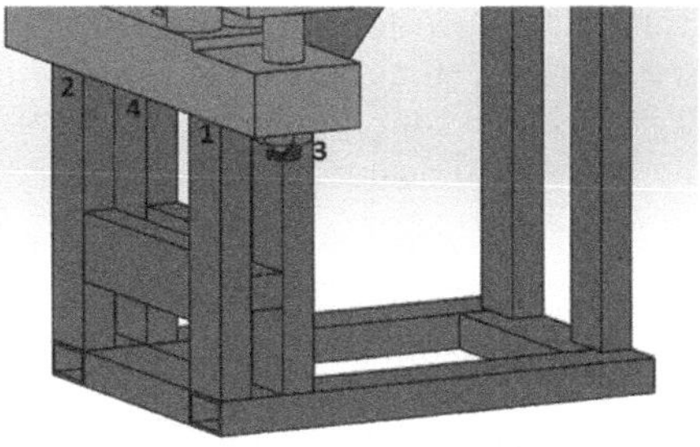

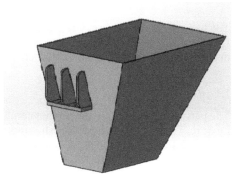

Figure 4.   Frame support rectangular bar.

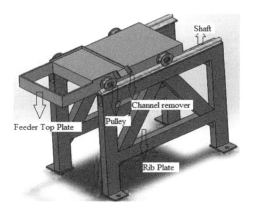

Figure 5.   Feeder assembly.

Figure 6.   Hopper.

ease of welding. Quad mesh is used in the top and bottom channels, and standard mesh in the guide pillar. The maximum element size is 10 mm. Fixed support is given at the end of the plate with the channel stand. A reaction force of 200 kN is generated at the top plate by the cylinder. A maximum von Mises stress of 173.55 MPa is generated between the top plate and the nut. Here, the generated stress is lower than the permissible stress limit.

2.6   Feeder

The feeder is a component unit which is used for feeding the brick material to the mold for making bricks. The operation of the feeder takes place after the mixed material falls onto the feeder plate from the hopper. The material used in the design of the feeder is EN 24 mild steel. The principle of the feeder is that it feeds the material to the main machine frame mold cavity plate from the feeder. From the options for single-stroke motors, a hydraulic cylinder of bore diameter 50 mm and piston rod diameter 28 mm are selected. The important components that are used in the feeder are the feeder top plate, feeder bottom plate, shaft, pulley, rib plate and hydraulic cylinder. The structure of the feeder assembly is as shown in Figure 5.

2.7   Hopper

The hopper is a component used for spraying and distributing the supplied mixture. It is a form of frustum which is described as being a truncated square pyramid (see Figure 6). The material which is mixed in the hopper is fly ash, sand, cement and water. Taking the cycle time for each stroke as 10 min, for single stroke and the machine runs for 5 min on a full single cycle. Thus, on the basis of the composition and volumes of the material: total volume required for 90 bricks, $V_b = 23 \times 10^7$ mm^3; a = inlet length = 540 mm; b = outlet length = 360 mm; h = height = 720 mm. The final volume required for the hopper is: $V_h = 23.2 \times 10^7$ mm^3. Here, the volume of the hopper is greater than the volume of 90 bricks, and thus the design is safe.

2.8   *Conveyor idler arm pulley*

The belt will sag due to the weight of the belt and the conveyed material. Normally, the idler pulley is fitted at 300–350 mm distance from the conveyor at the angle of inclination of the conveyor. The total number of idlers for a length of 2,250 mm is seven. Three idler pulleys are fitted in a flattened V-shape to one another at an angle of 35°. These three idler pulleys are fixed on the idler stand, which consists of a pulley stand and arm.

3   SELECTION OF MIXER AND CONVEYOR

The pan mixer and conveyor are the important material handling equipment of the machine. Initially, the material gets deposited on the pan mixer and after the crushing operation the material reaches the hopper with the help of a flatbed incline conveyor.

3.1   *Pan mixer*

The pan mixer is a circular device that combines various ingredients or materials to form the mixture. These inputs include fly ash, sand, cement, lime and water, and the composition of the material is 35% fly ash, 45% sand, 12% lime and 8% cement.

The total weight of material in the pan mixer is derived from the formula:

$$M_{PM} = M_F + M_S + M_C + M_L \qquad (1)$$

where M_F is mass of fly ash in the pan mixer, M_S is mass of sand in the pan mixer, M_C is mass of cement in the pan mixer, and M_L is mass of lime in the pan mixer. Thus, in our case the total weight of material is:

$$M_{PM} = 125.83 + 118.22 + 17.03 + 25.61 = 286.69 \text{ kg.}$$

The pan mixer drum dimension and capacity are selected from Table 1. Given a total weight of material in the pan mixer of 286.69 kg, we select a drum of capacity 500 kg with a 1,500 mm diameter.

3.2   *Conveyor*

A belt conveyor consists of an endless belt of a resilient material, connected between two pulleys and moved by rotating one of the pulleys through a drive unit gear box connected to an electric motor. The angle of inclination for conveying the material should not exceed 35°. After leaving the pan mixer, the material enters the hopper conveyor, which is on the tail side;

Table 1.   Selection of pan mixer (Revomac, 2016).

Capacity	Motor (hp)	Muller size (inch)	RPM	Diameter (mm)
50 kg	2	16 × 10	2428	600
100 kg	5	12 × 7.5	2428	900
150 kg	5	12 × 7.5	2428	1000
250 kg	7.5	16 × 7.5	2428	1200
500 kg	15	16 × 10	2428	1500

then the operation of the conveyor proceeds. The material then falls into the material feeder from the conveyor. The height of the machine frame is 1,600 mm and the angle of inclination is 35 degrees.

To find the length from the machine frame to the initial point of the conveyor (Haideri, 2012):

$$\tan\alpha = h/L_h \tag{2}$$

where $L_h = 2285$ mm.

Pulley width is calculated thus:

$$W = B + 2S = 650 \text{ mm} \tag{3}$$

Standard-size pulley widths are 200, 250, 315, 400, 500, 630, 800, 1,000 and 1,250 mm. From these, a pulley of diameter 630 mm is selected. Speed of pulley, $N_p = 45.77$ rpm; speed of motor, $N_M = 1,440$ rpm. Hence the reduction ratio or gear ratio, N_G, is:

$$N_G = N_m/N_p = 31:1 \tag{4}$$

Hence, the mounting type of gear in the conveyor is a flange-type mounting.

4 CONCLUSIONS

Fly ash bricks produced by using this machine are relatively economical and affordable for those in rural areas and for low-income earners. In the initial phase, brick material and size are selected as per government norms, which gives the best physical properties at low cost compared to conventional clay bricks. Here, the machine is designed for three cavities per stroke. The design of the mold plate, punch, frame structure, hopper, and feeder are conducted according to strength criteria. The selection of the pan mixer and belt conveyor are conducted to fulfill the machine requirements. This gives the most affordable design with good productivity. This machine is highly affordable for small-scale industries.

REFERENCES

BIS. (1992). *IS 1077: Common burnt clay building bricks – Specification*. New Delhi, India: Bureau of Indian Standards.

Haideri, F. (2012). *Design of material equipment engineering*. Mumbai, India: Nirali Publications.

High Peak Steels. (2016). Black engineering bar: EN8. Retrieved from http://www.highpeaksteels.com/stock/black/black-080m40/

Kaushike, P. & Maitra, R. (2015). A case study on the fly ash brick manufacturing plant for uplifting the deprived human resource of the society by TATA Power: A step toward sustainable business. In *Twelfth AIMS International conference on management, 2–5 January 2015* (pp. 2052–2055). Retrieved from http://www.aims-international.org/aims12/12 A-CD/PDF/K748-final.pdf

Nataatmadja, A. (2008). Development of low-cost fly ash bricks. In R. Haigh & D. Amaratunga (Eds.), *Building resilience: Proceedings from international conference on Building Education and Research (BEAR)* (pp. 831–843). Salford, UK: School of the Built Environment, University of Salford.

Revomac. (2016). Concrete pan mixer. Retrieved from http://www.revomac.net/pan-mixer.html

Technology Drivers: Engine for Growth – Mahajan, Modi & Patel (Eds)
© 2018 Taylor & Francis Group, London, ISBN 978-1-138-56042-0

Design and development of thermal actuator for refocusing mechanism of secondary mirror of a space telescope

Pratik Patel
Department of Mechanical Engineering, Institute of Technology, Nirma University, Ahmedabad, India

Yeshpal Yadav
Space Application Centre, ISRO, Ahmedabad, India

K.M. Patel
Department of Mechanical Engineering, Institute of Technology, Nirma University, Ahmedabad, India

Haresh Jani
Space Application Centre, ISRO , Ahmedabad, India

ABSTRACT: A Satellite telescope consists of various components such as primary mirror, secondary mirror, detector unit and metering structure etc. Generally, space borne telescopes are subjected to orbital loads like temperature excursion, absences of gravity, moisture release of metering structure and due to this, there is a change in focal length and misalignment of optical system takes place which ultimately reduces the quality of images. So to get the perfect focal point, it is necessary to reset the distance between primary and secondary mirror by means of re-focusing mechanism. This re-focusing mechanism is placed behind the secondary mirror in order to compensate the change in distance between primary and secondary mirror. In this paper, the design and development of thermal actuator based refocusing mechanism have been proposed. The refocusing mechanism has been designed for ± 10 micron displacement in both forward and backward directions by means of heating only. It also provides six degrees of movement to a secondary mirror. Material for thermal actuator is selected and amount of temperature variation required for specified movement of the mechanism is studied. Experiments are performed in thermo-vacuum chambers and results are validated with structural and thermal analysis using commercially available software.

Keywords: Primary Mirror, Secondary Mirror, Focal Length, Refocusing, Thermal Actuator, Co-efficient of Thermal Expansion. Inner cylinder (IC), Outer cylinder (OC)

1 INTRODUCTION

The objective of satellite telescope is to capture good quality of image throughout its life cycle since it is used for both civilian and defense users. The actual location of focal point is very important for high quality images. Very precise actuating mechanism is required for resetting the distance between primary and secondary mirror so as to get the perfect focal point.

Previously, piezoelectric actuator based mechanisms were widely used for refocusing purpose. But due to the concern of its survivability against launch load and power and temperature limitation, use of a piezoelectric based mechanism has been limited [1]. Raval [2] had designed single cylinder thermal actuator based refocusing mechanism for both forward and backward strokes. Thermal actuator with double stage actuation is more reliable and less time consuming as both positive and negative translations are achieved by heating inner and outer cylinder respectively [3].

Generalized ray diagram of optics for cassegrain focus type telescope is shown in Figure 1. Folding and diverging effect of the secondary mirror creates a telescope with a long focal length while having short tube length [4].

2 DESIGN OF REFOCUSING MECHANISM

2.1 Working principle of thermal actuator

The working principle of thermal actuator can be demonstrated with the help of Joule's law of heating and differential thermal expansion which is expressed by:

$$Q = j^2\rho = I^2R \tag{1}$$

where, j is the current density vector, ρ is the specific electric resistivity, I is the current passing through the materials, R is the materials resistance and Q is the heat generated. This heat gained and stored by thermal actuator structure and is given by:

$$Q = mc_p\Delta T/t \tag{2}$$

A general thermal expansion is given by:

$$\Delta l = l_o \alpha\Delta T \tag{3}$$

where, l_o is the initial length, α is the coefficient of thermal expansion and ΔT is the temperature change. From Equation 3, it can be seen that Joule heating in electro-thermal actuators is heavily dependent on the properties of materials. So careful material selection is important.

Table 1. Performance requirements parameters.

SM Refocusing Mechanism Performance Requirements	
Parameter	Value (µm)
Total stroke	±30
Elongation accuracy	± 0.5
Minimum actuation step (i.e. Resolution)	0.5
Tilt in arc sec	±15
Decentre	±10

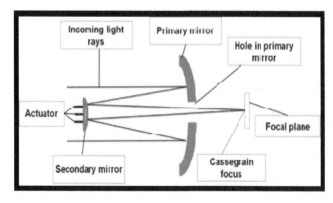

Figure 1.   Generalized ray diagram of optics [4].

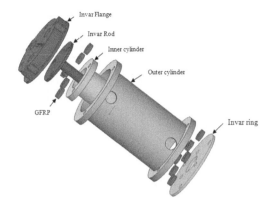

Figure 2a. Exploded view of thermal actuator.

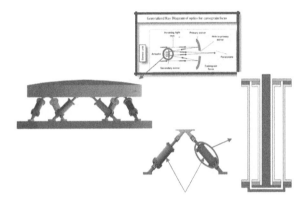

Figure 2b. Thermal actuator.

2.2 *Performance requirements*

Performance requirement parameters for the refocusing of secondary mirror are shown in the table below. Stroke, accuracy and resolution requirements are the main parameters which are to be satisfied in any orbital environment condition. Other functional requirements are also to be satisfied in order to get the six degree of freedom flexibility.

2.3 *Design description*

3D model of the thermal actuator prototype is shown in Figure 2. Mass of the whole actuator assembly is 0.0421 kg. Length of the invar rod is 61.5 mm and diameter is 4 mm. Thickness of both inner and outer cylinders is 1 mm whereas inner diameters are 21 mm and 10 mm respectively. Thickness selected for invar ring is 1.6 mm. Initially it was 1 mm thick which was subsequently increased to 1.6 mm to avoid bending.

2.4 *Design description*

Secondary mirror is mounted on three pairs of actuators (six actuators) 120 degree apart to each other. Each actuator is connected to the MFDs on both sides. GFRP washers are provided in between inner and outer cylinder for isolation purpose. Here one end of thermal actuator is connected to interfacing ring while other end is connected to secondary mirror through flexure arrangement.

While selecting the material, the main objective is to achieve high thermal expansion at minimum temperature gradient and minimum power requirement.

Aluminum 6061 T6 material is selected for inner and outer cylinder because of good coefficient of thermal expansion, High thermal conductivity and low density. GFRP material is used for spacer as it has low weight, high insulating properties, good heat resistance, low thermal conductivity and comparatively low in cost.

Invar material is used as it has high strength, low coefficient of thermal expansion, high stiffness and high young's modulus.

3   TEST SETUP

3.1   *Experimental test setup*

Here, Clamp is used to fix the displacement measuring probe and the material used for probe clamp is invar. Any displacement in micron can be measured with the help of LVDT.

Table 2.   Material selection for actuator components.

Sr. No.	Component Name	Material
1	Inner cylinder	Al 6061T6
2	Outer cylinder	Al 6061T6
3	Invar rod	Invar
4	Spacer	GFRP
5	Invar ring	Invar
6	Invar flange	Invar

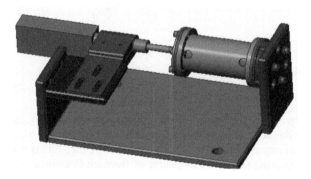

Figure 3.   Experimental test setup.

Table 3.   Actuation with to respect temperature gradient.

Temp (°C)	Inner cylinder (μm)	Outer cylinder (μm)
1	1.02	−0.97
2	2.05	−1.93
3	3.08	−2.90
4	4.11	−3.86
5	5.13	−4.83
6	6.16	−5.79
7	7.19	−6.76
8	8.22	−7.72
9	9.24	−8.69
10	10.27	−9.65

212

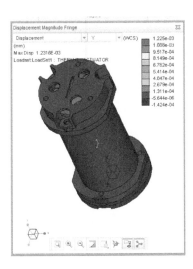

Figure 4. IC expansion.

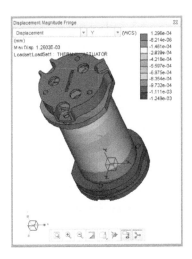

Figure 5. OC expansion.

In order to reduce the radiative interaction with atmosphere, low emissive coating is applied on outer cylinder. Kapton foil heater having Wattage capacity of 0.5 W is wrapped around inner and outer cylinders. Two PT100 temperature sensors which are placed on the single cylinder are used to measure the temperature of cylinders.

The whole actuator assembly is wrapped by MLI and low emissivity coating. To maintain temperature of a base and an actuator plate at 20°C, three copper strips are attached to invar flange, invar ring and invar rod which throws the heat outside on radiator and provide cooling.

4 RESULTS AND DISCUSSION

4.1 *Displacement analysis for single actuator*

In order to check whether the actuation of thermal actuator with respect to change in temperature is uniform or not, displacement analysis is carried out and the results are shown below.

Analysis is carried out by fixing one end of the thermal actuator and measuring the displacement at the free end. Temperature load of 1°C to 10°C is applied to both inner and outer cylinders of the thermal actuator respectively.

4.2 *Results comparison*

Heat is applied on inner and outer cylinders respectively through the heater placed on the circumference of the cylinder. Comparison between theoretical and experimental displacements for inner and outer cylinder heating is shown in Figure 6 and Figure 7 respectively.

Graph of displacements for both experimental results and structural analysis is plotted and is shown in Figure 6. It shows that both experimental and analytical results are identical to each other and slope of the graph is almost same.

As shown in Figure 7, linearity between temperature gradient and displacement is achieved. The slope for experimental results is not same as that of simulated results due to the heat convection that takes place on outer cylinder. It can be corrected by performing experiment in vacuum chamber.

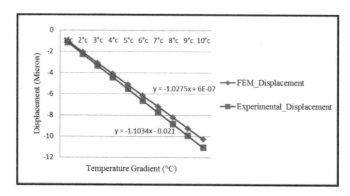

Figure 6.   Graph for inner cylinder heating.

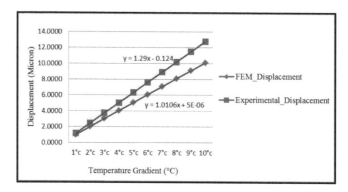

Figure 7.   Graph for outer cylinder heating.

5   CONCLUSION

Working principle of single thermal actuator has been successfully demonstrated. The selection of best material considering both thermal and structural point of view is carried out. Finite element analysis has been carried out for the whole thermal actuator assembly. This mechanism provides 6 degrees of freedom flexibility to any type of axi-symmetric mirror. Design, modeling and analysis has been carried out for ±10 micron displacement in both forward and backward directions. Thermal expansion has also been checked experimentally and compared with simulation results. Experiments are performed in thermo-vacuum chambers for both electro thermal and thermo-mechanical analysis and results are validated with structural analysis in software. Careful evaluation of test results is carried out and it shows that concept of thermal actuator for secondary mirror refocusing is successful and highly reliable.

REFERENCES

[1] Alberto Franzoso and RaffaeleD'Imporzano, "A sub-micrometric thermal refocusing mechanism for high Resolution EO telescopes", international conference on environmental systems, Bellevue, Washington. ICES-2015-66 12–16 July 2015.
[2] Apurv Raval, "A Detail Review of Optical Misalignment and Corrected by Refocusing Mechanism", IJSRD—International Journal for Scientific Research & Development| Vol. 3, Issue 4, 2015 | ISSN (online): 2321–0613.
[3] Patrice damilano, *Pleiades high resolution satellite: a solution for military and civilian needs in metric class optical observation,* 15th annual/USU conference on small satellite (2012).
[4] Jean-Luc Lamard, Catherine Gaudin-Delrieu, David Valentini, Christophe Renard, *Design of the high resolution optical instrument for the Pleiades HR earth observation satellite,* proceedings of the ICSO 2004.

Identification of faulty condition of rolling element bearing with outer race defect using time and frequency domain parameters of vibration signature

Shrey Trivedi
Department of Mechanical Engineering, Ahmedabad Institute of Technology, Gujarat Technological University, Ahmedabad, Gujarat, India

D.V. Patel, V.M. Bhojawala & K.M. Patel
Department of Mechanical Engineering, Institute of Technology, Nirma University, Ahmedabad, Gujarat, India

ABSTRACT: In the present work, various vibration parameters are used for fault identification in a rolling element bearing. Different sizes of artificial defect were seeded on the outer race of a rolling element bearing. Time domain parameters such as Root Mean Square (RMS), peak to peak, crest factor and kurtosis have been measured for different sizes of defect and are compared with a healthy bearing. In order to identify the defect, a location frequency domain study has been done. As the defect size decreases, the vibration signals obtained become very weak and it is difficult to identify the defect frequency because of external noise. Envelope analysis is used to remove this high-frequency noise from the low-frequency fault signature.

1   INTRODUCTION

Rolling element bearings are widely used in most rotating machinery. Damage occurring in a bearing can damage the entire system. Therefore, the diagnosis of a bearing is required. In the present work, different sizes of defects, such as 2 mm, 1 mm and 0.3 mm, are seeded on the outer race of a bearing using wire-cut Electrical Discharge Machining (EDM) (Karacay & Akturk, 2009). Scalar measures such as Root Mean Square (RMS), kurtosis, and crest factor are employed to identify the defect in the rolling element bearing (Dyer & Stewart, 1978). It was observed that kurtosis can be used to identify the defect in the bearing, and determined that a healthy bearing should have a kurtosis value of 3. Mathew and Alfredson (1984) used spectral and statistical parameters to check the condition of a bearing. Durkhure and Lodwal (2014) used statistical and spectral methods for identifying defects localized on the inner race, outer race and balls. Wu (2016) used a time domain parameter to find defects in a rolling element bearing and also discussed the effect of this time domain parameter for the healthy and faulty bearing. Randall and Antoni (2011) evaluated the usefulness of different methods, such as envelope analysis, minimum entropy deconvolution and spectral kurtosis, to identify defects in a bearing. The effectiveness of these parameters is also explained in the fault diagnostics of the rolling element bearing. Ho and Randall (2000) discussed the use of squared envelope and higher power for the detection of a defect frequency.

2   EXPERIMENTAL SETUP AND EXPERIMENTAL WORK

An experimental test rig is designed for fault detection in a rolling element bearing, as shown in Figure 1. The test rig is designed in such a way that it can rotate a defective bearing with the application of the load. In the test setup, an electric motor is rotating at 1,440 rpm and a shaft is connected to the electric motor through a flexible coupling. The shaft is supported by two

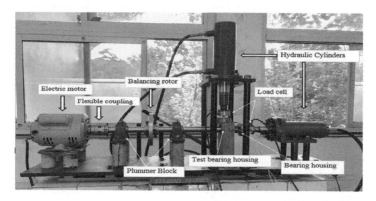

Figure 1. Experimental setup.

Figure 2. Defects of 2 mm, 1 mm and 0.3 mm.

Table 1. Loading conditions and sampling rates.

Load (kg)	0	100		200	300		400
Sampling rate (Hz)	800	1,024	2,048	5,120	12,800	25,600	51,200

plummer blocks. One bearing housing is designed in which the test bearing is fitted. Load is applied on the test bearing through hydraulic cylinders. The capacity of the hydraulic cylinder is 10 tons. Two hydraulic cylinders are used, one for radial load and the other for axial load. The applied load is measured with the help of a pancake-type load cell, having capacity of 1 ton.

A 6205 bearing with a polyamide cage has been used, with artificially seeded 2 mm, 1 mm and 0.3 mm defects on the outer race using wire-cut EDM. Figure 2 shows the artificially seeded defects and their actual size measured with a USB-connected digital microscope.

The vibration signals for the faulty bearings are captured with the help of a four-channel Fast Fourier Transform (FFT) analyzer. Readings were taken for different loading conditions and different sample rates. The vibration signatures obtained were analyzed in MATLAB software for time domain and frequency domain. Table 1 shows the magnitudes of the radial load and sampling rates for the current work.

3 RESULTS AND DISCUSSION

A time domain study of a healthy bearing has been performed to compare it with the defective bearings. Measurement of the vibration signal has been conducted in terms of variation of acceleration 'g' with respect to time.

3.1 *Time domain study of a 2 mm defect on the outer race*

For the 2 mm defect on the outer race, similar experimentation has been performed as described in the preceding section. Various statistical parameters were measured and compared with those of a healthy bearing. Figure 4 shows the variation in statistical parameters for 2 mm defects on an outer race.

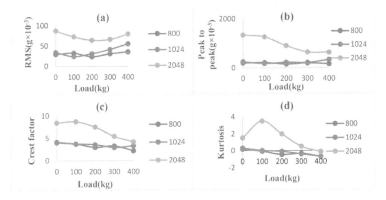

Figure 3. Variation in statistical parameter of a healthy bearing: (a) RMS; (b) peak to peak; (c) crest factor; (d) kurtosis.

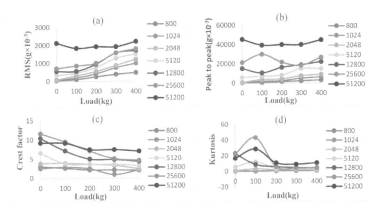

Figure 4. Variation in statistical parameter for a 2 mm defect on the outer race: (a) RMS; (b) peak to peak; (c) crest factor; (d) kurtosis.

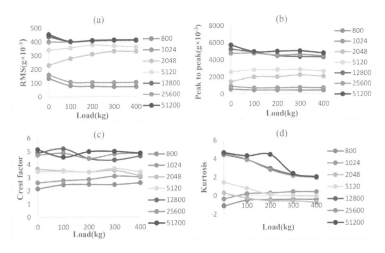

Figure 5. Variation in statistical parameters for a 1 mm defect on the outer race: (a) RMS; (b) peak to peak; (c) crest factor; (d) kurtosis.

217

A comparison of Figures 3 and 4 shows that with an increase in sampling rate for vibration signals, the values of all the statistical parameters increase. Statistical parameters for the 2 mm defect size shows that all four statistical parameter values are higher than for the healthy bearing, clearly indicating the presence of a defect.

3.2   Time domain study for a 1 mm defect on the outer race

Statistical parameters were measured for the 1 mm defect on the outer race with the same loading condition. Figure 5 shows the variation in statistical parameters.

A comparison of Figures 3 and 5 shows that with an increase in sampling rate for vibration signals, the values of all the statistical parameters increase. For the 1 mm defect size, RMS and peak to peak have higher values when compared to a healthy bearing, while the other parameters—crest factor and kurtosis—have the same values as in a healthy bearing.

3.3   Time domain study for a 0.3 mm defect on the outer race

Statistical parameters were measured for the 0.3 mm defect on the outer race with the same loading condition. Figure 6 shows the variation in statistical parameters.

A comparison of Figures 3 and 6 shows that with an increase in sample rate for vibration, the signal values of all the statistical parameters increase. For the 0.3 mm defect size, RMS and peak to peak have higher values when compared to a healthy bearing, while the other parameters—crest factor and kurtosis—have the same values as in a healthy bearing.

3.4   Frequency domain analysis

For a bearing with an outer race defect, a FFT has been performed using the MATLAB signal processing toolbox. The obtained frequency data has been analyzed for the presence of a fault frequency corresponding to an outer race defect.

3.4.1   Calculation of theoretical outer race fault frequency

As the balls in the bearing pass a defect in the outer race they generate a frequency associated with the defect. This Ball Pass Frequency Outer (BPFO) is calculated thus:

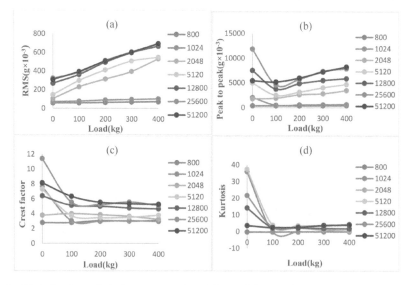

Figure 6.   Variation in statistical parameters for a 0.3 mm defect on the outer race: (a) RMS; (b) peak to peak; (c) crest factor; (d) kurtosis.

$$BPFO = \frac{N_b}{2} S \left(1 - \frac{B_d}{P_d} \right) \cos \phi = 87 Hz \qquad (1)$$

where N_b = number of balls, S = shaft speed (sec), B_d = ball diameter (mm), P_d = pitch diameter (mm), and ϕ = contact angle.

3.4.2 *Frequency spectra for a 2 mm defect on the outer race*
Figure 7 shows the frequency spectrum of a 2 mm size localized defect on the outer race, loaded with 400 kg and with a sampling rate of 51,200 Hz. Figure 7 shows the fundamental of the BPFO race and its harmonics present in the raw frequency spectrum obtained through FFT.

Figure 7 shows that for a 2 mm defect on the outer race the defect frequency is clearly identified. By setting the MATLAB variable *NFFT* to be the FFT window size, more defect frequencies can be clearly identified and deviation in theoretical and experimental fundamental defect frequency is lowered, hence further analysis is not required.

3.4.3 *Frequency domain analysis for a 1 mm defect on the outer race*
Figure 8 shows the frequency domain spectra for the 1 mm defect on the outer race. Vibration signals are taken at 400 kg radial load and 51,200 Hz sampling rate.

From a comparison of Figures 7 and 8, it can be clearly seen that as the defect size becomes smaller, the associated vibration signature becomes very weak and may not be detected if it is buried within the external noise present. Envelope analysis is used to remove high-frequency noise from the low-frequency fault signature. The method of performing envelope analysis is to apply the Hilbert transform on the time domain data. The transformed time domain vibration signature is converted into a frequency domain signature with conventional FFT analysis. Figure 9a shows the original time domain signal of the faulty bearings and Figure 9b shows the enveloped signal in the time domain.

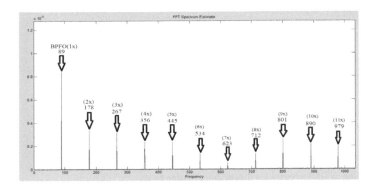

Figure 7.   FFT spectra for a 2 mm defect on the outer race.

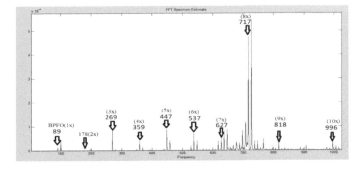

Figure 8.   FFT spectra for a 1 mm defect on the outer race.

219

The FFT of the enveloped signal will give the defect frequency. Figure 10 shows the FFT spectra required defect frequency.

3.4.4 *Frequency domain analysis for a 0.3 mm defect on the outer race*
Figure 11 shows the frequency domain spectra for the 0.3 mm defect on the outer race. A sample of the vibration signal is taken at 400 kg radial load and 51,200 Hz sampling rate.

The smallest size of defect created is 0.3 mm. For such a small size it is very difficult to identify the defect frequency, as shown in Figure 11. Hence, the radial load has been increased to 600, 700 and 800 kg, and the readings were taken with sampling rates of 12,800, 25,600 and 51,200 Hz. Figure 12 shows the FFT spectra for the radial load of 800 kg and sampling rate 51,200 Hz, and indicates that the vibration signature becomes more prominent at a higher radial load. Thus, no further analysis is required.

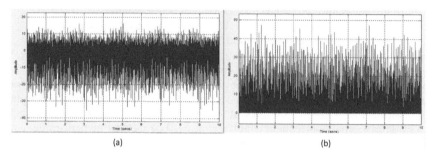

Figure 9.　(a) Original signal; (b) enveloped signal.

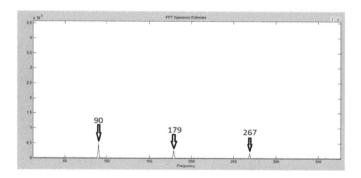

Figure 10.　Filtered FFT spectra for 1 mm defect outer race.

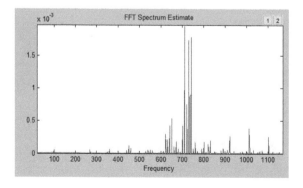

Figure 11.　FFT spectra for a 0.3 mm defect on the outer race.

220

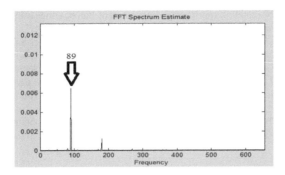

Figure 12.   FFT spectra for a 0.3 mm defect at higher load.

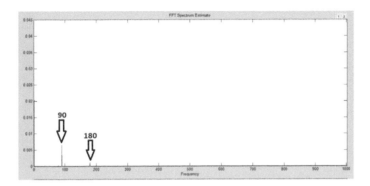

Figure 13.   Filtered FFT spectra for a 0.3 mm defect at lower load.

To check the effectiveness of the envelope, the analysis signal of a 0.3 mm defect size at 400 kg load is filtered using the Hilbert transform. Figure 13 shows the FFT spectra with the required defect frequency.

4   CONCLUSION

The present work is focused on the use of time domain and frequency domain parameters for the evaluation of defects present in an outer race of a rolling element bearing. A bearing with a localized outer race defect subjected to pure radial loading shows that RMS and peak to peak values are higher for all sizes of defect compared to the healthy bearing. As the defect size decreases, other parameters such as crest factor and kurtosis value do not give significant indications of the fault. A frequency spectrum of the outer race defect in pure radial loading shows the corresponding values of BPFO and its harmonics. However, for smaller defect sizes operating under lower loads, the Hilbert transform is shown to be a useful tool for obtaining the required fault information.

REFERENCES

Durkhure, P. & Lodwal, A. (2014). Fault diagnosis of ball bearing using Time Domain Analysis and Fast Fourier Transformation. *International Journal of Engineering Science and Research Technology*, 3(7), 711–715.

Dyer, D. & Stewart, R.M. (1978). Detection of rolling element bearing damage by statistical vibration analysis. *Journal of Mechanical Design*, 100(2), 229–235.

Ho, D. & Randall, R.B. (2000). Optimization of bearing diagnostic techniques using simulated and actual bearing faults. *Mechanical Systems and Signal Processing*, 14(5), 763–788.

Karacay, T. & Akturk, N. (2009). Experimental diagnostics of ball bearings using statistical and spectral methods. *Tribology International, 42*(6), 836–843.

Lyons, R.G. (2010). *Understanding digital signal processing* (3rd ed.). Upper Saddle River, NJ: Prentice Hall.

Mathew, J. & Alfredson, R.J. (1984). The condition monitoring of rolling element bearings using vibration analysis. *Journal of Vibration, Acoustics, Stress, and Reliability in Design, 106*(3), 447–453.

Randall, R.B. & Antoni, J. (2011). Rolling element bearing diagnostics—A tutorial. *Mechanical Systems and Signal Processing, 25*(2), 485–520.

Wu, Z. (2016). Rolling bearing fault evolution based on vibration time-domain parameters. *Key Engineering Materials, 693*, 1412–1418.

Technology Drivers: Engine for Growth – Mahajan, Modi & Patel (Eds)
© 2018 Taylor & Francis Group, London, ISBN 978-1-138-56042-0

Identification of faulty condition of rolling element bearing with inner race defect using time and frequency domain parameters of vibration signature

Shrey Trivedi

Department of Mechanical Engineering, Ahmedabad Institute of Technology, Gujarat Technological University, Ahmedabad, Gujarat, India

D.V. Patel, V.M. Bhojawala & K.M. Patel

Department of Mechanical Engineering, Institute of Technology, Nirma University, Ahmedabad, Gujarat, India

ABSTRACT: Machinery fault diagnostics is an essential part of condition-based maintenance activity in industry. Many techniques are available, and among these vibration analysis
is widely used. In this work, different sizes of artificial defect were seeded on the inner race of
a rolling element bearing. The vibration signatures obtained were analyzed and comparisons
made between healthy and faulty bearings. All the readings were taken with different loading
conditions and different sampling rates. Bearings are loaded with hydraulic cylinders. Vibration signatures were analyzed for pure radial loading condition. The vibration signals obtained
were analyzed in the time domain and the frequency domain. Various statistical parameters
were measured, namely Root Mean Square (RMS), peak to peak, kurtosis and crest factor. As
the fault is present on the inner race, vibration signatures are combined with the side bands
and it becomes difficult to identify the defect frequency. In this work, side bands are due to the
rotation of the shaft with rotating inner race. In order to remove the side bands, a band-pass
filter is used, and a kurtogram is used to select the proper bandwidth. Further, an envelope
analysis technique using the Hilbert transform is used for removing side bands.

1 INTRODUCTION

Bearings are the most important element in any rotating machinery. If any type of fault occurs
in the bearing, it will lead to catastrophic failure, and that in turn will lead to machine downtime.
In order to avoid such failures, monitoring of the bearing condition is required. Bearing defects
mainly occur on the inner race, outer race and balls. Defects are mainly of two types, localized and
distributed. In this project work, different sizes of defects, such as 1 mm and 2 mm, are generated
on the inner race of the bearing with the help of wire-cut Electrical Discharge Machining (EDM).

Mathew and Alfredson (1984) incorporated spectral and statistical methods to check the
condition of a bearing. Karacay and Akturk (2009) studied time domain parameters such as
Root Mean Square (RMS), kurtosis, and crest factor to identify a defect in a rolling element
bearing. Tandon (1994) used statistical parameters and the Shock Pulse Method to detect
faults in a bearing. Patel et al. (2012) studied envelope analysis and used the Duffing oscillator method to identify the faults in a rolling element bearing. Tandon et al. (2007) compared different condition monitoring techniques for identifying the fault in an induction
motor ball bearing. Randall and Antoni (2011) studied envelope analysis, minimum entropy
deconvolution and spectral kurtosis to identify the defects in a bearing and also discussed
the effectiveness of these parameters in fault diagnostics of rolling element bearings. Ho and
Randall (2000) described envelope analysis with the Hilbert transform and also discussed
the use of squared envelope and higher power for the detection of a defect frequency.

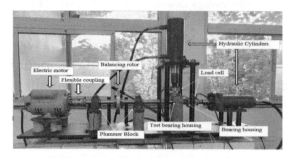

Figure 1.   Experimental setup.

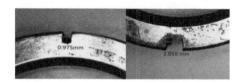

Figure 2.   Defects of 1 mm and 2 mm.

Table 1.   Loading conditions and sampling rates.

Load (kg)	0	100		200	300		400
Sampling rate (Hz)	800	1,024	2,048	5,120	12,800	25,600	51,200

2   EXPERIMENTAL SETUP AND EXPERIMENTAL WORK

To identify the defects in a rolling element bearing, an experimental setup is designed, as shown in Figure 1. An electric motor is rotated at 1,440 rpm. A shaft is connected to the motor through a flexible coupling. This shaft is supported by two plummer blocks. The test bearing housing is then mounted as shown in Figure 1. Load is applied on the test bearing through hydraulic cylinders having a capacity of 10 tons. Applied load is measured with the help of a pancake-type load cell having a capacity of 1 ton, mounted between the hydraulic cylinder and test bearing.

A 6205 bearing with a polyamide cage has been used, with artificially seeded 2 mm and 1 mm defects made on the inner race using wire-cut EDM. Figure 2 shows the artificially seeded defects and their actual size measured with a USB-connected digital microscope.

Vibration signals for the faulty bearing are then captured with the help of a four-channel Fast Fourier Transform (FFT) analyzer. Readings were taken for different loading conditions and different sampling rates. The vibration signatures obtained were analyzed in MATLAB software for time domain and frequency domain. Table 1 shows the magnitudes of the radial load and the sampling rates used for the current work.

3   RESULTS AND DISCUSSION

To differentiate between healthy and faulty bearings, a time domain study has been carried out under pure radial load. Vibration data has been obtained in terms of variation of acceleration 'g' with respect to time during the measurement.

3.1   *Time domain study of a 2 mm defect on the inner race*

For a 2 mm defect on the inner race, similar experimentation has been performed as discussed in the preceding section. Various statistical parameters were measured and compared with those of a healthy bearing. Figure 4 shows the variation in statistical parameters for a 2 mm defect on the inner race.

A comparison of Figures 3 and 4 shows that with an increase in sampling rate for vibration signals, the values of all the statistical parameters increase. For a defect size of 2 mm, all the statistical parameters have a higher value compared to those of a healthy bearing; hence the presence of fault is clearly confirmed. However, the kurtosis has an unrealistic value, which could be misleading in fault detection.

3.2 *Time domain study for a 1 mm defect on the inner race*

Statistical parameters were measured for a 1 mm defect on the inner race with the same loading condition. Figure 5 shows the variation in statistical parameters. A comparison of Figures 3 and 5 shows that with an increase in sampling rate for vibration signals, the values of all the statistical parameters increase. For the 1 mm defect size, all the statistical parameters have a higher value as compared to a healthy bearing; hence the presence of fault is clearly confirmed.

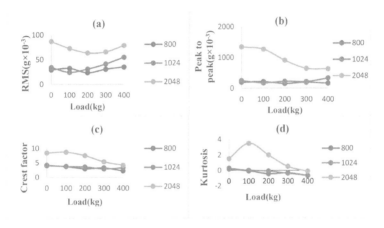

Figure 3. Variation in statistical parameter of healthy bearing: (a) RMS; (b) peak to peak; (c) crest factor; (d) kurtosis.

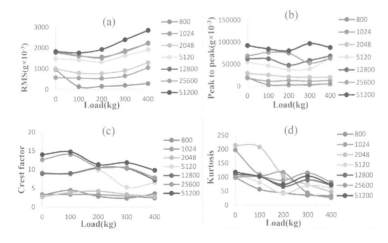

Figure 4. Variation in statistical parameters for a 2 mm defect on the inner race: (a) RMS; (b) peak to peak; (c) crest factor; (d) kurtosis.

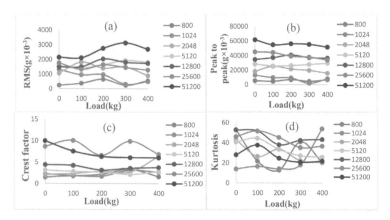

Figure 5. Variation in statistical parameters for a 1 mm defect on the inner race: (a) RMS; (b) peak to peak; (c) crest factor; (d) kurtosis.

3.3 *Frequency domain analysis*

For the inner race defect bearing, an FFT has been performed using the MATLAB signal processing toolbox. The obtained frequency data has been analyzed for the presence of a fault frequency corresponding to an inner race defect.

3.3.1 *Calculation of theoretical inner race fault frequency*
As the balls in the bearing pass a defect in the inner race they generate a frequency associated with the defect. This Ball Pass Frequency Inner (BPFI) is calculated thus:

$$BPFO = \frac{N_b}{2} S \left(1 + \frac{B_d}{P_d} \right) \cos \phi = 130 Hz \tag{1}$$

where N_b = number of balls, S = shaft speed (sec), B_d = ball diameter (mm), P_d = pitch diameter (mm), and ϕ = contact angle.

3.3.2 *Frequency spectra for a 1 mm defect on the inner race*
Figure 6 shows the frequency spectrum of a 1 mm size localized defect, loaded with 400 kg and with a sampling rate of 51,200 Hz. Figure 6 shows the fundamental of the BPFI race and its harmonics present in the raw frequency spectrum obtained through FFT.

As shown in Figure 6, the frequency spectrum contains a number of side bands alongside the fault frequency and shaft rotation frequency as a result of signal modulation. Ho and Randall (2000) have proposed a methodology for carrying out envelope analysis of such a modulated signal. The steps are as follows:

- identify the frequency band containing resonant frequencies of the bearing and structure;
- band-pass the signal with the above identified frequency;
- shift the frequency signal which is band-passed;
- perform an inverse FFT to obtain the signal in the time domain;
- perform the Hilbert transform of the obtained time domain signal;
- perform a FFT of the Hilbert transformed signal.

In the methodology described above, the difficult part is to find the parameter for the band-pass filter (i.e. center frequency and bandwidth). Randall and Antoni (2011) have suggested a methodology to identify the band of frequency where the presence of the impact is prominent using a fast kurtogram. Figure 7a shows the kurtogram plot for the signal discussed in the present section. From the kurtogram plot, it can be said that the frequencies having a bandwidth of 6,400 Hz and a center frequency of 16,000 Hz contain the resonance frequency resulting from impact. This parameter has been used to band-pass the signal. The output of the band-passed signal is shifted and is shown in Figure 7b.

An inverse FFT has been performed for the shifted signal to obtain a new time domain signal. This time domain signal is further transformed using the Hilbert transform technique. Figure 8 shows the resulting envelope signal.

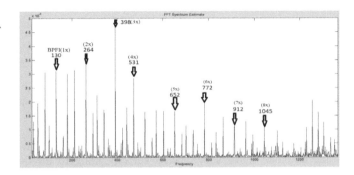

Figure 6.   FFT spectra for a 1 mm defect on the inner race.

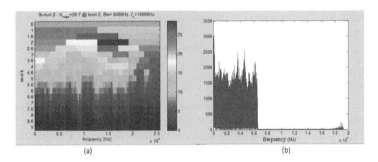

Figure 7.   (a) Kurtogram plot; (b) frequency shift band.

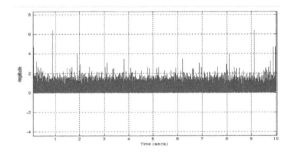

Figure 8.   Envelope signal for a 1 mm defect on the inner race.

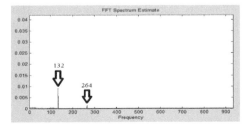

Figure 9.   Filtered FFT spectra with defect frequency.

227

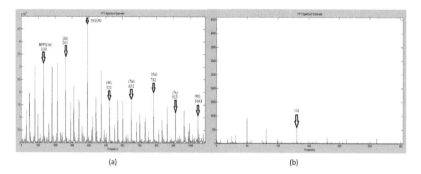

Figure 10. (a) FFT spectra for a 2 mm defect on the inner race; (b) filtered FFT spectra with defect frequency.

The FFT of the envelope signal shows peaks corresponding to the defect frequency. Figure 9 shows the filtered FFT spectra.

3.3.3 *Frequency domain analysis for a 2 mm defect on the inner race*

The same methodology is used to identify the defect frequency for the 2 mm defect present on the inner race (see Figure 10).

4 CONCLUSION

For a bearing with a localized inner race defect subjected to pure radial loading, all statistical parameters of the vibration signatures have higher values when compared to those of a healthy bearing. Hence the presence of a fault can be easily detected. The frequency signature of the inner race fault under pure radial load shows fault frequencies with side bands and shaft rotation frequencies. To distinguish between the defect frequencies and the other non-relevant frequencies, a methodology proposed by Ho and Randall (2000) has been applied, which gives clear peaks for the inner race defect.

REFERENCES

Ho, D. & Randall, R.B. (2000). Optimization of bearing diagnostics techniques using simulated and actual bearing faults. *Mechanical Systems and Signal Processing, 14*(5), 763–788.

Karacay, T. & Akturk, N. (2009). Experimental diagnostics of ball bearings using statistical and spectral methods. *Tribology International, 42*(6), 836–843.

Lyons, R.G. (2010). *Understanding digital signal processing* (3rd ed.). Upper Saddle River, NJ: Prentice Hall.

Mathew, J. & Alfredson, R.J. (1984). The condition monitoring of rolling element bearings using vibration analysis. *Journal of Vibration, Acoustics, Stress, and Reliability in Design, 106*(3), 447–453.

Patel, V.N., Tandon, N. & Pandey, R.K. (2012). Defect detection in deep groove ball bearing in presence of external vibration using envelope analysis and Duffing oscillator. *Measurement, 45*(5), 960–970.

Randall, R.B. & Antoni, J. (2011). Rolling element bearing diagnostics—A tutorial. *Mechanical Systems and Signal Processing, 25*(2), 485–520.

Tandon, N. (1994). A comparison of some vibration parameters for the condition monitoring of rolling element bearings. *Measurement, 12*(3), 285–289.

Tandon, N., Yadava, G.S. & Ramakrishna, K.M. (2007). A comparison of some condition monitoring techniques for the detection of defect in induction motor ball bearings. *Mechanical Systems and Signal Processing, 21*(1), 244–256.

Technology Drivers: Engine for Growth – Mahajan, Modi & Patel (Eds)
© *2018 Taylor & Francis Group, London, ISBN 978-1-138-56042-0*

Performance investigation of a pulse-tube refrigerator with different regenerative materials

Krunal R. Parikh

Department of Mechanical Engineering, Indus Institute of Technology and Engineering, Indus University, Gujarat, India

Balkrushna A. Shah

Department of Mechanical Engineering, Institute of Technology, Nirma University, Gujarat, India

ABSTRACT: A cryorefrigerator is a device which is used to produce very low temperature by compressing and expanding a gas. As a pulse-tube refrigerator has no moving parts in its cold section, it has the potential to achieve lower vibration and higher reliability than other cryorefrigerators. The performance of a pulse-tube refrigerator is mainly dependent on its configuration and pulse rate. In addition, the performance also depends on the regenerative material, size and pressure drop through the regenerator. The main objectives of this research were to obtain the no-load characteristic of a pulse-tube refrigerator with different regenerative materials such as stainless steel and phosphor bronze, and also with different wire mesh sizes for basic pulse-tube refrigerators, orifice pulse-tube refrigerators and double-inlet pulse-tube refrigerators. It was observed that a 150 mesh stainless steel regenerative wire mesh resulted in a lower temperature than a 150 mesh phosphor bronze one.

Keywords: Basic pulse-tube refrigerator, orifice pulse-tube refrigerator, double-inlet pulse-tube refrigerator, wire mesh

1 INTRODUCTION

A cryocooler is a refrigeration system used for cooling temperature below 123 K. Cryocoolers have special applications, such as liquefaction of gases, cryogenic catheters and cryosurgery, storage of biological cells and specimens, cooling of infrared sensors, and satellites. Therefore, cryocooler requires less weight and vibration reliable, long life. To fulfill such requirements, a pulse-tube refrigerator is used. A compressor, wave generation device, regenerator, pulse tube and two heat exchangers are the main components of the pulse-tube refrigerator [1]. The compressor compresses the working gas that flows through a pulse (wave) generation device to the regenerator, which absorbs heat from the cold gas leaving the pulse tube and returns it to the warm gas entering the pulse tube in the next cycle. The pulse tube is a hollow tube which has some residual gases that are compressed by the high pressure gas. Residual gas temperature increases due to compression which is produced by a water jacket at the closed end of the pulse tube. Because of the switching of flow in and out of the pulse tube, the gas is compressed and then expanded to a low pressure and temperature, ultimately being used to produce a cooling effect at the cold end. The cooling effect is further absorbed in the regenerator and the gas leaves at about atmospheric pressure. This cycle repeats and achieves the cold end-temperature (Gifford & Longsworth, 1964). The cool-down effect mainly depends upon the regenerator. An efficient regenerator must have a low thermal conductivity and high specific heat (Mikulin et al., 1984).

2   LITERATURE REVIEW

The history of Pulse-Tube Refrigerators (PTRs) began with the research of Gifford and Longsworth (1964). Their cryocooler system was without moving parts and could reach a temperature of about −123 °C. The Orifice Pulse-Tube Refrigerator (OPTR) was introduced by Mikulin et al. (1984), in which an orifice and reservoir were added at the hot end of a basic pulse tube to improve performance. A temperature of about 60 K was achieved by Radebaugh et al. (1986) using an OPTR. The Double-Inlet PTR (DIPTR) was introduced by Zhu et al. (1990), in which another orifice was introduced between the hot end of the regenerator and that of the pulse tube to achieve a lower temperature than the OPTR. The thermodynamic model of a basic pulse-tube refrigerator was developed by De Boer (1994) and his main focus was on the behavior of the gas in the cooling and heating processes and the temperature profiles generated. He then worked on the thermodynamic analysis of Basic Pulse-Tube Refrigerators (BPTRs) with hot and cold ends (De Boer, 1996). In the present research work, an experimental study was carried out to investigate the effect of regenerative material and size on the performance of a BPTR, OPTR and DIPTR. Woven wire meshes of different size were used as the regenerative material. Experimental results show that the cooling performance with stainless steel woven wire screens was better than that with phosphor bronze screens.

3   EXPERIMENTAL WORK

The experimental setup of the single-stage pulse-tube refrigerator is shown in Figure 1. A 2.2 kW capacity commercial air compressor was used and the required pressure waves achieved by a 3/2 solenoid valve. The discharge pressure of the compressor can go up to 10 bar. A cyclic timer was used to vary the cycle time, so that the frequency of the pressure wave can be varied smoothly. Needle valves were used for the bypass and orifice valves. The regenerator consists of about 1,500 disks of 150, 200, 250, 300 wire meshes of stainless steel or a 150 wire mesh with phosphor bronze inside a stainless steel (S.S.) seamless tube of length 200 mm,16 mm outside diameter (o.d.), and 1 mm wall thickness. The pulse tube is made up of a S.S. tube of 14.9 mm o.d., 0.9 mm wall thickness, and 300 mm length. The water circulation was at the hot end through water jacket is used for heat removal. The pulse tube and regenerator assembly with platinum resistance temperature sensors (PT100) were kept inside the vacuum enclosure. The outlet from this unit forms the common inlet for both main and bypass lines. The buffer (reservoir) volume is almost 20 times the pulse tube volume. The vacuum vessel was connected to the rotary pump system. A rubber gasket with vacuum grease was used at the top cover of the vacuum vessel to maintain a vacuum within the vessel.

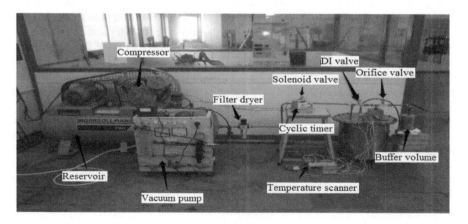

Figure 1.   Experimental set up of the pulse-tube refrigerator.

4 RESULTS AND DISCUSSION

The experimental results were analyzed and the data summarized in graphical form for the lowest end-temperature by comparing different materials and different wire mesh sizes. From the experimental data, a comparison of lowest end-temperature for different materials—stainless steel and phosphor bronze—of 150 mesh size, as well as different stainless steel wire mesh sizes – 150, 200, 250 and 300 – was plotted for a time interval of five minutes for the different pulse tube configurations of **BPTR**, **OPTR** and **DIPTR**.

4.1 *Effect of S.S. 150 and phosphor bronze 150 wire meshes on cool-down characteristics of a BPTR*

Table 1 shows the effect of regenerative materials in the form of S.S. and phosphor bronze (PB) 150 wire meshes for a **BPTR**. Based on this table, Figure 2 was plotted.

4.2 *Effect of S.S. 150 and phosphor bronze 150 wire meshes on cool-down characteristics of an OPTR*

Table 2 shows the effect of regenerative materials in the form of S.S. and phosphor bronze (PB) 150 wire meshes for an **OPTR**. Based on this table, Figure 3 was plotted.

Table 1. Experimental results of BPTR with S.S. 150 and phosphor bronze 150 wire meshes.

Time (min)	Cool-down temp. for S.S. 150 (°C)	Cool-down temp. for PB 150 (°C)
0	24.9	24.9
5	11.4	11.3
10	5.9	7.6
15	−2	4.5
20	−3.9	2.3
25	−5.3	0.5
30	−5.7	−1

Table 2. Experimental results of OPTR with S.S. 150 and phosphor bronze 150 wire meshes.

Time (min)	Cool-down temp. for S.S. 150 (°C)	Cool-down temp. for PB 150 (°C)
0	24.9	23.8
5	12.2	7.8
10	9.8	3.2
15	6.2	−1.1
20	−3.5	−4
25	−6.3	−5.4
30	−8.5	−6.7
35	−8.7	−7.5

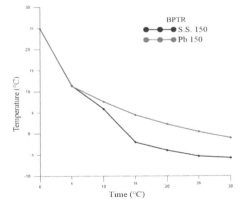

Figure 2. Graph of cool-down characteristics of BPTR with S.S. 150 and phosphor bronze 150 wire meshes.

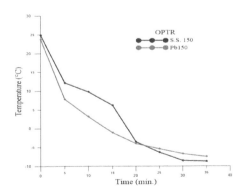

Figure 3. Graph of cool-down characteristics of OPTR with S.S. 150 and phosphor bronze 150 wire meshes.

231

4.3 Effect of S.S. 150 and phosphor bronze 150 wire meshes on cool-down characteristics of a DIPTR

Table 3 shows the effect of regenerative materials in the form of S.S. and phosphor bronze (PB) 150 wire meshes for a DIPTR. Based on this table, Figure 4 was plotted.

4.4 Effect of S.S. 150, 200, 250 and 300 wire mesh on cool-down characteristics of a BPTR

The experimental results for regenerative material in the form of S.S. wire mesh of sizes 150, 200, 250 and 300 for a BPTR are shown in Table 4. Based on this table, Figure 5 was plotted.

4.5 Effect of 150, 200, 250 and 300 S.S. wire meshes on cool-down characteristics of an OPTR and a DIPTR

The results of using S.S. regenerative material in the form of 150, 200, 250 and 300 wire meshes for an OPTR and a DIPTR are shown in Tables 5 and 6, respectively. The results show that the lowest temperature achieved was in the DIPTR arrangement with a S.S. 150 wire mesh.

Table 3. Experimental results of DIPTR with S.S. 150 and phosphor bronze 150 wire meshes.

Time (min)	Cool-down temp. for S.S. 150 (°C)	Cool-down temp. for PB 150 (°C)
0	24.9	24.9
5	12.9	11.1
10	−8.1	3.2
15	−10.1	0.5
20	−14.3	−1.1
25	−15.1	−2.6
30	−15.8	−3.6
35	−16.1	−5.7
40	−16.4	−6.2

Table 4. Experimental results of BPTR with S.S. 150, 200, 250 and 300 wire meshes.

Time (min)	Cool-down temp. for S.S. 150 (°C)	Cool-down temp. for S.S. 200 (°C)	Cool-down temp. for S.S. 250 (°C)	Cool-down temp. for S.S. 300 (°C)
0	24.9	26.3	25.7	22.4
5	11.4	12.8	9.8	11.2
10	5.9	7.3	7.6	6.4
15	−2	3.1	4.5	2.2
20	−3.9	1.8	2.5	−2.4
25	−5.3	1.6	1.5	−3.6
30	−5.7	1.4	0.5	−3.8

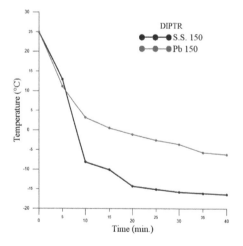

Figure 4. Graph of cool-down characteristics of DIPTR with S.S. 150 and phosphor bronze 150 wire meshes.

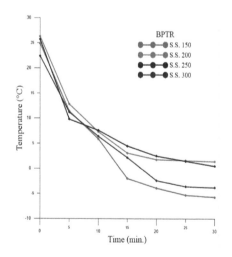

Figure 5. Graph of cool-down characteristics of BPTR with S.S. 150, 200, 250 and 300 wire meshes.

Table 5.	Effect of S.S. 150, 200, 250 and 300 on cool-down characteristics of an OPTR.			
Time (min)	Cool-down temp. for S.S. 150 (°C)	Cool-down temp. for S.S. 200 (°C)	Cool-down temp. for S.S. 250 (°C)	Cool-down temp. for S.S. 300 (°C)
0	24.9	25.1	24.3	23.6
5	12.2	6.5	4.5	11.2
10	9.8	2.9	2.4	6.1
15	6.2	0.4	−0.5	1.8
20	−3.5	−1.4	−2.3	−3.7
25	−6.3	−2.4	−3.4	−4
30	−8.5	−2.8	−4	−4

Table 6.	Effect of S.S. 150, 200, 250 and 300 wire wire meshes meshes on cool-down characteristics of a DIPTR.			
Time (min)	Cool-down temp. for S.S. 150 (°C)	Cool-down temp. for S.S. 200 (°C)	Cool-down temp. for S.S. 250 (°C)	Cool-down temp. for S.S. 300 (°C)
0	24.4	24.7	24	23
5	12.9	5.1	4.4	8
10	−8.1	0.2	−1.7	5.4
15	−10.1	−2.6	−3.5	3.9
20	−14.3	−4	−5.8	2.7
25	−15.1	−5.5	−6.2	−4.6
30	−15.8	−6.1	−7.2	−3

5 CONCLUSION

An experimental investigation was carried out to observe the difference in performance of the regenerator in terms of material and wire mesh size for a BPTR, OPTR and DIPTR. A series of tests were conducted using different pulse rates. After comparison and analysis of the results, the following conclusions were drawn:

1. It was found that for the same regenerative material, the DIPTR gives better cooling than the OPTR, and the OPTR gives better cooling than the BPTR.
2. It was found that when comparing different regenerative materials for the same wire mesh size of 150, S.S. gives better cooling than phosphor bronze, because of the higher specific heat of S.S. compared to phosphor bronze.
3. By comparing different S.S. wire mesh sizes of 150, 200, 250 and 300, it was found that 150 gave the best cooling for the same regenerative material.

REFERENCES

De Boer, P.C.T. (1994). Thermodynamic analysis of the basic pulse-tube refrigerator. *Cryogenics, 34*, 699–711.
De Boer, P.C.T. (1996). Analysis of basic pulse-tube refrigerator with regenerator. *Cryogenics, 36*, 547–553.
Gifford, W.E. & Longsworth, R.C. (1964). Pulse-tube refrigeration. *Journal of Engineering for Industry, 86*(3),264–268.
Mikulin, E.I., Tarasov, A.A. & Shkrebyonock, M.P. (1984). Low-temperature expansion pulse tubes. In R.W. Fast (Ed.), *Advances in cryogenic engineering* (Vol. 29, pp. 629–637). Boston, MA: Springer.
Radebaugh, R., Zimmerman, J., Smith, D.R. & Louie, B. (1986). Comparison of three types of pulse tube refrigerators: New method for reaching 60 K. In R.W. Fast (Ed.), *Advances in cryogenic engineering* (Vol. 31, pp. 779–789). Boston, MA: Springer.
Zhu, S., Wu, P. & Chen, Z. (1990). Double inlet pulse tube refrigerators: An important improvement. *Cryogenics, 30*, 514–520.

Technology Drivers: Engine for Growth – Mahajan, Modi & Patel (Eds)
© *2018 Taylor & Francis Group, London, ISBN 978-1-138-56042-0*

Effect of various geometrical parameters on the thermal performance of an axially grooved heat pipe

Akshay Desai
Thermal Engineering, Institute of Technology, Nirma University, Ahmedabad, Gujarat, India

V.K. Singh & R.R. Bhavsar
Thermal Engineering Division, Space Application Center—ISRO, Ahmedabad, Gujarat, India

R.N. Patel
Thermal Engineering, Institute of Technology, Nirma University, Ahmedabad, Gujarat, India

ABSTRACT: Heat pipes have become essential components in thermal systems for a variety of applications including electronics cooling, nuclear reactors, cryogenic systems, and so on. Axially grooved heat pipes are widely used for spacecraft thermal management due to their ability to transport heat over a large distance. A mathematical model was developed and solved using the MATLAB software package of MathWorks® to study the effect of various geometrical parameters of a trapezoidal-shaped ammonia-charged axially grooved heat pipe. It was found that heat pipe performance was highly influenced by wetting contact angle, groove inclination angle and groove depth. An increase in wetting contact angle reduces the heat transportation capability of the heat pipe. A decrease in groove inclination angle reduces heat transportation capacity and thermal resistance. An increase in groove depth leads to an initial increase in heat transportation capacity, which decreases further after attaining its maximum. Thermal resistance increases almost linearly by increasing groove depth. The current study is helpful in designing the optimal shape and tolerance level of the geometrical parameters of a trapezoidal-shaped axially grooved heat pipe.

1 INTRODUCTION

Heat pipes are highly efficient heat transfer devices which use evaporation and condensation of working fluid to transfer a large amount of heat with very little temperature gradient between the heat source (hot end) and heat sink (cold end) within a closed metallic tube having a capillary wick structure on the internal periphery for circulation of the working fluid. For spacecraft thermal control, axially grooved heat pipes are used as a part of honeycombed structural panels and act as isothermalizers (Rassamakin et al., 1997). Wire mesh and sintered heat pipes perform excellently in microgravity or terrestrial conditions for a short distance but they are not viable for long-distance heat transportation (>1 meter) due to the large pressure drop experienced in liquid (Engelhardt, 2008; Kempers et al., 2006). Axially grooved heat pipes have long-distance heat transportation capability, with a wide range of operation, flexibility, reliability, low manufacturing cost and less pressure drop in liquid over wire mesh and sintered powder heat pipes (Hoa et al., 2003).

Kim et al. (2003) developed an analytical model for the operational characteristics of miniature grooved heat pipes considering the effect of shear stress at the liquid vapor interface and fluid inventory. Using the analytical model of Kim et al. (2003), Arab and Abbas (2014) developed a model-based approach for analysis of working fluid in heat pipes. Chen et al. (2009) developed a mathematical model for maximum heat transportation capacity and temperature gradient in a heat pipe structure having wicking 'Ω'-shaped axial microgrooves.

Optimization of the geometrical parameters of a heat pipe with trapezoidal-shaped axial grooves may lead to an increase in its maximum heat transportation capacity and reduction in thermal resistance. To the best of the authors' knowledge, the trapezoidal groove's geometrical parameters, such as groove inclination angle, groove depth and wetting contact angle, are not reported in open literature. This paper is an attempt to study the effect of the geometrical parameters of trapezoidal-shaped axially grooved heat pipes. To this end, previous mathematical models of axially grooved heat pipes reported in open literature (Anand et al., 2008; Arab & Abbas, 2014; Kim et al., 2003) are modified and developed in MATLAB code to account for the effect of liquid–vapor interfacial shear stress, wetting contact angle, and thin-film resistance due to evaporation and condensation.

2 PHYSICAL DESCRIPTION OF AXIALLY GROOVED HEAT PIPES

In the present study, ammonia-charged axially grooved aluminum heat piping having trapezoidal-shaped grooves was used for parametric analysis. Ammonia was chosen as the working fluid because of its higher merit number and good compatibility with aluminum (Faghri, 2014; Zohuri, 2011). The reference heat pipe cross section chosen for study is shown schematically in Figure 1. It has 28 trapezoidal-shaped grooves and the total length of the heat pipe is 1.0 m. The inner diameter of the heat pipe (D_i) is 10.5 mm, vapor core diameter (D_v) is 7.9 mm, depth of groove (h_g) is 1.3 mm, groove inclination angle (2v) is 84°, evaporator and condenser length is 300 mm. and ω_1 and ω_2 are 0.5 mm and 0.79 mm, respectively, for the reference heat pipe. The performance parameters of the heat pipe reported in this paper are at C operating condition, horizontal orientation, and constant heat flux boundary condition.

3 NUMERICAL MODEL

3.1 *Assumptions*

- Steady fluid flow and heat transfer
- Laminar and incompressible flow (Dunn & Reay, 1978; Zohuri, 2011)
- Properties of fluid remain constant (Chen et al., 2009)
- Wick is fully saturated (Anand et al., 2008; Arab & Abbas, 2014; Kim et al., 2003)

3.2 *Maximum heat transport capacity*

Given the assumptions and model previously developed in the open literature (Anand et al., 2008; Kim et al., 2003), the numerical model was developed using the MATLAB software package of MathWorks® to determine the performance characteristics of the heat pipe. As the heat pipe starts functioning, the working fluid recedes into the grooves and generates a radius of curvature which varies along the axial direction of the heat pipe due to surface tension forces. The curvature between liquid and vapor interface shows a pressure differ-

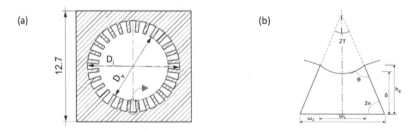

Figure 1.   Heat pipe geometry and nomenclature: (a) cross section of heat pipe; (b) detailed view of groove.

ence between the two phases, which can be determined using the well-known Young–Laplace equation:

$$P_c = P_v - P_l = \sigma\left(\frac{1}{r_{cr}} + \frac{1}{r_{ca}}\right) \tag{1}$$

where r_{cr} is radial capillary radius, and r_{ca} is axial capillary radius. Here, $r_{ca} \gg r_{cr}$, so we can assume $r_{ca} \to \infty$ and $r_{cr} \approx r_c(x)$. Thus, a differential equation can be formed from Equation 1:

$$\frac{dp_v}{dx} - \frac{dp_l}{dx} = -\frac{\sigma}{r_c^2(x)}\frac{dr_c(x)}{dx} \tag{2}$$

For incompressible and laminar flow of liquid and vapor, the pressure drop along the length of the heat pipe can be derived from:

$$\frac{dp}{dx} = -\frac{2\mu(f\,\mathrm{Re})\bar{V}}{D^2} \tag{3}$$

where the mean velocity can be determined as $\bar{V} = \frac{\dot{m}(x)}{A\rho}$ and $\dot{m}_l(x) = -\dot{m}_v(x) = -\frac{Q(x)}{h_{fg}}$

For a closed circular flow passage, the Poiseuille number generally has a value of 16. Here, the trapezoidal groove is not enclosed but open to the vapor phase from one side. Therefore, liquid–vapor interfacial shear stress plays an important role in the performance of the heat pipe. The Poiseuille numbers ($f\,\mathrm{Re}$) for specific groove geometries, considering the effect of liquid–vapor interfacial shear stress, are reported in the open literature (Schneider & DeVos, 1980).

The capillary radius varies non-linearly from evaporator to condenser, having its minimum at the evaporator and its maximum at the condenser. For a given heat load, if the capillary radius at the evaporator end reaches the minimum meniscus radius for that wick geometry, the given heat load is known as the maximum heat transport capacity of the heat pipe. The methodology to obtain maximum heat transport capacity by running the numerical model in MATLAB is shown in Figure 3a. As per the logical sequence, the program runs with an assumption of the lowest possible heat load and calculates the capillary radius variation along the length before comparing it with the minimum capillary radius for that specific geometry. If the convergence criteria are satisfactory, then the assumed heat load is called the maximum heat transport capacity; otherwise, the program runs in a continuous loop by increasing the heat load. For a trapezoidal groove geometry, the minimum meniscus radius is determined as:

$$r_{c,min} = \frac{\omega_2}{2\left[\cos(\theta - \gamma) - \tan(-\gamma)\{1 - \sin(\theta - \gamma)\}\right]} \tag{4}$$

3.3 Total thermal resistance

The thermal resistance of a heat pipe is the second important parameter of study, which is evaluated by making an equivalent resistance network which shows the relationship between different possible heat flow paths. The resistance network at evaporator and condenser sections of a single-groove geometry of heat pipe are shown in Figure 2. Equations 5, 6 and 7 are used to calculate the equivalent total thermal resistance for each section, namely evaporator, adiabatic and condenser.

The detailed calculation of all resistances, such as R1 to R10, is shown in open literature (Desai, 2017). The major contributor to thermal resistance is liquid–vapor interfacial resistance due to evaporation and condensation. Total thermal resistance is found by the series parallel method (Desai, 2017).

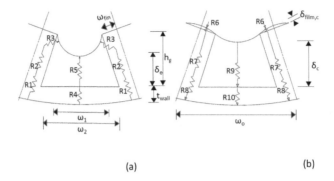

(a)   (b)

Figure 2.   Equivalent resistance network for axially grooved heat pipe: (a) evaporator; (b) condenser section.

Table 1.   Various resistances in evaporator and condenser sections.

Resistances	Description
R1 & R8	Resistance through metal container in evaporator and condenser, respectively
R2 & R7	Resistance due to metal fins in evaporator and condenser, respectively
R3 & R6	Resistance through very thin liquid film region in evaporator and condenser, respectively
R4 & R10	Resistance due to metal wall thickness in evaporator and condenser, respectively
R5 & R9	Resistance due to existence of liquid phase of working fluid in evaporator and condenser, respectively

Resistances at evaporator, adiabatic and condenser sections can be derived, respectively, as:

$$R_e = \frac{\left(R_1 + R_2 + R_3 \right)\left(R_4 + R_5 \right)}{L_e N_g \left(R_1 + R_2 + R_3 + 2\left(R_4 + R_5 \right) \right)} \tag{5}$$

$$R_a = \frac{T_{v,e} T_{v,c} R \log \left(\dfrac{P_{v,z=0}}{P_{v,z=l}} \right)}{h_{fg} Q_{in}} \tag{6}$$

$$R_c = \frac{\left(R_6 + R_7 + R_8 \right)\left(R_9 + R_{10} \right)}{L_c N_g \left(R_6 + R_7 + R_8 + 2\left(R_9 + R_{10} \right) \right)} \tag{7}$$

$$R_{Total} = R_e + R_a + R_c \tag{8}$$

Hence, the temperature difference across the length is $\Delta T = Q_{in} \times R_{Total}$

4   SENSITIVITY OF GEOMETRIC PARAMETERS

After validating the current model with previously reported experimental results (Anand et al., 2008; Kim et al., 2003) as shown in Figure 3b, the model was used to determine the effect of different geometrical parameters of groove on the performance of the heat pipe. It has been proven from the parametric study that any modification in geometrical parameters gradually affects the performance of the heat pipe. The main focus lies on the maximum heat transport capacity and total thermal resistance.

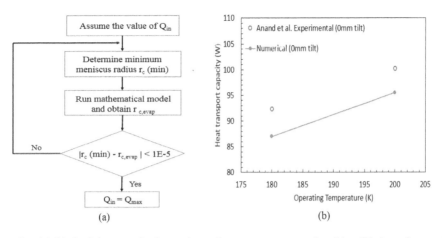

(a)

(b)

Figure 3. (a) Methodology to obtain maximum heat transport capacity; (b) validation of numerical model with experimental results.

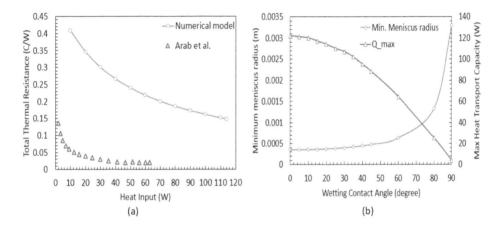

(a)

(b)

Figure 4. (a) Total thermal resistance at different heat inputs; (b) effect of different wetting contact angles on maximum heat transport capacity.

Geometrical nomenclatures are shown in Figure 1. The input geometrical parameters are taken as reference for parametric study. As shown in Figure 4a, an increase in heat load causes the rate of evaporation to be increased, which reduces liquid volume at the evaporator section. Therefore, the thermal resistance due to the liquid thickness reduces, which decreases the overall thermal resistance. The trend of the graph shows good agreement with previously reported experimental results (Arab & Abbas, 2014), which were obtained for a wire mesh wick structured heat pipe having a maximum heat transport capacity of 65 W.

As shown in Figure 4b, the wetting contact angle, which mainly depends on the surface roughness of the groove, has a great influence on maximum heat transport capacity. A rise in contact angle (from perfectly wetting to partially non-wetting) increases the minimum meniscus radius which becomes nearly flat at a 90° angle. Therefore, the liquid is incapable of generating sufficient capillary force to reach the evaporator section, which reduces the maximum heat transport capacity. The variation in wetting contact angle may be due to either insufficient cleaning of the grooves or an ineffective machining process.

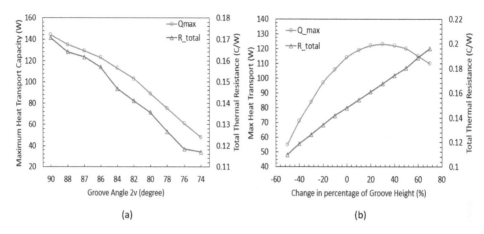

Figure 5.   Effect of different parameters on maximum heat transport capacity and total thermal resistance: (a) groove inclination angle; (b) groove depth (height).

It is shown in Figure 5a that as the groove inclination angle reduces from 90° (rectangular shape) to 74° (trapezoidal shape), the maximum heat transport capacity also decreases due to the reduction in cross-sectional area of liquid flow. The leading factors that affect the total thermal resistance are liquid thickness variation in the groove and on the wall of the groove. Due to a reduction in inclination angle, liquid volume decreases, leading to a decrease in total thermal resistance.

As shown in Figure 5b, total thermal resistance increases with an increase in groove depth from 0.65 mm to 2.21 mm. The percentage increase and decrease in depth of groove is taken from a specific reference depth of 1.3 mm. With an increase in depth, the liquid cross-sectional area expands, which increases the total thermal resistance. The liquid returning through the microgroove experiences greater frictional forces, which results in a pressure drop across the length. Therefore, an increase in groove depth raises the maximum heat transport capacity until the pressure drop is within the limits. As the depth goes beyond a certain limit, the pressure drop increases in such a way that the liquid stops flowing.

5   CONCLUSION

In the present study, a numerical model was developed and validated with that previously reported in open literature to analyze the thermal performance of an axially grooved heat pipe. This model was used to carry out a sensitivity analysis of geometrical parameters on maximum heat transport capacity and total thermal resistance, which are the main indicators of heat pipe performance. The increase in the wetting contact angle occurred for several reasons including a rough surface due to inefficient machining, chemicals present on the surface because of insufficient cleaning, impurities in the working fluid and so on, which eventually reduced the performance of the heat pipe. It decreased the heat carrying capacity of the heat pipe. A decrease in groove inclination angle also affected the thermal performance of the heat pipe. It is advantageous to decrease the groove inclination angle as far as the total thermal resistance is concerned. However, it also decreases the maximum heat carrying capacity due to the decrease in liquid volume in the groove. Groove depth also follows the same phenomena. An increase in groove depth leads to an increase in liquid volume, resulting in less conductivity. Consequently, the overall resistance of the heat pipe increases at the same time as the heat transport capacity also increases up to its maximum. The increase in groove depth eventually ends in a greater pressure drop of liquid, which decreases the maximum heat transport capacity again. Therefore, all the parameters should be chosen optimally and wisely as per the application and design criteria.

REFERENCES

Anand, A.R., Vedamurthy, A.J., Chikkala, S.R., Kumar, S., Kumar, D. & Gupta, P.P. (2008). Analytical and experimental investigations on axially grooved aluminum-ethane heat pipe. *Heat Transfer Engineering*, *29*(4), 410–416. doi:10.1080/01457630701825846.

Arab, M. & Abbas, A. (2014). A model-based approach for analysis of working fluids in heat pipes. *Applied Thermal Engineering*, *73*(1), 749–761. doi:10.1016/j.applthermaleng.2014.08.001.

Chen, Y., Zhang, C., Shi, M., Wu, J. & Peterson, G.P. (2009). Study on flow and heat transfer characteristics of heat pipe with axial "Ω"-shaped microgrooves. *International Journal of Heat and Mass Transfer*, *52*(3–4), 636–643.

Desai, A. (2017). *Numerical and experimental investigations of axial groove and wire mesh heat pipe for spacecraft payload thermal management.* Nirma University.

Dunn, P. & Reay, D.A. (1978). *Heat pipes* (2nd ed.). Oxford, UK: Pergamon Press.

Engelhardt, A. (2008). Pushing the boundaries of heat pipe operation. *Electronics Cooling*, *14*(4), 1–7. Retrieved from https://www.electronics-cooling.com/2008/11/pushing-the-boundaries-of-heat-pipe-operation/.

Faghri, A. (2014). Heat pipes: Review, opportunities and challenges. *Frontiers in Heat Pipes*, *5*(1). doi:10.5098/fhp.5.1

Hoa, C., Demolder, B. & Alexandre, A. (2003). Roadmap for developing heat pipes for ALCATEL SPACE's satellites. *Applied Thermal Engineering*, *23*(9), 1099–1108. doi:10.1016/S1359-4311(03)00039-5.

Kempers, R., Ewing, D. & Ching, C.Y. (2006). Effect of number of mesh layers and fluid loading on the performance of screen mesh wicked heat pipes. *Applied Thermal Engineering*, *26*(5–6), 589–595. doi:10.1016/j.applthermaleng.2005.07.004.

Kim, S.J., Seo, J.K. & Do, K.H. (2003). Analytical and experimental investigation on the operational characteristics and the thermal optimization of a miniature heat pipe with a grooved wick structure. *International Journal of Heat and Mass Transfer*, *46*(11), 2051–2063. doi:10.1016/S0017-9310(02)00504-5.

Rassamakin, B.M., M.G. Semena, M.G., Badayev, S., Khairnasov, S., Tarasov, G. & Rassamakin, A. (1997). High effective aluminium heat pipes in heat control systems of honeycomb panel platform of the Ukrainian space vehicle.

Schneider, G. & DeVos, R. (1980). Non-dimensional analysis for the heat transport capability of axially grooved heat pipes including liquid/vapor interaction. In *18th Aerospace Sciences Meeting, Pasadena, CA, 14–16 January 1980*. Reston, VA: American Institute of Aeronautics and Astronautics. doi:10.2514/6.1980-214.

Zohuri, B. (2011). *Heat pipe design and technology: Modern applications for practical thermal management.* Boca Raton, FL: CRC Press.

Technology Drivers: Engine for Growth – Mahajan, Modi & Patel (Eds)
© 2018 Taylor & Francis Group, London, ISBN 978-1-138-56042-0

Development of activated TIG welding technology for low alloy steels: A step towards sustainable manufacturing

Ashutosh Naik, Darshan Kundal, Sagar Suthar, Jay Vora, Vivek Patel,
Subhash Das & Ritesh Patel
Pandit Deendayal Petrolium University, PDPU, School of Technology, Gandhinagar, Gujarat, India

ABSTRACT: The present study is an attempt to investigate the effect of five different single component oxide fluxes such as MoO_3, NiO, SiO_2, TiO_2, and ZnO on Activated TIG (A-TIG) welding of Cr-Mo-V steels, which are widely used low alloy steels. Effects of these fluxes on surface appearance and weld attributes of 6 mm thick Cr-Mo-V steel weldments were studied. A novel melting pool technique was proposed for A-TIG welding in the given study. Different weld morphological features such as depth of penetration (DOP) and bead width (BW) was analyzed by developing the weld macrostructures. Experimental results indicated that activated TIG (A-TIG) process could increase the joint penetration An increase in the penetration was conceived to make the process a sustainable technology as it increases the weld depth penetration and allow high thickness plates to be welded in more economical and environmental friendly way.

1 INTRODUCTION

Low alloy steels are used to fabricate the components for pressure vessels, nuclear vessels, power systems, chemical and food processing to name a few. However, the steels has a comparatively restricted weldability and hence limited welding process has been developed for the welding of these steels (Rao and Kalyankar, 2013). Activated TIG (A-TIG) welding process was invented at Paton institute of electric welding in 1960 at Ukraine (Vora and Badheka, 2017, Vora and Badheka, 2015). A-TIG welding is a variant of TIG welding process which employs a single component or mixture of chemical powder/s (termed as flux) in TIG welding process. The flux is essentially converted to paste form by adding alcoholic reagents (such as acetone, methanol etc.) and applied on the upper surface of the plate. Standard autogenous TIG welding process is then carried out with the flux layer in between welding arc and plate. An increase of about 300% in the weld penetration has been reported by researchers. Thus, the A-TIG welding technology is considered as an upcoming "sustainable" and "green welding technology" wherein plates of higher thickness can be welded with reduced welding power, harmful gases and increased welding speed and increased penetration at similar welding parameters. The A-TIG welding process has been exhaustively carried out for welding of stainless steels and its alloys along with limited reported studies on HSLA and alloy steels. Authors Tathgir et al. (2015) used oxide flux TiO_2 in A-TIG welding of different AISI steels and duplex 2205 alloy. Increase in penetration up to 70% was reported in the study. Arc constriction mechanism was depicted as a responsible mechanism for the increase in penetration. Similarly, authors Tseng and Hsu (2011) and Tseng and Chen (2016) carried out an in depth analysis of the A-TIG welding of stainless steels and an increase in penetration was reported. Similarly, authors Arunkumar et al. (Kumar and Sathiya, 2015) studied the effect of SiO_2 and ZnO fluxes on Incoloy 800 H and reported an increase in penetration compared to TIG welding without flux. There are available literatures reporting the use of oxide fluxes (Shyu et al., 2008, Leconte et al., 2006) in A-TIG welding. Liu et al. (Liu et al., 2007) analyzed the effect of oxide fluxes on TIG welding of magnesium alloys and Vora et al. (Vora and Badheka, 2016b,

Vora and Badheka, 2016a) analyzed the use of oxide fluxes on LAFM steel. Common oxide fluxes such as Fe_2O_3, CaO, Cr_2O_3 etc. were used by both researchers and enhanced penetration was achieved. Hence oxide fluxes have a promising effect which makes them candidate compounds for A-TIG welding.

Thus Present study address the investigations on A-TIG welding 6 mm thick 2.25 Cr–1.0 Mo–0.25 V steel using five different types of single component oxide fluxes. Effect on weld attributes such as weld depth of penetration (DOP) and weld bead width (BW) in addition to depicting the exact mechanism responsible for deeper penetration is discussed. A novel technique known as melting pool technique has been suggested in this study to make A-TIG welding more effective and authors firmly believe it can give a new direction to the welding of low alloy steels and particularly establishment of A-TIG welding technology as a main stream fabrication process.

2 EXPERIMENTAL WORK

The base material used in present study was low alloy steel grade SA 542 type D Class 4a with chemical composition as shown in Table 1. Five different oxide fluxes such as MoO_3, NiO, SiO_2, TiO_2, and ZnO were used for the study. 6 mm strips of dimensions 175 mm length and 30 mm width were cut and machined.

The thickness of flux layer was maintained approximately as 0.15 mm by taking requisite amount of flux. The A-TIG welding methodology as shown in Figure 1 was followed. Extensive Bead-on-plate (BOP) trials were taken at 200 A welding current and 100 mm/min travel speed. A Miller make dynasty 280 model TIG welding machine coupled with special purpose system for the automated welding was used for carrying out the welding trials. The welding trials were carried out in a way that arc remains at the center of the plate and width of applied flux.

After welding trials, metallographic samples were prepared following generic procedures and 2% Nital as an etchant to develop macrostructures. The macrostructures were then observed under Olympus inverted stereomicroscope to depict and measure different weld bead attributes such as DOP, BW and HAZ. The authors firmly believe that the present study shall serve as a vital direction for the further research in establishing the A-TIG technology for main stream fabrication industries.

Table 1. Chemical composition of base material.

Elements	C	Mn	Cr	Mo	Si	Cu	Ni	V
%	0.15	0.48	2.27	1.0	0.09	0.06	0.22	0.30

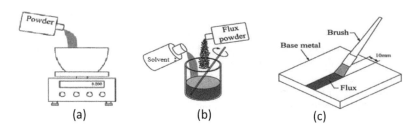

Figure 1. A-TIG welding methodology (a) Weighing of flux (b) Mixing of flux (c) Flux application.

3 RESULTS AND DISCUSSIONS

Five different single component oxide fluxes were used for carrying out the bead-on-plate trials with A-TIG welding process. Several important results were obtained as explained further the base material used in present study was low alloy steel grade SA 542 type D Class 4a.

3.1 *Melting pool technique for welding*

In the present study, a novel method for welding named as "Melting Pool (MP) Technique" has been proposed for A-TIG welding.

In this technique instead of making the torch travel immediately along the welding line the travel is started after 2 seconds. This technique allows the arc to form a considerable amount of molten weld pool as shown in Figure 2. 6 mm thick Cr-Mo-V steel plate was welded by employing the normal and MP technique. The Figure 3 shows the macrostructure of the weld bead obtained with and without melting pool technique. The trials were taken at same welding parameters and TiO_2 flux. It can be observed that, at same welding parameters and conditions an evident increase in the DOP is achieved on applying the melting pool technique. Thus, it can be proposed that the technique increases the possible reversed Marangoni effect and hence deeper penetration can be achieved at lower welding current.

3.2 *Effect of activating fluxes on macrostructure*

The weld macrostructures obtained by TIG welding with different oxide fluxes as well as without flux is a shown in Figure 4. It can be observed that the TIG welded macrostructure without flux showed a shallow penetration and characteristic wide weld bead. The resultant weld bead is an inherent nature of the autogenous TIG welding technique which was the prime reason for inventing the A-TIG welding. It can further be observed that on application of the single component oxide fluxes, there was a change in the weld bead shape. Primarily the technique address the increase in the weld DOP. Analogous results were obtained as seen from Figure 5 that shows that an evident increase in penetration and decrease in the weld bead width has been achieved with the use of fluxes MoO_3, NiO, SiO_2 and ZnO. However, it is worth noting that, none of the above fluxes were able to give complete penetration (> = plate thickness). With the use of flux TiO_2 it can be observed that the penetration obtained was more than the plate thickness (6 mm) which was evident in form of a lump at the bottom of the plate. Above all, it is worth noting that all the trials were taken at same welding parameters/conditions.

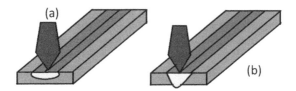

Figure 2. (a) Normal vs. (b) Melting pool Technique.

Figure 3. Macrostructure of welded plate (a) Normal (b) Melting pool Technique.

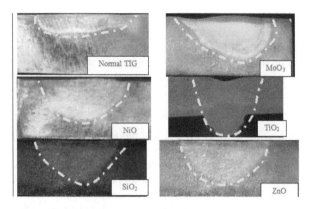

Figure 4.   Effect of activating fluxes on weld.

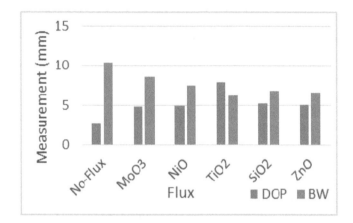

Figure 5.   Effect of activating fluxes on DOP and BW macrostructures.

3.3   *Effect of activating fluxes on weld attributes*

The resultant weld dimensions such as DOP and BW are as shown in Figure 5. It can be observed that the weld DOP is increased with the use of fluxes MoO_3, NiO, SiO_2 and ZnO in A-TIG welding as compared to TIG welding without flux. However, full and secure weld penetration (> = Plate thickness) was achieved with the use of flux TiO_2 in A-TIG welding. he weld macrostructures obtained by TIG welding with different oxide fluxes as well as without flux is a shown in Figure 4. It can be observed that the TIG welded macrostructure without flux showed a shallow penetration and characteristic.

The maximum penetration achieved was 7.9 mm. Similarly, it can also be noticed that the weld bead width is decreased with the used of fluxes MoO_3, NiO, SiO_2 and ZnO in A-TIG welding as compared to TIG welding without flux and lowest BW is achieved with flux TiO_2 as 6.2 mm. The reason for the drastic change in the weld bead attributes can be attributed to the different depth-enhancing mechanisms dominant during the A-TIG welding.

4   CONCLUSIONS

The effect of different single component oxide fluxes on 6 mm thick Cr-Mo-V steel has been carried out in present study and following noticeable conclusions can be carved out. The melting pool technique proposed in the study is proved to be successful in further enhancing the penetration of the A-TIG welding process. The technique can be further harnesses to achieve increased penetration at much lower welding currents. An evident change in the weld bead shape was observed with the use of oxide fluxes in A-TIG welding as compared to TIG welding without flux, which confirms the effectiveness of the process. The selected welding parameters and flux TiO_2 imparted full and secure penetration (> 6 mm) in single pass operation. The characteristic shape having reduced bead width and deep penetration was achieved with TiO_2. With flux TiO_2 almost 200% increase in the weld DOP was achieved as compared to TIG welding without flux.

REFERENCES

Kumar, S.A. & Sathiya, P. 2015. Experimental investigation of the A-TIG welding process of Incoloy 800 H. *Materials and Manufacturing Processes,* 30, 1154–1159.

Leconte, S., Paillard, P., Chapelle, P., Henrion, G. & Saindrenan, J. 2006. Effect of oxide fluxes on activation mechanisms of tungsten inert gas process. *Science and Technology of Welding & Joining,* 11, 389–397.

Liu, L., Zhang, Z., Song, G. & Wang, L. 2007. Mechanism and microstructure of oxide fluxes for gas tungsten arc welding of magnesium alloy. *Metallurgical and materials transactions A,* 38, 649–658.

Rao, R.V. & Kalyankar, V. 2013. Experimental investigation on submerged arc welding of Cr–Mo–V steel. *The International Journal of Advanced Manufacturing Technology,* 69, 93–106.

Shyu, S., Huang, H., Tseng, K. & Chou, C. 2008. Study of the performance of stainless steel A-TIG welds. *Journal of Materials Engineering and Performance,* 17, 193–201.

Tathgir, S., Bhattacharya, A. & Bera, T.K. 2015. Influence of current and shielding gas in TiO_2 flux activated TIG welding on different graded steels. *Materials and Manufacturing Processes,* 30, 1115–1123.

Tseng, K-H. & Chen, P.Y. 2016. Effect of TiO2 crystalline phase on performance of flux assisted GTA welds. *Materials and Manufacturing Processes,* 31, 359–365.

Tseng, K.-H. & Hsu, C.-Y. 2011. Performance of activated TIG process in austenitic stainless steel welds. *Journal of Materials Processing Technology,* 211, 503–512.

Vora, J.J. & Badheka, V.J. 2015. Experimental investigation on mechanism and weld morphology of activated TIG welded bead-on-plate weldments of reduced activation ferritic/martensitic steel using oxide fluxes. *Journal of Manufacturing Processes,* 20, 224–233.

Vora, J.J. & Badheka, V.J. 2016a. Experimental Investigation on Effects of Carrier Solvent and Oxide Fluxes in Activated TIG Welding of Reduced Activation Ferritic/Martensitic Steel. *International Journal of Advances in Mechanical & Automobile Engineering,* 3, 75–79.

Vora, J.J. & Badheka, V.J. 2016b. Improved Penetration with the Use of Oxide Fluxes in Activated TIG Welding of Low Activation Ferritic/Martensitic Steel. *Transactions of the Indian Institute of Metals,* 69, 1755–1764.

Vora, J.J. & Badheka, V.J. 2017. Experimental investigation on microstructure and mechanical properties of activated TIG welded reduced activation ferritic/martensitic steel joints. *Journal of Manufacturing Processes,* 25, 85–93.

Technology Drivers: Engine for Growth – Mahajan, Modi & Patel (Eds)
© *2018 Taylor & Francis Group, London, ISBN 978-1-138-56042-0*

Effect of surface roughness on working of electromagnets

Chirag Swamy & Bimal Mawandiya
Department of Mechanical Engineering, Nirma University, Ahmedabad, Gujarat, India

ABSTRACT: With the rise in the applications of electromechanical devices, magnetic mate-rials now play a key role in their proper functioning. Soft ferromagnetic materials are widely used in all the electromechanical devices. They have property of instant magnetization as well as demagnetization in the presence and absence of current flowing in coil surrounding it respec-tively. The behaviour of electromagnetism depends on the parameters during manufacturing it. In the present research, effect of surface roughness achieved on the mating surface of an electromagnet is studied with respect to its behaviour during movement of the electromagnets, i.e. attraction and repulsion between fixed electromagnet and the moving electromagnet or spring loaded plunger. In some electromechanical devices, it is observed that even when coil is discharged, two electromagnets do not get pulled away from each other. To resolve this prob-lem, DMAIC methodology is used to investigate prime root causes and implement solutions. During the experimentation, two grinding parameters i.e. depth of cut and grit of the grinding wheel were varied. With set of the eight trials, obtained surface roughness is recorded with the help of Mitutoyo surface tester. Further, with the combination of four sets of Ra value bands for both the electromagnets, i.e. fixed and moving *magnet*, sixteen combinations of trials are undertaken to find the nature of attraction and repulsion between the two electromagnetic magnets during magnetization and demagnetization. With the use of box-plot charts, suitable parameters are decided with respect to the required surface roughness band for proper func-tioning of electromechanical device. Minitab Software is used for the analysis of the data.

Keywords: DMAIC, Electrical Non-trip, Electromagnets, Surface Roughness, Soft ferro-magnetic material

1 INTRODUCTION

In the recent development of integrated technologies and automation, electromechanical devices play a key role to sense and actuate the system. Electromechanical devices convert electrical energy to mechanical energy and vice versa. Most of the electromechanical devices consist of stationary electric circuit elements and one or more moving elements made up of soft magnetic material or permanent magnet (Furlani, 2001). Moving magnet actuator is one of the most effective assemblies of electromechanical devices such as switchgears, auto-lock-ing devices, etc. The process of latching and unlatching is controlled by the functioning of moving magnet actuators. In the present research paper, study of the behaviour of magnets in an electromechanical device is studied. In the device under consideration, a fixed magnet and a spring loaded moving magnet are kept in the assembly surrounded by an electric coil. Both the magnets are made of soft ferromagnetic material. In no-supply condition or rest position, plunger/moving magnet is suspended on a spring above a fixed core as shown in Figure 1.

When the coil is energised, the plunger magnet gets attracted to the fixed magnet as shown in Figure 1 (energised position) and when the supply in the coil is stopped, the plunger mag-net gets pulled back to its rest position with the help of stored energy of suspended spring. Although in some cases, when the coil is de-energised, plunger magnet is not pulled back to its rest position. This phenomenon is known as Electrical Non-Trip (ENT). With the application

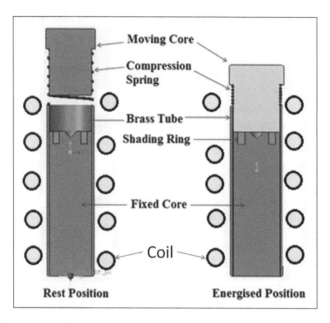

Figure 1.   Model comparing rest position and energised position of moving magnet actuator assembly.

of DMAIC methodology, the root causes of ENT are identified and the behaviour of magnets with different surface roughness (Ra value) on mating surfaces is studied in order to resolve the occurrence of ENT during the working of moving magnet actuator assembly.

2   LITERATURE REVIEW

– Cheng *et al* (2002) determined the relation between roughness on the pull-off force and adhesive contact for smooth as well as rough surfaces. They comparatively simulated the relations to know the attributes of surface roughness in subsistence of molecular adhesion. Modelling of adhesion for smooth surfaces was developed and used to generate a mathematical theory to determine effects of surface roughness. The study concluded that by enlarging the surface roughness, pull-off force reduces significantly and vice-versa. Also, subsequently increasing the Ra value, the tendency for the microsphere to snap-on and snap-off decreases.

– Another research survey on magnetic domains and their relation with surface roughness by Zhao and Gamache (2001) gave a glimpse on the effects of surface roughness over magnetic domain size, domain wall thickness, as well as coercivity of thin magnetic films. Their observation showed that with decrease in surface roughness, the domain wall movement is affected and coercivity is increased. With the above study, surface roughness value (Ra) of the mating surfaces was considered as one of the important factors affecting the functioning of magnetic actuators.

– Various experiments and trials to observe the effect of surface grinding parameters on surface roughness were also studied by Padda et al. (2015). Trials were conducted in order to analyse the effect of varying depth of cut, wheel speed and wheel grain size on surface roughness observed that with increasing grain size of wheel and keeping depth of cut constant, Ra was decreased. Also, with rise in depth of cut, less cutting and more rubbing was observed due to wheel wear and complete break-off of wheel grains. Therefore, in grinding process, optimum value of parameters has to be fixed depending upon nature of wheel used and the surface roughness Ra value needed.

– Demir and Gullu (2010) also conducted an experiment to observe the influence of grain size and other grinding parameters on Ra value in grinding operation. They found that with the increasing grit of grinding wheel, surface finish becomes better while decreasing the Ra value. Also when depth of cut was increased, surface irregularities were observed with naked eyes affecting the end application of job.

– In an investigation for optimizing the parameters in surface grinding process for steel, Mahajan and Nikalje (2015) used Taguchi method to determine the relation between metal removal rate, surface roughness, wheel speed, depth of cut and the grain size. They concluded that the table speed has a major effect on surface finish of the job while material removal rate is affected by the depth of cut. They also generated a relation between grinding wheel parameters, table speed, spindle speed and the depth of cut using ANOVA method and gave optimum parameters for achieving required surface finish for the job.

3 MEASUREMENT AND ANALYSIS PHASE

To identify the root causes for the Electrical Non-Trip defect in moving magnet actuators, cause and effect diagram of ENT is created as shown in Figure 2.
Following critical observations were made from the above diagram:

1. *Mating surface between fixed magnet/core and moving magnet/core*: Friction created between the mating surfaces of both the magnets is dependent upon their respective surface roughness. Ra value of mating surface of magnet may be less than the specified range resulting into more contact surface area which would produce ENT defect.
2. *Dirt Particles*: Oil or dirt particles on the mating surface may arise due to heating of magnets after successive workings. Oil may keep the mating surfaces attracted to each other during de-energising condition of the coil.
3. *Compressive Spring*: With less compressive load on the spring moving core may not be pulled up as effectively as designed. Moreover, restricted movement of the spring may also limit the pulling up of moving magnet.

Range of surface roughness (Ra) on the mating surfaces of both the magnets/cores is considered as one of the most important causes of ENT. Further set of experiments with different grinding parameters undertaken to obtain optimum Ra value for fixed as well as moving magnets.

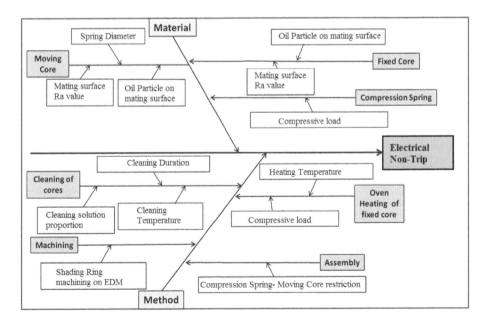

Figure 2.   Cause and effect diagram of electrical non-trip defect.

4 IMPROVEMENT PHASE

From review of the literature, it is observed that when the surface roughness (Ra) of the mating surfaces is increased then the adhesion force is decreased and pull-off force is increased. Various experiments are undertaken to increase the surface roughness (Ra) of moving magnet and fixed magnet. Two grinding parameters, wheel grain size and depth of cut are selected and eight sets of trial combinations are undertaken for the grinding process. Roughness values for the ten samples of each trial are measured and main effect graph are plotted for both moving core as well as fixed core. Surface Roughness (Ra) value is measured on Mitutoyo SJ-410 roughness measuring instrument. Further, to set a standardized surface roughness range, various roughness ranges are selected in both the cores and a matrix of sixteen sets of trials are programmed to compare the effect of roughness band on occurrence of ENT defect. From the result of this trial, roughness band is selected. To achieve the selected roughness band, grinding parameters were picked from the box-plot graph depicting the surface roughness of moving cores and fixed cores with various grinding parameters. Validation of the selected roughness band is carried out to check the occurrence of ENT defect.

4.1 Change in grinding process parameter for magnets/cores

Set of grinding parameters are:
   Wheel Grit: 46, 60; Depth of Cut (mm): 0.02, 0.03, 0.04 and 0.05.
   Using the combinations of considered parameters, eight sets of trials are programmed. The sample size for each trial is 10. Obtained surface roughness values for fixed magnet and moving magnet are displayed in Table 1 and Table 2 respectively.
   Main effect plot for surface roughness (Ra) of fixed magnet and moving magnet are shown in Figure 3 and Figure 4 respectively.

Table 1.   Surface roughness values for fixed magnet after trial.

Wheel grain size	Depth of cut (mm)	Ra value for fixed magnet										Avg. Ra (μm)
		1	2	3	4	5	6	7	8	9	10	
46	0.02	0.68	0.67	0.68	0.67	0.64	0.70	0.71	0.75	0.68	0.76	0.69
46	0.03	0.76	0.67	0.60	0.69	0.71	0.82	0.89	0.71	0.80	0.82	0.75
46	0.04	0.86	0.92	1.05	1.12	1.04	1.16	1.20	1.23	0.97	1.10	1.07
46	0.05	0.99	1.56	1.21	1.26	1.65	1.83	1.38	1.42	1.28	1.15	1.37
60	0.02	0.60	0.58	0.46	0.52	0.45	0.32	0.38	0.42	0.44	0.35	0.45
60	0.03	0.35	0.48	0.32	0.52	0.41	0.42	0.46	0.38	0.45	0.46	0.43
60	0.04	0.51	0.38	0.48	0.46	0.42	0.50	0.48	0.46	0.52	0.58	0.48
60	0.05	0.62	0.71	0.75	0.68	0.58	0.75	0.62	0.65	0.72	0.48	0.66

Table 2.   Surface roughness values for moving magnet after trial.

Wheel grain size	Depth of cut (mm)	Ra value for moving magnet										Avg. Ra (μm)
		1	2	3	4	5	6	7	8	9	10	
46	0.02	0.67	0.88	0.77	0.80	0.71	0.86	0.80	0.83	0.65	0.85	0.78
46	0.03	0.92	0.95	0.61	0.79	0.86	0.86	0.86	0.92	1.11	0.98	0.88
46	0.04	1.11	1.15	1.08	1.06	1.30	1.15	0.98	1.12	1.07	1.11	1.11
46	0.05	1.24	1.56	1.21	1.26	1.65	1.83	1.38	1.42	1.28	1.48	1.43
60	0.02	0.35	0.48	0.32	0.52	0.41	0.38	0.42	0.35	0.48	0.45	0.42
60	0.03	0.51	0.38	0.48	0.46	0.42	0.52	0.48	0.42	0.40	0.42	0.45
60	0.04	0.60	0.58	0.46	0.52	0.45	0.55	0.51	0.42	0.45	00.46	0.50
60	0.05	0.63	0.60	0.70	0.73	0.66	0.55	0.75	0.69	0.65	0.59	0.65

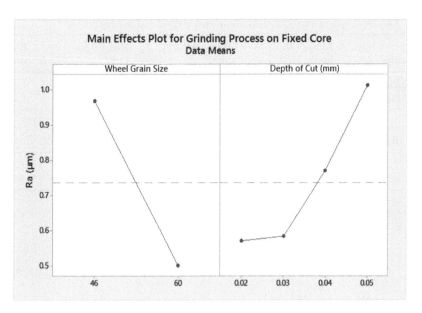

Figure 3.　Main effect plot for grinding process on fixed magnet/core.

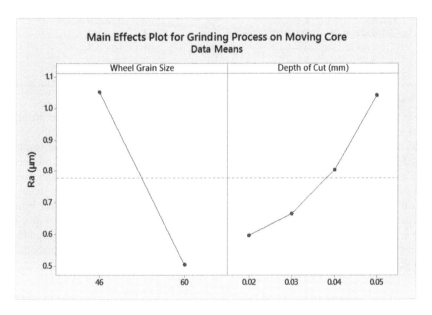

Figure 4.　Main effect plot for grinding process on moving magnet/core.

It is observed that Ra value on mating, surface of both the magnets reduces with the use of grinding wheel having higher grain size/grit. Moreover, surface roughness (Ra value) of both the magnets increases with the increase in depth of cut while keeping the grinding wheel same.

4.2　Testing of roughness band matrix

To obtain the optimum band of surface roughness for fixed and moving magnet for the proper functioning of moving magnet actuator mechanism, various ranges of roughness in both the cores are selected and tested with each other. Table 3 lists down the four different bands of Ra value for both the magnets.

Table 3. Various ranges of surface roughness in fixed magnet and moving magnet.

Component	Range 1 Ra (μm)	Range 2 Ra (μm)	Range 3 Ra (μm)	Range 4 Ra (μm)
Fixed Magnet	0.3–0.6	0.6–0.8	0.8–1.1	1.1–1.45
Moving Magnet	0.3–0.6	0.6–0.8	0.8–1.1	1.1–1.45

Table 4. Behaviour of magnets with combinations of various surface roughness bands.

Sr. No.	Ra of fixed magnet (μm)	Ra of moving magnet (μm)	Testing samples	Ok samples	ENT samples	Chattering samples
1	0.3–0.6	0.3–0.6	10	7	3	0
2	0.3–0.6	0.6–0.8	10	8	2	0
3	0.3–0.6	0.8–1.1	10	9	0	1
4	0.3–0.6	1.1–1.45	10	8	1	1
5	0.6–0.8	0.3–0.6	10	8	2	0
6	0.6–0.8	0.6–0.8	10	8	2	0
7	0.6–0.8	0.8–1.1	10	8	2	0
8	0.6–0.8	1.1–1.45	10	9	1	0
9	0.8–1.1	0.3–0.6	10	9	0	1
10	0.8–1.1	0.6–0.8	10	10	0	0
11	0.8–1.1	0.8–1.1	10	10	0	0
12	0.8–1.1	1.1–1.45	10	9	0	1
13	1.1–1.45	0.3–0.6	10	7	2	1
14	1.1–1.45	0.6–0.8	10	8	1	1
15	1.1–1.45	0.8–1.1	10	9	0	1
16	1.1–1.45	1.1–1.45	10	8	0	2

Sixteen combinations of trials with fixed magnet and moving magnet are programmed with different surface roughness ranges as shown in Table 4.

It was observed that magnets with Ra value range of 0.8–1.1 μm results into no occurrence of ENT defect as well as chattering. Chattering is a loud noise due to improper mating of fixed and moving magnets. When Ra value in both the magnets ranges below 0.8 μm, contact surface area is increased resulting in the occurrence of ENT. Whereas, when surface roughness values in both the cores ranges above 1.1 μm, chattering is observed. When mating surface of fixed magnet and moving magnet have different ranges of Ra value, moving magnet actuator functions improperly with occurrence of ENT and chattering. Hence, Ra value range of 0.8–1.1 μm is finalised on the mating surfaces of both the magnets in moving magnet actuator assembly.

4.3 Selection of grinding parameters

To select the parameters to achieve the required surface roughness (Ra) values, box-plots depicting various roughness ranges for fixed magnet and moving magnet produced during surface grinding are plotted in Figure 5 and Figure 6.

It is observed that for fixed core, Ra 0.75–1.15 μm is obtained when depth of cut is 0.04 mm while for moving core Ra 0.75–1.15 μm is obtained with depth of cut 0.03 mm. Wheel grit for both the cores should be selected as 46.

4.4 Process capability of grinding process

To validate the selection of grinding parameters and Ra ranges for both fixed and moving magnet to obtain a correct functioning of moving magnet actuator assembly, 150 nos. of assemblies are taken into consideration. Process capability of grinding process in fixed

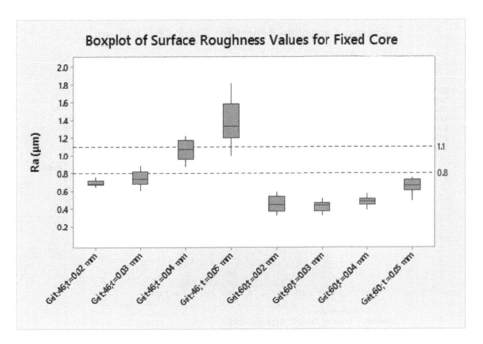

Figure 5.   Box-plot of surface roughness values for fixed core/magnet (*t: depth of cut*).

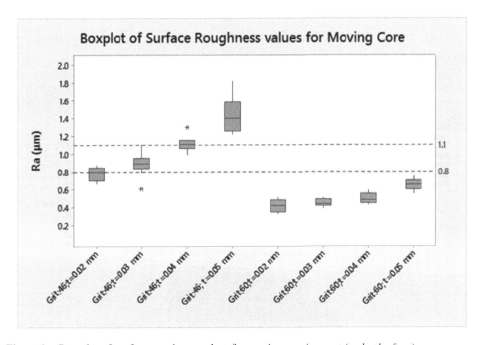

Figure 6.   Box-plot of surface roughness values for moving core/magnet (*t: depth of cut*).

magnet and moving magnet are shown in Figure 7 and Figure 8 respectively. Cpk values for the machining of both the magnets are between 0.4–0.5. It shows that steps to achieve consistency in grinding process for both the magnets have to be undertaken in order to improve process capability of grinding process.

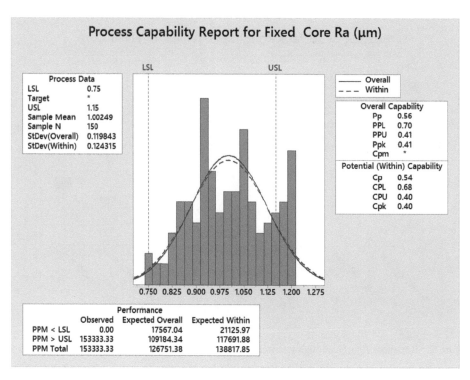

Figure 7. Process capability report for fixed core/magnet.

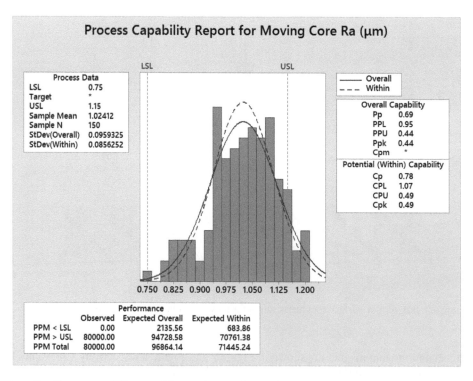

Figure 8. Process capability report for moving core/magnet.

4.5 Validation of surface roughness band

Testing of moving magnet actuator assemblies is also undertaken and it is observed that the occurrence of ENT defect is nil (nil out of 150 assemblies) when magnets with mating surface have the surface roughness band as 0.8–1.1 μm. Although chattering was detected in some of the assemblies because of the Ra value above 1.15 μm in fixed and moving magnets or with the presence of dirt the particles between the mating surfaces during working condition.

With this study it was observed that the working of moving magnet actuator assembly or electromechanical device is dependent on the machining parameters for individual magnets present in them. Grinding parameters play a key role in determining the surface roughness of mating surface of magnets which controls the functioning of the moving magnet actuator assembly.

5 CONCLUSION

This paper presents a study of electrical non-trip defect found in moving magnet actuator assembly used in electromechanical devices. The results can be summarized as below:

– During the analysis phase, main root causes for Electrical Non-Trip (ENT) defect are observed as surface roughness on mating surface of fixed and moving magnets, presence of oil particles on the mating surfaces and restrictive movement of spring. Improper range of surface roughness for magnets generated with the grinding operation is considered as the major root cause of ENT defect.
– During a set of grinding operation, it is observed that with increase in depth of cut surface roughness Ra value becomes higher when grinding wheel with grain size 46 is used compared to grinding wheel with grain size 60. Further, it is observed that when magnets with Ra 0.8–1.1 μm are used in moving magnet actuator assembly, presence of Electrical Non-Trip defect is nil.
– With the help of box-plot chart, grinding parameters for both fixed and moving magnets are set for surface roughness range of Ra 0.8–1.1 μm.
– It is observed that the behaviour of two magnets in moving magnet actuator assembly gets changed with respect to the surface roughness values of their mating surfaces.

REFERENCES

Amandeep Singh Padda, Satish Kumar, Aishna Mahajan (2015), *Effect of Varying Surface Grinding Parameters on the Surface Roughness of Stainless Steel*, International Journal of Engineering Re-search and General Science, Vol. 3, Issue 6, pp. 314–319.

Cheng, W., F. Dunn, M. Brach (2002), *Surface roughness effects on micro-particle adhesion*, The Journal of Adhesion, Vol. 78, pp. 929–965.

Edward P. Furlani (2001), *Permanent Magnet and Electromechanical Devices*, Academic Press, pp. 335–336.

Halil Demir, Abdulkadir Gullu (2010), An Investigation into the influences of Grain Size and Grinding Parameters on Surface Roughness and Grinding Forces when Grinding, *Journal of Mechanical Engineering, Vol. 56, pp. 447–454.*

Mahajan, T.V., A.M. Nikalje, (2015) *Optimization of Surface Grinding Process Parameters for AISI D2 Steel*, International Journal of Engineering Sciences & Research Technology, Vol. 4, Issue 7, pp. 945–949.

Zhao, Y.-P., R.M. Gamache (2001), Effect of surface roughness on magnetic domain wall thickness, domain size, and coercivity, *Journal of Applied Physics, Vol. 89, pp. 1325–1330.*

Technology Drivers: Engine for Growth – Mahajan, Modi & Patel (Eds)
© *2018 Taylor & Francis Group, London, ISBN 978-1-138-56042-0*

Numerical investigation of a high-voltage circuit breaker

P. Kumar
Department of Mechanical Engineering, Aadishwar College of Technology-Venus, Gujarat, India

S. Jain
Department of Mechanical Engineering, Institute of Technology, Nirma University, Gujarat, India

ABSTRACT: The current interruption process in circuit breakers is quite complex and involves interaction of the electric arc, gas flow, ablation and radiation. It is very important to carry out a thorough analysis of circuit breakers for the safety of energy-consuming devices. In the present study, a numerical analysis of a 145 kV high-voltage circuit breaker was carried out using the commercial code FLUENT. A dynamic mesh model was used for the moving boundaries and mesh deformation. Unsteady state simulations were carried out using unsteady Reynolds-averaged Navier–Stokes equations along with a standard k-ε turbulence model and discrete ordinate radiation model. To consider the effects of moving boundaries, valve movement, arc generation and variation in properties of SF_6, different user-defined functions were written and incorporated in FLUENT. From the analysis, a temperature of around 2000 K was observed in puffer and self-blast volumes. The maximum pressure and temperature near the arc zone were found to be 31 bar and 5000 K, respectively. The study of path lines showed vortex formation and turbulence resulting in energy dissipation in different parts of the circuit breaker. The computational fluid dynamics results were compared with the results provided by the manufacturer and the literature and were found to be in very good agreement.

1 INTRODUCTION

A circuit breaker is used for the coupling of transformers, bus bars, transmission lines and so on. The most important objective of a circuit breaker is to interrupt fault currents and thus protect electric and electronic equipment. A circuit breaker is an apparatus in a system that should be capable of switching from being an ideal conductor to an ideal insulator and vice versa in the shortest possible time (Dilawer et al., 2013). The current interruption process in such a device is quite complex and involves the interaction of the electric arc, gas flow, ablation and radiation. All these factors make the design and analysis of circuit breakers very important in view of the safety of various energy-consuming devices in numerous applications. In typical self-blast high-voltage circuit breakers (> 50 kV), a plasma arc is ignited after arcing contacts separate from each other and the breaker uses arc energy to create pressure inside a chamber and to blow the arc (Ye & Dhotre, 2012). The working principle of a self-blast circuit breaker is shown in Figure 1.

Ye and Dhotre (2012) reported that a density-based explicit flow solver algorithm gives a better result than the SIMPLEC-based (Semi-Implicit Method for Pressure Linked Equations—Consistent) pressure-correction Computational Fluid Dynamics (CFD) solver in predicting dielectric breakdown voltages inside a circuit breaker. Lee (2012) evaluated the whole arcing history of sulfur hexafluoride (SF_6) thermal plasma inside a puffer-based self-blast chamber of high-voltage switchgears and found that the k-ε model and Prandtl's mixing-length model can give good predictions of the thermal-flow characteristics of SF_6 inside the chamber. Srikanth and Bhasker (2009) carried out compressible flow simulation in a puffer-type chamber comprised of moving contacts, fixed electrodes, inlet and exit locations and so on, using a numerical approach. With the displacement of the moving contact, they found indications of swirl flow between the fixed electrodes and the exit.

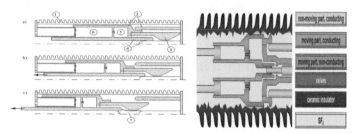

Figure 1. Working principle of self-blast circuit breaker (left); stationary and moving parts (right). Key – 1: porcelain insulation; 2: main contacts; 3: Teflon nozzle; 4: arcing contacts; 5: self-blast volume; 6: puffer volume; a) closed position; b) main contact open; c) arcing contact open (Nielsen, 2001).

Ye and Dhotre (2012) reported that despite the latest computational resources, research in understanding the electro-fluid dynamics of quenching the arc has achieved limited success. In the present study, the numerical analysis of a 145 kV high-voltage circuit breaker was carried out, for better understanding of energy utilization, using the commercial code FLUENT. A 2D axisymmetric computational model was created in Gambit software and a dynamic mesh model was used for the moving boundaries. The unsteady state simulations were carried out using a standard k-ε turbulence model. The Discrete Ordinate Method (DOM) was applied for arc radiation energy transfer. User-Defined Functions (UDFs) were written for moving boundaries, valve movements, arc generation, and for temperature-based variation in the properties of SF_6 gas.

2 GOVERNING EQUATIONS

2.1 Dynamic mesh model

The dynamic mesh model was used for the moving boundaries. It was applied after describing the starting volume mesh and the motion of moving zones using UDFs. The integral form of the generalized conservation equation for a scalar, φ, on an arbitrary control volume, whose boundary is moving, is given in the FLUENT 6.3 User's Guide (FLUENT, 2006):

$$\frac{d}{dt}\int \rho \varnothing dV + \int \rho \varnothing \left(\vec{u} - \vec{u_g} \right) \cdot d\vec{A} = \int \Gamma \nabla \varnothing \cdot d\vec{A} + \int S_\varnothing dV \qquad (1)$$

where ρ is the fluid density, is the velocity vector, u_g is the grid velocity of the moving mesh, Γ is the diffusion coefficient, and S_ϕ is the source term of φ.

2.2 Discrete ordinates model

For arc modeling, a discrete ordinates model was used. In the context of the operation of a breaker device and thermal plasmas in general, scattering is usually neglected. This model solves the Radiation Transfer Equation (RTE) in the direction as shown below (FLUENT, 2006):

$$\nabla \cdot \left(I(\vec{r},\vec{s}) \cdot \vec{s} \right) + \left(a + \sigma_s \right) \times I(\vec{r},\vec{s}) = an^2 \frac{\sigma T^4}{\pi} + \frac{\sigma_s}{4\pi} \int_0^{4\pi} (\vec{r},\vec{s}) \varnothing \left(\vec{s} \cdot \vec{s} \right) d\Omega' \qquad (2)$$

where is del operator, I is radiation intensity, is position vector, a is absorption coefficient, σ_s is scattering coefficient, n is refractive index, σ is Stefan-Boltzmann constant, T is temperature, φ is phase function, and Ω' is solid angle.

3 NUMERICAL METHODOLOGY

3.1 Model and meshing

The detailed specifications and drawing of the circuit breaker were obtained from the manufacturing company located in Vadodara, India. Because of the symmetry of the circuit

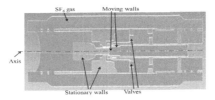

Figure 2. Grid generation for computational domain.

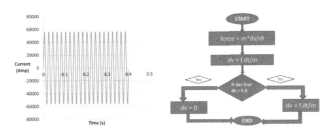

Figure 3. Variation of current with time (left); flow chart for valve movements (right).

breaker, the analysis of the 2D axisymmetric model was carried out as shown in Figure 2. In view of the complex geometry, an unstructured grid consisting of triangular elements was used. The flow domain was discretized into 7,54,586 elements using the pave scheme with a grid size of 0.5–1 mm. Mesh quality was examined for equisize skewness and found to be within the limits. A finer mesh was generated near the moving contact of the circuit breaker to accommodate the higher gradients predicted in this region.

3.2 Solution technique

A dynamic mesh model was used for the moving boundaries and mesh deformation. Inside the circuit breaker, a pressure of 6 bar was specified inside the computational domain. The transient axisymmetric simulations were performed using the Finite Volume Method (FVM) based on governing equations using absolute velocity formulation. For turbulence and radiation modeling, standard k-ε and discrete ordinate models were used, respectively. A second order upwind scheme was used for turbulent kinetic energy and dissipation rate. SF_6 gas was defined as a fluid in the circuit breaker and UDFs were applied to consider variation in density, viscosity, thermal conductivity and specific heat. User-defined functions were also applied for the moving boundaries, valve movements and arc generation. In the circuit breaker, the valve movement took place due to a pressure difference. The sinusoidal variation of current with time (for arc generation) and the flow chart for the valve movements (for the UDF) are shown in Figure 3 (Kumar, 2013).

4 RESULTS AND DISCUSSION

To study the effects of grid size on numerical results, a grid independence test was carried out. The number of grid elements varied from 12,000 to 774,000. Above around 197,000 grid elements, the variations in temperature and pressure were found to be within 1%; hence, further simulations were done with 197,000 grid elements.

The temperature contours at different time intervals are shown in Figure 4 (Kumar, 2013). At 19 ms, the arcing contacts were separated and the plasma arc was generated, so the temperature of the SF_6 gas increased. The temperature near the main contact and self-blast volume was found to be greater than 2000 K. At 30 ms, under the effect of the plasma arc, there is a rapid rise in temperature inside the circuit breaker.

The variation of temperature with position at 19 ms is shown in Figure 5. It can be seen that as the distance from the main contact increased, the temperature decreased due to sinusoidal variation of current and power with respect to time. The maximum temperature in the circuit breaker was found to be 5000 K (near the arcing zone), at the time of opening of both the contacts at 19 ms, whereas in other regions it was around 2000 K to 3000 K. Lee (2012), Cho et al. (2013) and Claessens et al. (2006) also reported similar temperature variations in circuit breakers.

The static pressure variation along the length at 28 ms is shown in Figure 5. In the absence of an arc source, when half the puffer volume was compressed, the maximum pressure was found as 12 bar. When a plasma arc was generated (at 19 ms), the pressure increased in the arcing zone as well as in the self-blast volume. The maximum pressure in the circuit breaker was found to be 31 bar. Lee (2012) and Cho et al. (2013) also found similar pressure variations in circuit breakers.

The velocity contour (at 19 ms) and velocity vectors in the bottom part of the circuit breaker are shown in Figure 6. It can be seen that the maximum velocity occurred in the self-blast volume region. Due to the complicated shape and transient state, high turbulence was observed inside the circuit breaker.

Figure 4.   Temperature contours at 19 ms and 30 ms.

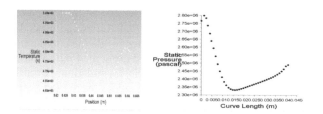

Figure 5.   Temperature variation at 19 ms (left); pressure variation at 28 ms (right).

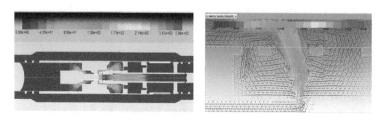

Figure 6.   Velocity contours at 19 ms (left); velocity vectors in bottom of circuit breaker (right).

5 CONCLUSION

In the present study, a numerical analysis of a 145 kV high-voltage circuit breaker was carried out using the commercial code FLUENT. The unsteady state simulations were carried out using a standard k-ε turbulence model and discrete ordinate radiation model. UDFs were written for moving boundaries, valve movements and arc generation, as well as for properties of the SF_6 gas. The maximum pressure and temperature near the arc zone were found to be 31 bar and 5000 K, respectively. In puffer and self-blast volumes, a temperature around 2000 K was found and corresponded to the value of 2000 K provided by the circuit breaker manufacturer, which indicates very good agreement. The study of velocity vectors revealed that vortices were formed around the self-blast volume and bottom part of the circuit breaker. The study of path lines inside the circuit breaker showed a vortex formation and turbulence effect in different parts of the circuit breaker. The results were compared with the literature and found to be in very good agreement. The analysis presented may help researchers working in a similar area, thereby improving the safety aspects of energy-consuming devices.

REFERENCES

Cho, Y.C., Kim, H.K., Chong, J.K. & Lee, W.Y. (2013). Investigation of SLF interruption capability of gas circuit breaker with CFD and a mathematical arc model. *Journal of Electrical Engineering & Technology, 8*(2), 354–358.

Claessens, M.S., Drews, L., Govindarajan, R., Holstein, M., Lohrberg, H. & Jouan, P.R. (2006). Advanced modelling methods for circuit breakers. In *International Council on Large Electric Systems, CIGRE 2006, 29 August 2006, Paris.*

Dilawer, S.I., Junaidi, A.R., Samad, M.A. & Mohinoddin, M. (2013). Steady state thermal analysis and design of air circuit breaker. *International Journal of Engineering Research & Technology, 2*(11), 705–715.

FLUENT. (2006). *FLUENT 6.3 user's guide.* Canonsburg, PA: Ansys, Inc. Retrieved from https://www.sharcnet.ca/Software/Fluent6/html/ug/main_pre.htm.

Kumar, P. (2013). *CFD analysis of high-voltage circuit breaker* (Master's thesis, Nirma University, Gujrat, India).

Lee, J.C. (2012). Numerical study of SF_6 thermal plasma inside a puffer-assisted self-blast chamber. *Physics Procedia, 32*, 822–830.

Nielsen, T. (2001). *Electric arc-contact interaction in high current gasblast circuit breakers.* Stockholm, Sweden: Royal Institute of Technology. Retrieved from https://www.mech.kth.se/thesis/2001/lic/lic_2001_torbjorn_nielsen.pdf.

Srikanth, C. & Bhasker, C. (2009). Flow analysis in valve with moving grids through CFD techniques. *Advances in Engineering Software, 40*, 193–201.

Ye, X. & Dhotre, M. (2012). CFD simulation of transonic flow in high-voltage circuit breaker. *International Journal of Chemical Engineering, 2012*, 609486. doi:10.1155/2012/609486.

Technology Drivers: Engine for Growth – Mahajan, Modi & Patel (Eds)
© 2018 Taylor & Francis Group, London, ISBN 978-1-138-56042-0

Modeling and analysis of surface roughness in a turning operation using minimum-quantity solid lubrication

Anand S. Patel, K.M. Patel & Mayur A. Makhesana
Department of Mechanical Engineering, Institute of Technology, Nirma University, Gujarat, India

ABSTRACT: Machining is a fundamental manufacturing process used for removing additional material from workpiece surfaces and achieving desired dimensional accuracy. Metal working fluids are also known as cutting oils, lubricants and coolants. They are used to reduce friction and wear between the workpiece and cutting tool, to remove heat from the cutting zone, and to remove chips. However, the application of solid lubricants in the form of Minimum-Quantity Solid Lubrication (MQSL) is a novel approach in machining. An attempt was made to analyze the effect of the MQSL approach during turning operations. Experiments were conducted based on the design of a model to predict surface roughness during the turning operation. The feed rate is the predominant process parameter among those considered, directly affecting the surface roughness.

1 INTRODUCTION

Machining is a manufacturing process in which material is removed through plastic deformation of workpieces by way of a shearing action. Lots of friction occurs in the contact area of the cutting tool and workpiece, resulting in heat generation. The heat generation mainly depends on the strength of the cutting tool and workpiece, and other parameters such as the cutting speed during the machining process. Excessive heat generation adversely affects the surface finish, dimensional accuracy, tool life and properties of the machined surface. Therefore, heat generation needs to be minimized to improve the overall performance of the machining process. Various alternatives commonly used are cutting fluid, air cooling, Minimum-Quantity Lubrication (MQL), and cryogenic cooling.

2 LITERATURE REVIEW

The performance of graphite and boric acid powder mixed in SAE 40 oil has been evaluated in turning operations by Krishna and Rao (2008). The workpiece selected was made of EN8 steel and the cutting tool used was cemented carbide. The solid lubricant mixture used was graphite and boric acid powder in different proportions by weight, mixed with SAE 40 oil. The researchers reported that the solid lubricant assisted machining better than a dry and wet machining environment and 20% of boric acid in SAE 40 oil assisted machining, providing better results.

Similar work has been reported in the turning of an alloy steel (EN31) workpiece under different process parameters by Abhang and Hameedullah (2014). The cutting speed, feed rate, depth of cut and tool nose radius were selected as input process parameters for investigation. The cutting fluid used was boric acid mixed with SAE 40 oil and the cutting tools were CNMA 120404, CNMA 120408 and CNMA 120412 carbide turning inserts. It was observed that surface finish worsened with an increase in cutting speed and tool nose radius, and improved with an increase in feed rate and depth of cut. The investigation concluded that turning with boric acid mixed with SAE 40 oil within an MQL environment is better than dry and wet turning.

Amrita et al. (2014) evaluated the effect of using a nanographite-based cutting fluid as a coolant during the turning operation. The experiments were conducted using dry, flood (water-soluble oil as coolant) and mist-based cutting fluid (nanographite mixed with soluble oil) applied through an MQL technique. The cutting conditions selected were a cutting velocity of 40 m/min, feed rate of 0.14 mm/rev and depth of cut of 1 mm for different cutting environments. In the case of a nanographite-based cutting fluid, a reduction in cutting forces due to better lubricity was observed, together with a reduction in cutting temperature due to better penetration of cutting fluid, a reduction in flank wear of tool, and an enhanced surface finish compared to other cooling techniques.

Reddy and Rao (2005) investigated the use of graphite powder as a solid lubricant in an end milling operation on AISI 1045 steel. The cutting speed, feed, radial rake angle and nose radius were taken for the study. Water-soluble oil was used (ratio 1:10) for wet machining, whereas graphite powder with a 2 μm average particle size was used as the solid lubricant. A considerable improvement in reduction of cutting force, improved surface finish, and reduction in chip thickness ratio were observed during graphite-assisted machining when compared to wet machining.

Rahim et al. (2015) investigated MQL as a sustainable cooling technique. AISI 1045 steel was turned with uncoated carbide inserts on a Computer Numerical Control (CNC) lathe machine. The experiments were performed at different cutting speeds and feed rates with a width of cut of 2 mm. Compared to dry conditions, the cutting temperature and cutting force were reduced by 10–30% and 5–28%, respectively, using the MQL technique.

Reddy and Rao (2006) investigated the effect of a solid lubricant on cutting forces, specific energy, chip thickness ratio and surface finish in end milling operations. A workpiece of AISI 1045 steel of dimensions $100 \times 75 \times 20$ mm^3 was selected with a four-flute solid coated carbide end mill cutter as the cutting tool. Graphite and molybdenum disulfide powder of 2 μm size were used as the solid lubricant. It was found that the radial rake angle was the most significant factor influencing cutting forces, and reported that the solid lubricants assisted machining, with a better surface finish compared to the wet cutting environment.

Abhang and Hameedullah (2010) evaluated the effect of nanoboric acid suspensions in SAE 40 oil as an MQL during a turning operation on EN31 steel alloy. Different values of cutting speed and feed rate were taken as input parameters for the experiment and their effects in terms of chip–tool interface temperature, cutting force, surface finish and chip thickness ratio were selected as output parameters. The researchers concluded that the MQL technique showed considerable improvements in all output parameters.

Dhar et al. (2006) studied the influence of the MQL technique on cutting temperature, dimensional accuracy and chip formation during a turning operation. Their experiments were conducted on AISI 1040 steel with uncoated carbide inserts during a plain turning operation. The process parameters selected for study were cutting speed, feed rate and depth of cut under three cutting conditions, namely, dry, flood and MQL. The dimensional accuracy improved because of a reduction in wear of the tool tip with the application of cutting fluid through the MQL technique.

Mukhopadhyay et al. (2007) investigated the use of a solid lubricant in a turning operation. AISI 1040 steel was turned with uncoated cemented carbide inserts and experiments were conducted with different tool geometries (approach angle and rake angle) under various cutting speeds and feed rates. Molybdenum disulfide powder of 2 μm size was used and supplied at a 6 g/sec flow rate. The factors considered for experimentation were cutting speed, feed rate, approach angle and rake angle. It was reported that the surface finish had improved by 5% to 30% in molybdenum disulfide assisted machining.

3 EXPERIMENTAL METHODOLOGY

3.1 Selection of workpiece material and cutting environment

The alloy/bearing steel, popularly known as EN31 and used to manufacture ball and roller bearings, bearing rings, ball screws, cams and pawls, gages, forming tools, punches and so on, was selected for the present investigation. Calcium fluoride powder was used as a solid lubricant, which was also used by Shaji and Radhakrishnan (2013) in a grinding operation. Calcium fluoride powder was mixed in SAE 40 oil by 15% weight concentration of powder and injected with 2 bar air pressure through the Minimum-Quantity Solid Lubrication (MQSL) setup. The nozzle tip diameter was 1 mm and kept at a 15 mm distance from the machining zone.

The MQSL setup was mounted on the guard of the lathe machine with the nozzle fixture kept on the saddle. Figure 1 shows the entire MQSL setup along with the stirrer as mounted on the lathe machine. In the present investigation, the cutting speed (70–150 m/min), feed rate (0.1–0.4 mm/rev) and depth of cut (0.5–1.5 mm) were selected as input parameters. The ranges of all three input parameters selected for study are based on the recommendations for combination of workpiece material and cutting tool material. A Box–Behnken design was used and 15 experiments were conducted for three input process parameters at three levels.

4 RESULTS AND DISCUSSION

An analysis of variance (ANOVA) was used to check the adequacy of the developed model. Table 1 shows the ANOVA (for the full quadratic model, Equation 1) for surface roughness (using MQSL) after elimination of insignificant parameters.

$$\text{Surface roughness} = (-0.519280) + 0.001931 \times \text{Cutting speed} + 1.664295 \times \text{Feed rate} + 1.456596 \times \text{Depth of cut} + 27.423077 \times (\text{Feed rate})^2 - 0.703923 \times (\text{Depth of cut})^2 \qquad (1)$$

Figure 1. MQSL setup with stirrer mounted on the lathe.

Table 1. ANOVA (for full quadratic model) for surface roughness (MQSL).

Source	SS	DF	MS	F value	P value Prob > F	
Model	44.2023	5	8.8405	898.5320	<0.0001	Significant
A – Cutting speed	0.0477	1	0.0477	4.8523	0.0551	
B – Feedrate	42.5549	1	42.5549	4325.2257	<0.0001	
C – Depth of cut	0.0048	1	0.0048	0.4831	0.5046	
B^2	1.4141	1	1.4141	143.7248	<0.0001	
C^2	0.1150	1	0.1150	11.6914	0.0076	
Residual	0.0885	9	0.0098			
Lack of fit	0.0738	7	0.0105	1.4309	0.4712	Not significant
Pure error	0.0147	2	0.0074			
Cor total	44.2908	14				
Std. deviation	0.0992		R^2	0.9980		
Mean	2.8111		Adj. R^2	0.9969		
CV%	3.5286		Pred. R^2	0.9942		

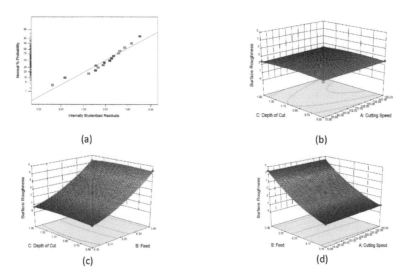

(a)   (b)

(c)   (d)

Figure 2. (a) Normal probability plot of residuals for the surface roughness model (using MQSL); (b), (c), (d) surface plots showing interactive effects of input variables on surface roughness.

Table 2. Details of confirmation experiments.

No.	Cutting speed (m/min)	Feed rate (mm/rev)	Depth of cut (mm)	Surface roughness	
				Experimental	Predicted
1	70	0.25	1.5	2.434	2.3470
2	150	0.4	1.0	5.664	5.6476
3	85	0.18	0.7	1.515	1.4853
4	121	0.2	1.5	1.811	1.7977

The ANOVA of the reduced model indicates that the model is significant because R^2 is 99.80%. The predicted R^2 of 0.9942 is in reasonable agreement with the adjusted R^2 of 0.9969. The lack of fit is not significant. Because the model's F value (898.53) is greater than F0.01, 5, 9 (6.06), there is a definite relationship between the response variable and the independent variables at the 99% confidence level. The lower value of coefficient of variation (CV%) of 3.53 indicates improved precision and reliability of the performed experiments.

Figure 2a displays the normal probability plot of residuals for the surface roughness model (using MQSL). It shows that the residuals lie reasonably close to a straight line.

The surface plots of interactive effects of the input variables on surface roughness are shown in Figures 2b, 2c and 2d. From the plots and analysis, it is evident that the feed rate has considerable influence on the surface finish, whereas the cutting speed and depth of cut only partially influence the surface finish by comparison.

In order to verify the adequacy of the model developed, four confirmation experiments reflecting the MQSL cutting environments were conducted. The responses measured are shown in Table 2.

5 CONCLUSION

Experiments were conducted with cutting parameters at different levels and their effects on the output response (surface roughness) were studied. Based on observations and analysis, it was found that feed rate is the main influencing factor on surface roughness for the MQSL environment. The surface roughness deteriorates with an increase in feed rate and improves with a decrease in feed rate. A mathematical model developed for surface roughness in relation to a range of cutting process parameters was found to be adequate. A 5% error was found between the predicted and experimental values.

REFERENCES

Abhang, L.B. & Hameedullah, M. (2010). Experimental investigation of minimum quantity lubricants in alloy steel turning. *International Journal of Engineering Science and Technology*, 2(7), 3045–3053.

Abhang, L.B. & Hameedullah, M. (2014). Parametric investigation of turning process on EN 31 steel. *Procedia Materials Science*, 6, 1516–1523.

Amrita, M., Srikant, R.R. & Sitaramaraju, A.V.(2014). Performance evaluation of nanographite-based cutting fluid in machining process. *Materials and Manufacturing Processes*, 29, 600–605.

Dhar, N.R., Islam, M.W., Islam, S. & Mithu, M.A.H. (2006). The influence of minimum quantity of lubrication (MQL) on cutting temperature, chip and dimensional accuracy in turning AISI – 1040 steel. *Journal of Materials Processing Technology*, 171, 93–99.

Krishna, P.V. & Rao, D.N. (2008). Performance evaluation of solid lubricants in terms of machining parameters in turning. *International Journal of Machine Tools & Manufacture*, 48, 1131–1137.

Mukhopadhyay, D., Banerjee, S. & Reddy, N.S.K. (2007). Investigation to study the applicability of solid lubricant in turning AISI 1040 steel. *Transactions of the ASME*, 129, 520–526.

Rahim, E.A., Ibrahim, M.R., Rahim, A.A., Aziz, S. & Mohid, Z. (2015). Experimental investigation of minimum quantity lubrication as a sustainable cooling technique. *Procedia CIRP*, 26, 351–354.

Reddy, N.S.K. & Rao, P.V. (2006). Experimental investigation to study the effect of solid lubricants on cutting forces and surface quality in end milling. *International Journal of Machine Tools and Manufacture*, 46, 189–198.

Reddy, N.S.K. & Rao, P.V. (2005). Performance improvement of end milling using graphite as a solid lubricant. *Materials and Manufacturing Processes*, 20, 673–686.

Shaji, S. & Radhakrishnan, V. (2013). A study on calcium fluoride as solid lubricant in grinding. *International Journal of Environmentally Conscious Design and Manufacturing*, 11, 29–36.

Technology Drivers: Engine for Growth – Mahajan, Modi & Patel (Eds)
© 2018 Taylor & Francis Group, London, ISBN 978-1-138-56042-0

Synthesis of aluminum-based Metal Matrix Composites (MMC) and experimental study of their properties

Arvind M. Sankhla, B.A. Modi & K.M. Patel
Department of Mechanical Engineering, Institute of Technology, Nirma University, Ahmedabad, Gujarat, India

ABSTRACT: Ongoing research and development of composite materials has revolutionized the existing trends of material selection and usage in almost in every engineering sector, such as the automotive sector, sporting goods industries and aerospace industries. Metal Matrix Composites (MMC), especially aluminum-based MMCs (Al-MMCs), have evolved as high-performance materials, light in weight and with a lot of scope for tailoring the properties of the composite material. It has been established that reinforcing the aluminum or aluminum alloy matrix with ceramic material such as SiC, Al_2O_3 and graphite can enhance properties such as hardness, compressive strength and tensile strength without much increase in density, making the MMC more promising on the basis of specific properties and thermal management, which is very useful in semiconductor-based instrument manufacturing. This paper presents an experimental study of the synthesis of Al-MMCs through powder metallurgy and subsequent analysis of properties such as density, hardness and compressive strength. Moreover, the microstructure of the composite is also obtained and studied to see the effect of proportions of SiC on the properties of the composites when added to the aluminum matrix in the range of 5–25% by weight.

1 INTRODUCTION

Composite materials are a class of material in which two or more materials are added together and processed in such a way that the newly developed material has the desired properties or we can say that the new material is superior to its individual constituents. Metal Matrix Composites (MMCs) are an upcoming newer family of such materials that are replacing conventional materials, especially in the automobile, sporting goods, aviation and recreation industries.

Compared to monolithic metals, MMCs have:

- Higher strength to density ratios
- Higher stiffness to density ratios
- Better fatigue resistance
- Better elevated temperature properties
- Higher strength
- Lower creep rate
- Lower coefficients of thermal expansion
- Better wear resistance.

The advantages of MMCs over polymer matrix composites are:

- Higher temperature capability
- Fire resistance
- Higher transverse stiffness and strength
- No moisture absorption
- Higher electrical and thermal conductivities
- Better radiation resistance.

Some of the disadvantages of MMCs compared to monolithic metals and polymer matrix composites are:

- Higher cost of some material systems
- Relatively immature technology
- Complex fabrication methods for fiber-reinforced systems (except for casting)
- Limited service experience.

2 PRODUCTION METHODS OF METAL MATRIX COMPOSITES

There are many processes available to fabricate MMCs; they can be classified as solid-state, liquid-state and deposition processes.

Solid-state processing includes the following methods:

- Powder blending and consolidation
- Diffusion bonding
- Physical vapor deposition.

Liquid-state processing includes the following methods:

- Stir casting
- Infiltration process
- Spray deposition
- In-situ processing (reactive processing).

In deposition processes, droplets of molten metal are sprayed together with the reinforcing phase and collected on a substrate where the metal solidification is completed. This technique has the main advantage that the matrix microstructure exhibits very fine grain sizes and low segregation but it has several drawbacks: the technique can only be used with discontinuous reinforcements; the costs are high; and the products are limited to the simple shapes that can be obtained by extrusion, rolling or forging.

3 POWDER METALLURGY FOR PROCESSING METAL MATRIX COMPOSITES

Powder Metallurgy (PM) is a well-known and established process of producing special components such as self-lubricating bearing and bushes, and also for processing materials which are difficult to process by other conventional processes. Transportation and automotive industries are the major sectors which use a large proportion of parts and components made via PM. PM is a material processing technology for creating new materials and parts by diffusing metal powders as raw ingredients through the sintering process. Applications of PM may be found in the manufacture of structural parts, tribological parts such as high contact pressure bearings, heat- and wear-resisting parts, magnetic parts, and other next-generation high-performance parts. The features of PM can be seen in the following areas:

1. Alloys created from high melting point metals including tungsten, molybdenum and tantalum.
2. Metal/non-metal composite materials as represented by cemented carbide, cermets and friction materials.
3. Creation of composites of metals that do not dissolve into each other, such as high thermal conductivity materials (W-Cu, Mo-Cu), high-density alloys and electrical contact materials (Ag-Cu, Cr-Cu).
4. Creation of porous materials such as oil-impregnated bearings and filters.
5. PM products that can be formed by pressing powders in molding tools, affording excellent economic efficiency.

Figure 1 shows a rotor core for the motor of a Hybrid Electrical Vehicle (HEV) produced by PM in which diffusion bonding is achieved by sintering. Its outer circumferences uses a

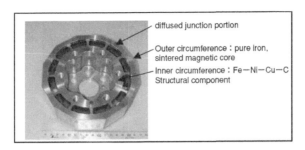

Figure 1.   Rotor core for motor of HEV formed by diffusion bonding at sintering (Tsutsui, 2012).

sintered magnetic core material made of pure iron. The interior requires a high degree of strength because the motor torque is directly transmitted to the shaft. It is composed of a Fe-Ni-Cu-C based material and is combined into a single body by diffusion bonding at sintering.

4   METHODOLOGY AND EXPERIMENTAL STUDY

For this experimental study, fine aluminum powder of 1000 grit size was used as a matrix material. This fine powder was of 99% purity and was of analytical grade, and is therefore also used for chemical analysis and synthesis of other compounds. Silicon carbide (SiC) in powder form of 1000 grit size was selected and used as a reinforcing element in the range of 5–25% in steps of 5%. A reinforcing element was added on a weight percentage basis; hence, the required quantity was obtained through carefully weighing the powder on a digital weighing machine that has a resolution of at least 0.01 g. Because blending of metal powder with a reinforcing element is important, a tumbling fixture was fabricated to ensure proper and uniform mixing. The powders were placed in a plastic container to achieve proper mixing. The container was held in a cubic cage of the tumbling fixture, which was assembled in such a way that it and the container rotate about an axis which passes through its diagonal. Because of this, a greater tumbling effect was obtained when the cage was rotated, which results in effective and uniform mixing of the two powders, one of aluminum and the other of the reinforcing element. After mixing well, the blended powder was packed into a die contour in amounts of 30 grams. The die and punch used in this case study are shown in Figure 3.3. Using a mounting press available in the laboratory, the powder was compacted while applying a pressure of 35 MPa (350 kg/cm²) and then specimens were ejected from the die. This is how ten samples of each composition, 5–25% (in steps of 5%) of SiC in an aluminum matrix were prepared. Subsequently, all the samples were sintered in a muffle furnace at 598°C for one hour. During sintering, an inert atmosphere was maintained in the muffle furnace by passing argon into the heating chamber at a pressure of 1.5 bars. This was done to reduce the reactivity of the aluminum, which is the base metal of the composite material. After sintering, specimens were finished for their end and their dimensions were measured to evaluate the experimental density. The hardness of each specimen was measured on the Rockwell B scale using a steel indenter of 1.587 mm and by applying a load of 100 kg. Moreover, compressive strength and indirect tensile strength were evaluated using a Universal Testing Machine (UTM). For the compression test, the length to diameter (l/d) ratio was kept at 1.25 and for determination of indirect tensile strength, the diameter to thickness (d/t) ratio was kept at 2. Microstructure examination was also carried out for one specimen from each composition. In one of the experiments, the blended powder was heated to 600°C for 20 minutes and then compacted in a die at the same pressure used during cold compaction of the powder: 35 MPa. This was done to fabricate the composite material through hot compaction of powder to see the effect of temperature on its properties. The specimens thus prepared are shown in Figure 2.

Figure 2. Specimen of Al-MMC prepared through PM.

5 RESULTS AND DISCUSSION

After successful preparation of Al-MMC in the form of cylindrical ingots, the following properties were analyzed:

1. Density
2. Hardness measurement
3. Indirect tensile strength
4. Compressive strength
5. Microstructure.

5.1 *Density analysis*

The theoretical and experimental density for each composition of Al-MMC was evaluated and the results in terms of variation in density are shown in Figure 3.

5.2 *Hardness measurement*

The hardness of Al-MMC was measured on the Rockwell B scale. It was found that reinforcing the aluminum matrix with SiC resulted in sufficient hardness.

5.3 *Indirect tensile strength*

Because it was not possible to prepare a specimen of sufficient length for tensile strength testing to be done on a UTM, the tensile strength was evaluated indirectly and is shown in Figure 5.

5.4 *Compressive strength*

As one of the objectives of this experimental study was to test mechanical properties, all the samples/specimens of different SiC weight proportions in the aluminum matrix were tested for compressive strength and the results are shown in Figure 6.

5.5 *Microstructure of AL-MMC*

In order to study the properties and material characterization, the microstructure of the composite material was studied through optical micrographic examination. To establish the microstructure of each type of composition of Al-MMC, specimens were prepared and each of the specimens polished using a disk-type polishing machine. The results are shown in Figure 7.

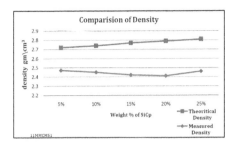

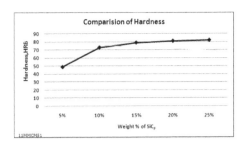

Figure 3. Variation in density of Al-MMC with increase of SiC content.

Figure 4. Hardness of Al-MMC with different proportions by weight of SiC.

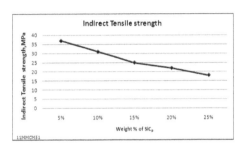

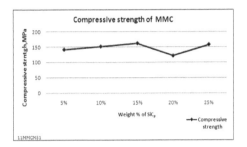

Figure 5. Indirect tensile strength of Al-MMC with different proportions by weight of SiC.

Figure 6. Compressive strength of Al-MMC with different proportions by weight of SiC.

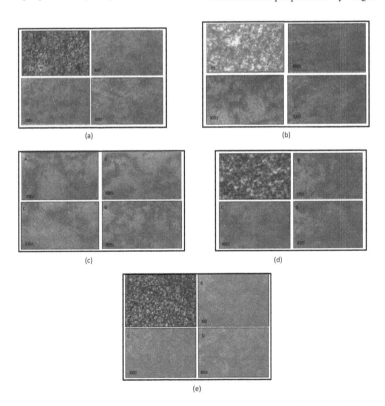

Figure 7. Micro structures of Al-MMC containing: (a) 5% SiC; (b) 10% SiC; (c) 15% SiC; (d) 20% SiC; (e) 25% SiC.

6 CONCLUSION

The following are the major conclusions of this experimental study:

- Powder metallurgy processes can be adopted for successful manufacturing of Al-MMCs
- Reinforcement of the aluminum matrix with SiC definitely has an impact on the mechanical properties of Al-MMCs
- The hardness of Al-MMCs is found to be around three to four times that of commercially pure aluminum
- Increasing SiC content in aluminum has little effect on the density of Al-MMC because the latter is very near that of base metal; hence, a better ratio of strength to density is possible
- Compressive strength was found to be nominal and to increase with an increase in the SiC content of the aluminum matrix; however, the compressive strength of the 20% SiC by weight composite subverts this trend for some reason.

The microstructures of Al-MMCs were also studied and it was found that local colonies of SiC particles were visible, which may be due to the fact that during the mixing of the two powders, local clusters of powder might have been formed which remained undispersed; prolonged mixing and incorporation of agitator-assisted mixing may bring more uniformity.

- Apart from the above, the remaining microstructures appear to show uniform distribution of SiC in the aluminum matrix
- Porosity still remains a considerable issue which definitely contributes greatly to the material characterization.

REFERENCES

Chawla, K.K. & Chawla, N. (2014). Metal matrix composites: Automotive applications. *Encyclopedia of Automotive Engineering.*

Das, S. (2004). Development of aluminium alloy composites for engineering applications. *Transactions of the Indian Institute of Metals, 57*(4), 325–334.

Nair, C.K. (2000). Research and development in metal matrix composites at HAL. In R.C. Prasad & P. Ramakrishnan (Eds.), *Composites: Science and technology* (pp. 20–30). New Delhi, India: New Age International.

Rosso, M. (2006). Ceramic and metal matrix composites: Routes and properties. *Journal of Materials Processing Technology, 175*(1), 364–375.

Singla, M., Dwivedi, D.D., Singh, L. & Chawla, V. (2009). Development of aluminium based silicon carbide particulate metal matrix composite. *Journal of Minerals and Materials Characterization and Engineering, 8*(6), 455–467.

Tsutsui, T. (2012). *Recent technology of powder metallurgy and applications.* Hitachi Chemical Review Paper 2. Retrieved from http://www.hitachi-chem.co.jp/english/report/054/54_sou2.pdf.

Yamaguchi, K., Takakura, N. & Imatani, S. (1997). Compaction and sintering characteristics of composite metal powders. *Journal of Materials Processing Technology, 63*(1–3), 364–369.

Technology Drivers: Engine for Growth – Mahajan, Modi & Patel (Eds)
© *2018 Taylor & Francis Group, London, ISBN 978-1-138-56042-0*

Mechanics of Cooling Lubricating Fluid (CLF) through chip deformation in eco-friendly machining of Inconel 718

Ganesh S. Kadam

Department of Mechanical Engineering, Dr. B.A.T.U., Lonere, Raigad, Maharashtra, India
Department of Mechanical Engineering, SIES GST, Nerul, Navi Mumbai, Maharashtra, India

Raju S. Pawade

Department of Mechanical Engineering, Dr. B.A.T.U., Lonere, Raigad, Maharashtra, India

ABSTRACT: This paper investigated the cutting mechanics aspects of the minutely explored eco-friendly Cooling and Lubricating Fluid (CLF), water vapor, in high-speed turning of Inconel 718 using coated carbide tools. Water vapor acts as a coolant and lubricant; however, the lubrication aspects need to be understood separately from the cooling. For this, the effect of water vapor CLF's parameters like nozzle diameter, stand-off distance, pressure and flow rate, as well as main machining parameters like cutting speed, feed rate and depth of cut in providing the necessary lubrication and/or cooling action were assessed in terms of the chip deformation coefficient, basically a prime indicator of lubrication status. The nozzle diameter and stand-off distance along with feed rate and depth of cut were found to be the most significant factors. Within the investigated regime, the optimal combination of process parameters for a lower chip deformation coefficient and hence, better lubrication, were a cutting speed of 80 m/min, feed rate of 0.20 mm/rev, depth of cut of 0.25 mm, nozzle diameter of 2 mm, stand-off distance of 20 mm and a pressure of two bar. Thus, by proper selection of optimal parameters, usage of water vapor for providing effective cooling and lubrication in machining can be a promising effort toward green manufacturing for the near future.

Keywords: Inconel 718, high-speed turning, water vapor, eco-friendly, chip deformation coefficient, green manufacturing

1 INTRODUCTION

Inconel 718, a nickel-based superalloy, is widely employed in the aerospace industry, marine equipment, nuclear reactors, petrochemical plants and food processing equipment due to its superior high-temperature strength, corrosion resistance and low thermal conductivity. However, Inconel 718 is known to be among the most difficult to cut materials due to properties that are responsible for poor machinability including rapid work hardening causing tool wear and poor thermal conductivity leading to high cutting temperatures (Pawade et al., 2007). Many problems caused during machining are due to heat generation and control over it for enhanced machining performance can be thus exercised by proper selection and application of Cooling and Lubricating Fluid (CLF). However, use of conventional CLFs cause problems such as high cost, pollution and hazards to the operator's health, and thus, have urged researchers to seek suitable eco-friendly alternatives. Apart from surface integrity in turning of Inconel 718, few researchers have also focused their attention on chip formation aspects (Wang & Rajurkar, 2000; Ezugwu & Bonney, 2005; Su et al., 2007). Furthermore, few research investigations have also involved water vapor as a coolant and lubricant in machining of steels (Liu et al., 2005; Junyan et al., 2010), titanium alloy (Pawade et al., 2013) and Inconel 718 (Kadam & Pawade, 2017). Hence, keeping the literature gaps in view, the present paper discusses an experimental study to analyse the effect of machining and mainly water vapor parameters on lubrication and cooling aspects through chip deformation in high-speed turning of Inconel 718.

2 EXPERIMENTAL WORK

High-speed turning experiments using water vapor as a CLF were carried out on standard Inconel 718 cylindrical bar specimens (ϕ25 mm × 200 mm length) by incorporating a L27 (3^7) Taguchi array (Phadke, 1989) under seven process variables (each having three levels) including nozzle diameter (2, 2.5, 3 mm), stand-off distance (20, 30, 40 mm), pressure (1, 1.5, 2 bar), valve position indicating relative flow rate (60, 120, 180°), cutting speed (80, 140, 200 m/min), feed rate (0.1, 0.15, 0.20 mm/rev) and depth of cut (0.25, 0.50, 0.75 mm). Iscar make, PVD coated carbide inserts WNMX080708-M4MW (grade IC806) clamped on a tool holder PWLNL 2525M-08X, were used as the cutting tools. The turning experiments were performed on a CNC lathe and the water vapor was supplied to the machining zone through the assistance of an external steam generation device. The response variable selected to assess the CLF's performance was the chip deformation coefficient considering a smaller-the-better quality characteristic. After the experiments, the chips were carefully collected and the chip thicknesses measured using a Mitutoyo TM505 toolmaker's microscope. Then, the chip deformation coefficient (ζ) was determined by taking the ratio of chip thickness after cut and uncut chip thickness chip.

3 RESULTS AND DISCUSSION

From previous studies on the effect of water vapor jets' lubrication–cooling aspects (Kadam, 2017), it was observed that a water vapor jet can be distinguished as having three zones as shown in Figure 1. Depending on the actual interaction of water vapor with the machining zone, we may get better lubrication or better direct cooling or even both. However, as the jet characteristics vary with supply pressure, nozzle orifice diameter and flow rate, the actual effects on machined surface quality can be found only through experimental analysis.

The chip deformation coefficient is one of the most important parameters to describe the lubrication status in machining (Liu, 2005; Junyan, 2010). A statistical analysis of the chip deformation coefficient was carried out and the results were analyzed by Analysis of Variance (ANOVA). The main effects plots (Figure 2) were plotted to see the effect of the process parameters on the chip deformation coefficient (ζ). The ANOVA revealed that the nozzle diameter, stand-off distance, feed rate and depth of cut are the crucial parameters which significantly influence chip deformation as their P values were 0.002, 0.015, 0.001 and 0.000, respectively, which are less than 0.05. The next subsections discuss the effect of each individual parameter on the chip deformation coefficient.

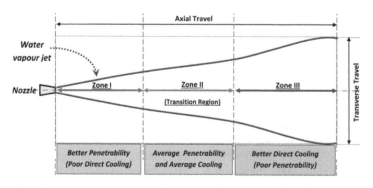

Figure 1.   Basic characteristics of water vapor jet flow (Kadam, 2017).

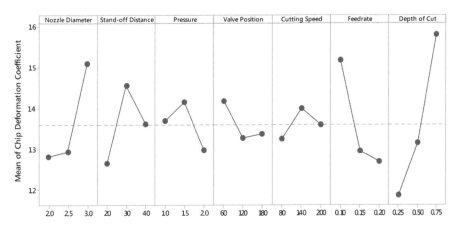

Figure 2.   Main effects plots for chip deformation coefficient.

3.1   *Effect of nozzle diameter (N_d)*

From Figure 2, it can be observed that as the nozzle diameter increased from 2 mm to 2.5 mm and then further to 3 mm, the chip deformation coefficient continually increased. This can mainly be attributed to the difference in interaction of the water vapor jet into the machining zone. At the lowest nozzle diameter of 2 mm, the water vapor jet coming out of the nozzle is narrow and hence, is able to penetrate the chip–tool and tool–work interface easily. As a result, the lubrication effect of the jet is better and thus, a lower value of chip deformation coefficient is observed. Also, thinner chip formation is favored as the lubrication effect allows the chips to glide easily over the tool face. With a further increase in nozzle diameter to 2.5 mm and then to 3 mm, there is an increase in the water vapor jet's axial and transverse length. Thus, as the water vapor jet's span becomes broad and non-focused, the previous dominant lubricating effect now gets shifted to a cooling effect. Due to reduced lubrication and enhanced direct cooling, the chips formed are thicker as they flow slowly and also get cooled quicker after formation. Thus, chip thickening leads to an increased chip deformation coefficient.

3.2   *Effect of stand-off distance (S_d)*

The stand-off distance is basically the distance between the nozzle tip and the tool–work interface. It was observed that with an increase in stand-off distance from 20 to 30 mm, the chip deformation coefficient increases. This is because at a low stand-off distance of 20 mm, the water vapor has better penetrability and hence, enhanced the lubrication effect and good chip flow leads to thinner chips. However, at a stand-off distance of 30 mm, sufficient distance exists between the nozzle tip and tool–work interface. This distance can be considered to be zone II or the transverse region (see Figure 1), which provides somewhat reduced jet penetrability and also some direct cooling effect. Thus, higher chip deformation coefficient values are encountered. With a further increase in stand-off distance from 30 to 40 mm, there is a reduction in chip deformation coefficient. At this distance of 40 mm, a direct cooling effect exists, however it is not that effective in cooling the faster flowing chips and thus, thin chip formation leads to a lower chip deformation coefficient.

3.3   *Effect of pressure (P)*

The effect of water vapor pressure on the chip deformation coefficient initially shows an increasing and then a decreasing trend. At a one bar and 1.5 bar pressure regime, the water vapor has just average penetrability and a direct cooling effect. However, as the pressure is increased to two bar, the chip deformation coefficient appreciably reduces. This can be attributed to the better water vapor jet's penetrability and thus, a better lubrication action leading to faster flowing thinner chips.

3.4 Effect of valve position (V_p)

The valve position basically corresponds to the flow rate. The larger the valve opening in terms of degrees, the greater will be the amount of water vapor flowing, resulting in an increase in flow rate and vice versa. It was observed that an increase in valve position from 60° to 120° led to a decrease in the chip deformation coefficient. This is because the greater the amount of water vapor flow will lead to better penetration and thus, a better lubrication effect. With a further increase in valve position from 120° to 180°, there was a very minute marginal increase in the chip deformation coefficient which was not that significant.

3.5 Effect of cutting speed (V_c)

It was seen that the effect of cutting speed was marginal as the chip deformation coefficient values lie nearer the mean line. It was observed that as the cutting speed increased from 80 m/min to 140 m/min, the chip deformation coefficient slightly increased. This can be mainly because an increase in cutting speed leads to a reduction in jet penetrability and thus, poor lubrication. Also, this agrees well with the fundamentals of metal cutting (Boothrod & Knight, 2006). However, with a further increase in cutting speed to 200 m/min, the chip deformation coefficient slightly decreased. This trend is also well in agreement with Pawade et al. (2013) regarding machining of titanium alloy using water vapor.

3.6 Effect of feed rate (f)

It was observed that as the feed rate increased from 0.10 mm/rev to 0.15 mm/rev and then to 0.20 mm/rev, the chip deformation coefficient continually decreases. This is well in agreement with the fundamentals of metal cutting (Boothrod & Knight, 2006). The chip thickness goes on decreasing with feed rate and hence, the chip deformation coefficient decreases. A similar trend was also reported by Pawade et al. (2013) regarding machining of titanium alloy using water vapor.

3.7 Effect of depth of cut (a_p)

The effect of depth of cut on the chip deformation coefficient was linear. It was observed that an increase in depth of cut led to an increase in chip deformation coefficient. This was because thicker chips were formed due to higher values of depth of cut. Hence, with an increase in chip thickness, the chip deformation coefficient also correspondingly increased.

4 CONCLUSIONS

The experimental investigation leads to the following conclusions:

– The machining parameters of feed rate and depth of cut, and the water vapor related parameters of nozzle diameter and stand-off distance, were found to be highly statistically significant.
– Lower values of nozzle diameter and stand-off distance, while higher values of pressure and flow rate are preferable for achieving better water vapor jet penetration and thus, for effective lubrication at the tool–work and tool–chip interfaces.
– A lower chip deformation coefficient and hence, better lubrication, can be obtained through the optimal combination of process parameters including a cutting speed of 80 m/min, feed rate of 0.20 mm/rev, depth of cut of 0.25 mm, nozzle diameter of 2 mm, stand-off distance of 20 mm, average flow rate with valve opening of 120° and a pressure of two bar.

– Thus, it is possible to achieve better lubrication and cooling by incorporating eco-friendly CLF water vapor and thus, machining using this option is a promising effort toward green manufacturing.

REFERENCES

Boothroyd, G. & Knight, W.A. (2006). *Fundamentals of machining and machine tools*. New York, NY: CRC Press-Taylor & Francis Group.

Ezugwu, E.O. & Bonney, J. (2005). Finish machining of Nickel-base Inconel 718 alloy with coated carbide tool under conventional and high-pressure coolant supplies. *Tribology Transactions, 48*, 76–81.

Junyan, Liu, Huanpeng, Liu, Rongdi, Han, Yang, Wang (2010). The study on lubrication action with water vapor as coolant and lubricant in ANSI 304 stainless steel. *International Journal of Machine Tools & Manufacture, 50*, 260–269.

Kadam, Ganesh S. & Pawade, Raju S. (2017). Surface integrity and sustainability assessment in high-speed machining of Inconel 718 – An eco-friendly green approach. *Journal of Cleaner Production, 147*, 273–283.

Kadam, Ganesh S. (2017). *Investigations on surface integrity in high-speed machining of Inconel 718 under different machining environments*, Unpublished Thesis Report, DBATU, Lonere, Raigad, M.S., India.

Liu, Junyan, Han, Rongdi, Sun, Yongfeng (2005). Research on experiments and action mechanism with water vapor as coolant and lubricant in Green cutting. *International Journal of Machine Tools & Manufacture, 45*, 687–694.

Pawade, R.S., Joshi, Suhas S., Brahmankar, P.K. & Rahman, M. (2007). An investigation of cutting forces and surface damage in high-speed turning of Inconel 718. *Journal of Materials Processing Technology, 192–193*, 139–146.

Pawade, Raju S., Reddy, D.S.N. & Kadam, Ganesh S. (2013). Chip segmentation behaviour and surface topography in high-speed turning of titanium alloy (Ti-6 Al-4V) with eco-friendly water vapour. *International Journal of Machining and Machinability of Materials, 13*(2/3), 113–137.

Phadke, M.S. (1989). *Quality engineering using robust design*. New Jersey: Prentice Hall Publications.

Su, Y., He, N., Li, L., Iqbal, A., Xiao, M.H., Xu, S. & Qiu, B.G. (2007). Refrigerated cooling air cutting of difficult-to-cut materials. *International Journal of Machine Tools & Manufacture, 47*, 927–933.

Wang, Z.Y. & Rajurkar, K.P. (2000). Cryogenic machining of hard-to-cut materials. *Wear, 239*, 168–175.

Technology Drivers: Engine for Growth – Mahajan, Modi & Patel (Eds)
© *2018 Taylor & Francis Group, London, ISBN 978-1-138-56042-0*

Design and analysis of centrifugal compressor

S. Dave, S. Shukla & S. Jain
Department of Mechanical Engineering, Institute of Technology, Nirma University, Gujarat, India

ABSTRACT: Centrifugal compressor is probably one of the most used compressors in the industry but at the same time the flow inside it is very complex, highly turbulent and three-dimensional in nature. It is very important to carry out energy-efficient design of centrifugal compressor followed by detailed analysis in view of energy saving opportunity. In the present study detailed design and analysis of centrifugal compressor is done. The input design parameters were obtained from the Compressor manufacturing company viz. pressure ratio of 3.5, mass flow rate as 4.2 kg/s, rotational speed as 14000 rpm. The design was done based on the standard process followed in the Industry. Iterative method was followed wherein the calculations were done for different mass flow rates and efficiencies were calculated. As the compressor was heavily not loaded radial curved vanes were selected in place of backward curved vanes. Pre whirl at the inlet was not recommended because of the low Mach number. After the design, outlet stagnation pressure was obtained as 352.14 kPa and the overall compressor efficiency was found as 84.72%. The results were compared with the Experimental data provided by Manufacturer and found in very good agreement.

1 INTRODUCTION

In order to protect environment and save energy, during last decades, governments around the world implemented extremely restrict regulations to reduce greenhouse gas emissions and energy consumption (Wan et al., 2017). Hence, it is very important to carry out energy-efficient design of energy consuming machines followed by detailed analysis. A centrifugal compressor is an energy consuming device which gives substantial rise in the pressure of a flowing gas. It consists of a rotating impeller followed by diffuser. Usually, in low-speed compressors, where simplicity and low cost is the criterion than the efficiency, the diffuser is not provided.

Centrifugal compressors found wide applications in refrigeration systems, heat pumps, turbochargers, jet engines, oil refineries, chemical and petrochemical plants, heat and power plants, natural gas processing plants etc. (Fang et al., 2014). The flow through a compressor stage is highly complicated, three-dimensional in nature and a full analysis is very difficult, time consuming and cumbersome. However, an approximate solution can be obtained by one-dimensional approach which assumes that the fluid conditions are uniform over certain flow cross-sections. The schematic diagram of centrifugal compressor with velocity triangles at inlet and outlet is shown in Figure 1(a).

Fang et al. (2014) carried out review of empirical models used for prediction of efficiency and mass flow rates of centrifugal compressors and concluded that the most models available for centrifugal compressors of vehicle engines and turbochargers are not satisfactory for refrigeration centrifugal compressors and thus accurate models need to be developed. (Shaaban, 2015) improved the centrifugal compressor performance by optimizing the design. The radial vane-less diffuser geometry was numerically optimized by minimizing the diffuser loss coefficient and maximizing the pressure coefficient. (Wan et al., 2017) developed an improved empirical parameter method for fuel cell powertrain system and obtained higher efficient stable operation range near low mass flow area compared to similar system with screw compressor.

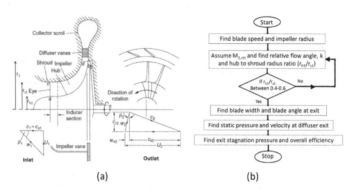

(a)   (b)

Figure 1. (a) Centrifugal compressor stage and velocity diagrams at impeller entry and exit. (Dixon & Hall, 2010) (b) Flow chart for design methodology.

In the present study, design of single stage centrifugal compressor is presented. The analysis of centrifugal compressor was done at different mass flow rates and results are discussed in terms of variations in head, power input and efficiency. The results are validated with the experimental data provided by the Manufacturer.

2 DESIGN OF CENTRIFUGAL COMPRESSOR

2.1 Input parameters

In the present study, the design of single stage centrifugal compressor was done correspond to pressure ratio of 3.5, mass flow rate ($\dot{m}$) as 4.2 kg/s, Power input (P) as 675 kW, rotational speed (N) as 14000 rpm, inlet stagnation temperature (T_{01}) and pressure (p_{01}) as 313 K and 100 kPa. Moreover; assumptions were made for number of radial curved impeller vanes (Z) as 24, number of diffuser vanes as 12, isentropic efficiency (η_i) as 92% and pre-whirl angle (α_1) as 0 (no pre-whirl). The flow chart for design methodology is shown in Figure 1(b).

2.2 Blade speed and impeller radius

Specific work of centrifugal compressor was calculated as per Eq. (1):

$$\Delta W = \frac{P}{\dot{m}} \tag{1}$$

Blade speed was calculated as $U_2 = (\Delta W/\sigma)^{0.5}$
Where, slip factor (σ) was found using Stanitz expression: $\sigma = 0.63\pi/Z$.
Impeller radius (r_2) was calculated as $r_2 = U_2/\omega$, where ω is angular velocity (rad/s).
Accordingly, U2 = 418.53 m/s and r2 = 0.285 m.

2.3 Design of impeller inlet

The inlet radius ratio was calculated by a suitable choice of the relative Mach number at the shroud ($M_{1, \text{rel}}$) using Eq. (2):

$$\frac{\dot{m}\omega^2}{\gamma\pi k p_{01}a_{01}} = \frac{M_{1,rel}^3 \sin^2\beta_{s1}\cos\beta_{s1}}{\left(1+\left(\dfrac{1}{2}\right)(\gamma-1)M_{1,rel}^2\cos^2\beta_{s1}\right)^{\frac{1}{(\gamma-1)+\frac{3}{2}}}} \tag{2}$$

In Eq. (2), by putting ratio of specific heat $\gamma = 1.4$ (for air),

$$f\left(M_{1,rel}\right) = \frac{\dot{m}\Omega^2}{1.4\pi k p_{01}a_{01}} = \frac{M_{1,rel}^3 \sin^2\beta_{s1}\cos\beta_{s1}}{\left(1+\left(\dfrac{1}{5}\right)M_{1,rel}^2\cos^2\beta_{s1}\right)^4} \qquad \text{where, } k = 1-\left(r_{h1}/r_{s1}\right) \qquad (3)$$

Here, a_{01} is stagnation velocity of sound at inlet; $\alpha_1 = 0$ and for a fixed value of $M_{1,rel}$, the optimum value of relative flow angle at shroud inlet (β_{s1}) occurs at maximum value of $f(M_{1,rel})$ as shown in Figure 2(a). By differentiating Eq. (3) it was found that this maximum occurs when,

$\cos^2\beta_{s1} = X - \sqrt{X^2 - 1/M_{1,rel}^2}$ where, $X = 0.7 + 1.5/M_{1,rel}^2$.

Using the given data several optimum values of k and hub-tip ratios were determined. To find the inlet dimensions, continuity equations was used i.e., $\dot{m} = \rho_1 A_1 c_{x1}$

$$\therefore r_{s1}^2 = \frac{\dot{m}}{\pi k \rho_1 c_{x1}} \qquad (4)$$

where, air density at inlet $\rho_1 = \dfrac{\rho_{01}}{\left[1+\frac{1}{5}M_1^2\right]^{2.5}}$ and axial velocity at inlet $c_{x1} = M_1 a_1$,

$M_1 = M_{1,rel}\cos\beta_{s1}$ and $a_1 = a_{01}/\left[1+\frac{1}{5}M_1^2\right]^{0.5}$

Accordingly, $r_{s1} = 0.131$ m and $r_{h1} = 0.083$ m.

2.4 Design of impeller exit

Assuming, radial velocity at impeller exit equal to axial velocity at impeller entry i.e. $c_{r2} = c_{x1}$. Hence,

$$\text{Velocity at exit } c_2 = \sqrt{c_{\theta 2}^2 + c_{r2}^2} \qquad (5)$$

And the flow angle $\alpha_2 = \tan^{-1}\left(c_{\theta 2}/c_{r2}\right)$

As stated, by assuming $\eta_i = 0.92$; T_{02}, T_2 and p_2 were determined as under.:

$$\eta_i = \frac{h_{02s} - h_{01}}{h_{02} - h_{01}} = \frac{\dfrac{T_{02s}}{T_{01}} - 1}{\dfrac{T_{02}}{T_{01}} - 1} \qquad (6)$$

$$\frac{T_{02}}{T_{01}} = \frac{\Delta W}{C_p T_{01}} + 1 \qquad (7)$$

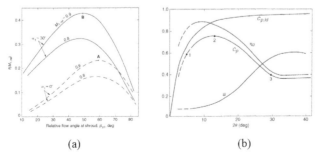

(a)   (b)

Figure 2.   (a) Variation of Mass Flow Function with βs1. (Dixon & Hall, 2010) (b) Diffuser perform-ance curves for a two-dimensional diffuser with L/W$_1$ = 8 (Kline et al., 1959).

$$T_2 = T_{02} - \frac{c_2^2}{2c_p} \quad \text{and} \quad p_2 = p_{02} / \left(\frac{T_{02}}{T_2} \right)^{\gamma(\gamma-1)}$$

From the continuity equation ($\dot{m} = \rho_2 A_2 c_{r2}$), area equation ($A_2 = 2\pi r_2 b_2$) and density at exit ($\rho_2 = p_2/R\,T_2$), the blade width at exit was determined from $b_2 = \dot{m}/2\pi r_2 \rho_2 c_{r2}$ and impeller exit the Mach number was found as $M_2 = c_2/a_2$.

Accordingly, $\alpha_2 = 73.02°$; $T_2 = 392.71$; $p_2 = 221.24\,kPa$; $p_{02} = 423.97\,kPa$ and $b_2 = 0.0101\,m$.

2.5   The vanned diffuser

Optimum diffuser channel length to width ratio (L/W₁) was taken as 8. From Figure 2(b), for a plate diffuser with $2\theta = 8°$, $C_p = 0.7$ and $C_{p,id} = 0.8$. Hence, the static pressure and velocity at diffuser exit were found as under:

$$p_3 = p_{2d} + C_p q_{2d} \tag{8}$$

$$c_3 = c_{2d} \left(1 - C_{p,id} \right)^{0.5}. \tag{9}$$

2.6   Exit stagnation pressure and overall compressor efficiency

Next, exit temperature ($T_3 = T_{03} - c_3^2 / sCp$) and exit density ($\rho_3 = p_3/R\,T_3$) were determined. The total stagnation pressure at the exit of compressor was found as $p_{03} = p_3 + (1/2)\rho_3 c_3^2$
The overall compressor efficiency was found using Eq. (10); where, $T_{03ss}/T_{01} = (p_{03}/p_{01})^{(\gamma-1/\gamma)}$

$$\eta_C = \frac{C_p T_{01} \left(\frac{T_{03ss}}{T_{01}} - 1 \right)}{\Delta W} \tag{10}$$

Accordingly, $p_{03} = 352.14\,kPa$ and $\eta_C = 84.72\%$.

3   RESULTS AND DISCUSSION

3.1   Validation of results

The results obtained from theoretical design were compared with the Manufacturer data for the Centrifugal compressor with similar specifications as summarized in Table 1. It can be seen that, the theoretical calculations were found in very good agreement with the Manufacturer data. From the design, the 3D model of impeller was developed as shown in Figure 3.

3.2   Performance analysis

The performance of centrifugal compressor was analysed at different mass flow rates. The comparison of performance curves obtained with theoretical calculations with the

Table 1.   Design output parameters and its comparison.

Output parameter	Theoretical	Manufacturer
Impeller radius, r_2 (mm)	285.4	280
Hub radius, r_{h1} (mm)	83.04	81
Shroud radius, r_{s1} (mm)	131.01	128
Flow angle at exit, α_2	73.02°	76.14°
Stagnation pressure ratio, p_{03}/p_{01}	3.52	3.5
Max. overall efficiency, η_C (%)	84.72	78

Figure 3.   Impeller model.

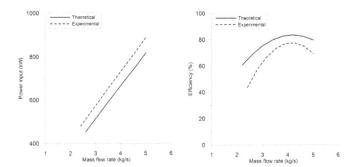

Figure 4.   Comparison of performance curves.

Manufacturer data is shown in Figure 4 (Shukla & Dave, 2017). It can be seen that, with increase in mass flow rate, output pressure decreased, power input increased; but, the efficiency first increased and then decreased. It is evident that the theoretical results are well in line with the experimental data. At design condition, the deviation was found as 7.93%; which shows very good agreement of the two approaches. The variations in the actual condition are due to the negligence of the frictional losses and heat transfer from the compressor walls; which resulted in increased power input and reduction in the efficiency than the theoretical calculations. The maximum efficiency was found as 84.72% at flow rate of 4.2 kg/s at design condition. However, at off-design conditions the relative velocity does not follow the inlet blade angle and leads to incidence (shock) losses and hence lower efficiency.

4   CONCLUSIONS

In this study a detailed design of centrifugal compressor is presented based on the input design parameters obtained from the Compressor manufacturing company viz. pressure ratio as 3.5 and mass flow rate as 4.2 kg/s. Iterative design approach was followed to ascertain the standard design procedure, safety of the equipment as well as an energy-efficient design. In view of low Mach number and lightly loaded conditions, radial curved vanes without pre whirl were recommended. From the design, the outlet stagnation pressure was obtained as 352.14 kPa, exit Mach number as 0.9 and the overall compressor efficiency as 84.72%. The results were compared with the experimental data provided by the Manufacturer and found in very good agreement. The results presented in the current study may be helpful to the designers working in the similar area in view of carrying out energy-efficient design of centrifugal compressors.

REFERENCES

Dixon, S.L. & Hall, C.A. (2010) *Fluid Mechanics and Thermodynamics of Turbomachinery*. Butterworth-Heinemann.

Fang, X., Chen, W., Zhou, Z. & Xu Y. (2014) Empirical models for efficiency and mass flow rate of centrifugal compressors. International Journal of Refrigeration, 41, 190–199.

Kline, S.J., Abbott, D.E., & Fox, R.W. (1959). Optimum design of straight-walled diffusers. Transactions of the American Society of Mechanical Engineers, Series D, 81.

Shaaban S. (2015) Design optimization of a centrifugal compressor vaneless diffuser. *International Journal of Refrigeration*, 60, 142–154.

Shukla, S. & Dave, S. (2017) *Design of a centrifugal compressor*. Nirma University, Project Report.

Wan, Y., Guan, J. & Xu S. (2017) Improved empirical parameters design method for centrifugal compressor in PEM fuel cell vehicle application. *International Journal of Hydrogen Energy*, 42, 5590–5605.

Technology Drivers: Engine for Growth – Mahajan, Modi & Patel (Eds)
© *2018 Taylor & Francis Group, London, ISBN 978-1-138-56042-0*

The effect of laser engraving on L-shaped Carbon Fiber Reinforced Polymer (CFRP) components during the curing process

Dhaval Shah, Shashikant Joshi, Kaushik Patel & Akash Sidhapura
Department of Mechanical Engineering, Institute of Technology, Nirma University, Gujarat, India

ABSTRACT: Carbon Fiber Reinforced Polymer (CFRP) is extensively used in space and aerospace applications due to its good specific properties over conventional metals and alloys. During the manufacturing process of laminated composite parts, whose dimensions do not match with the mold, that distortion is commonly referred to as spring-back or warpage. To measure the amount of spring-back or warpage, reference points are required on part as well as the mold surface. During the curing process, the mold and part are pressed at a high temperature in the presence of a release agent or release tape as simple marking will not work as a reference point.

In this research paper, the effect of laser engraving on L-shaped CFRP components during the curing process was examined. The various diameter markings were carried out with the help of at various depths on an aluminum mold. Different layers of twill prepreg were laid on a marked mold with the help of a release agent and release tape. The assembly of the mold and parts were cured using a hot air oven after applying consolidation. The impressions on cured CFRP parts were examined and it was found that marking can be easily seen with the naked eye in the case of a 0.5 mm diameter and 0.1 mm depth hole for a release agent compared to the release tape.

1   INTRODUCTION

A composite material is made by combining two or more materials to give a unique combination of properties. Composite material is widely used in aerospace, automotive and sporting goods industries due to its high strength to weight ratio in these industries (Mazumdar, 2003). A prepreg is a fiber pre-impregnated with a resin material and available as unidirectional tape, woven fabric tape and roving (Kaw, 2005). Many researchers have worked on the prediction of spring-back and warpage deformations for composite materials. Darrow and Smith (2002) considered three parameters, thickness cure shrinkage, mold expansion and fiber volume fraction gradients which affect spring-in. Fernlund and Poursartip (2004) demonstrated the effect of the tool surface condition and cure cycle on spring-back. Kaushik et al. (2010) studied the effect of different parameters including pressure, degree of cure and ramp rate on the coefficient of friction in the autoclave process. Twigg et al. (2004) determined warpage of flat laminate by considering part length and tool–part interaction and also measured the effect of release film and release agent on warpage. Stefniak et al. (2012) conducted experimental work to quantify the different mechanisms including property gradient and stress gradient mechanisms with respect to their relative contribution to composite part distortions. Kappel et al. (2013) considered the effect of essential parameters such as part thickness, layup, part radius and scattering of occurring distortion during the autoclave manufacturing process.

In this paper, the effect of laser engraving on L-shaped CFRP components during the curing process is examined. The various diameters at various depths of engraving were carried out with the help of laser on an aluminum mold. Different layers of twill prepreg were laid on a marked mold with the help of a release agent and release tape. The assembly of the mold and parts were cured using a hot air oven after applying consolidation. The impression on the parts was observed after removal from the mold.

Figure 1.   Cutting of carbon fiber prepreg.

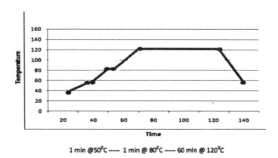

1 min @50°C ---- 1 min @ 80°C ---- 60 min @ 120°C

Figure 2.   Curing cycle for carbon fiber prepreg.

2   PREPARATION OF THE MOLD AND PART

Aluminum was selected as the mold material as it has a high Coefficient of Thermal Expansion (CTE) compared to the composite part so that a large measurable amount of spring-back can be achieved. The raw material size was 76 mm × 76 mm × 250 mm for the aluminum mold. The final dimensions of the mold achieved were 72 mm × 72 mm × 250 mm after machining with a corner radius of 4 mm in a vertical machining center machine. The surface roughness was measured as it affects the spring-in at different stages. The Ra value for the machined mold measured as 0.265 μm using a surface roughness tester instrument.

The twill configuration prepreg was selected for manufacturing the composite L-shaped part. During the manufacturing process, the curing temperature could be up to 120°C with a cure pressure of one bar based on the requirements of the prepreg supplier. The carbon fiber prepreg can be stored at −20°C to −18°C. The prepreg was marked using a marker and ruler scale and then cut into the required orientation using a cutter according to the required size as shown in Figure 1. The labeling of the degree of orientation on each strip of prepreg was carried out for further processes. The curing cycle for the CFRP prepreg was given by the supplier as shown in Figure 2.

3   EXPERIMENTAL PROCEDURE

Experiments were conducted to study the effect of engraving on the manufactured L-shaped CFRP components. Laser engraving was performed on two different locations of the aluminum mold. Different diameter holes with varying depth were engraved using laser. At one surface of the mold, various diameter sizes of 1 mm, 1.25 mm, 1.5 mm, 1.75 mm, 2 mm, 2.25 mm, 2.5 mm, 2.75 mm and 3 mm, with the depth varying from 0.5 mm, 0.6 mm, 0.7 mm, 0.8 mm, 0.9 mm and 1 mm were engraved as shown in Figure 3. Similar engraving was performed on a second location as shown in Figure 4. It had diameter sizes of 0.5 mm, 1 mm, 1.25 mm, 1.5 mm, 1.75 mm and 2 mm, with depth varying from 0.1 mm, 0.2 mm, 0.3 mm, 0.4 mm, 0.5 mm and 0.6 mm.

The cut prepreg layers were laid onto the mold for conducting the experiment. The release agent and PTFE release film were applied on the mold to check its impression on the composite part which was to be manufactured as shown in Figure 5. The release tape was used

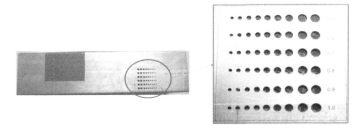

Figure 3. Laser marking on the mold at the first location.

Figure 4. Laser marking on the mold at the second location.

to prevent the adhesion of the composite layup with the tool during the curing process. The release agent was a liquid agent which was applied on the mold to prevent the adhesion of the composite layup with the tool during the curing process.

Different layers of twill prepregs were laid on four different locations (1), (2), (3) and (4) as shown in Figure 6. In Region 1, on one of the faces of laser marking was done and the size of the marking is shown in Figure 4. The tool surface beneath was treated with a release agent to avoid sticking of prepregs during the curing process. Two layers of prepregs were laid on this location. Using this configuration, the effect of laser engraving was observed. In Region 2, the release tape was applied over the mold. Two prepreg layers were laid on the mold. In this configuration, the effect of the release tape on the composite part of the surface was studied. In Region 3, the surface of mold was treated with a release agent and three layers of prepreg were laid. In this configuration, the effect of the release agent on the composite part of the surface was considered. In Region 4, laser marking was done on one of the faces and the size of the marking is shown in Figure 3. The mold surface beneath was treated with a release agent. Three layers of prepregs were laid on this location. Using this configuration, the effect of engraving was observed.

The vacuum bagging of the mold was performed as shown in Figure 7(a), followed by the prepreg layup on the mold, peel ply, breather cloth and so on. The whole assembly was kept in the hot air oven for curing as shown in Figure 7(b).

A temperature of 120°C was set in the oven and maintained for 140 minutes to complete the curing process. After that, the assembly was removed from the oven and kept for cooling to room temperature. The consolidation material was removed and composite parts were recovered from the aluminum mold as shown in Figure 8.

4 RESULTS AND DISCUSSION

The impressions on the composite were successfully obtained due to engraving at both locations as shown in Figure 9 (a) and (b). A different surface finish was obtained with the release agent and release film as shown in Figure 10 (a) and (b), respectively. The composite parts which were laid on the release tape have some waviness in the interacting surface and on the other hand, the composite part laid on the surface treated with a release agent has a smoother surface.

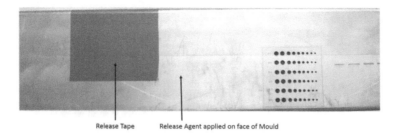

Release Tape   Release Agent applied on face of Mould

Figure 5.   Release tape and release agent on the mold.

Figure 6.   Prepreg layup laid on the mold.

Figure 7.   (a) Vacuum bagging, (b) Curing process in a hot air oven.

Figure 8.   CFRP L-shaped cured parts.

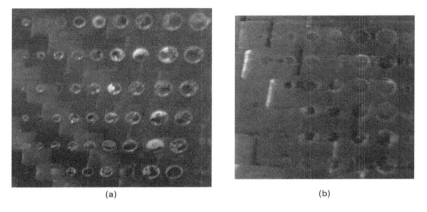

(a)   (b)

Figure 9.   Engraving impressions on manufactured composite parts.

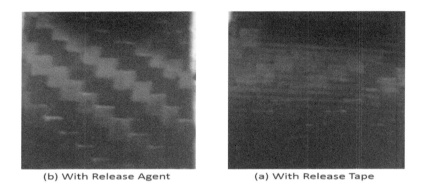

(b) With Release Agent   (a) With Release Tape

Figure 10.   Surface conditions of manufactured composites with release agent and release tape.

5   CONCLUSION

The CFRP composite L-shaped parts were manufactured using a hot air oven curing process with an aluminum mold. Laser engraving was performed on two surfaces of the mold to study its effect on the manufactured L-shaped component. A total of four prepreg layups were laid on the mold and some portion of the mold was treated with a release agent and the other with release film. Consolidation materials were arranged in proper sequence on the mold. The whole assembly was kept in a hot air oven for the curing purpose. L-shaped components manufactured with a release agent had a better surface finish than the components manufactured with release film. The Ra value obtained with the release agent was 1.019 μm and with the release film was 1.528 μm. The engraving effect was observed on the manufactured composite parts. The minimum size of laser engraving was of 0.5 mm diameter with 0.1 mm depth. The impression of this minimum size engraving could be observed with the naked eye on the part where the release agent was used compared to the area where the release tape was used.

REFERENCES

Darrow, D.A. & Smith, L.V. (2002). Isolating components of processing induced warpage in laminated composites. *Journal of Composite Materials*.

Fernlund, G. & Poursartip, A. (2004). *The effect of tooling material, cure cycle, and tool surface finish on springin of autoclave processed curved composite parts*. Department of Metals & Materials Engineering, University of Columbia.

Kappel, E., Stefaniak, D. & Hühne. C. (2013). Process distortions in prepreg manufacturing: An experimental study on CFRP L-profiles. *Composites Structures, 106*, 615–625.

Kaushik, V., Raghavan, J. & Zeng, X. (2010). Experimental study of tool–part interaction during autoclave processing of thermoset polymer composite structures. *Composites: Part A, 41*, 1210–1218.

Kaw, A.K. (2005). *Mechanics of composite material*, 2nd ed. CRC Taylor & Francis.

Mazumdar, S.K. (2001). *Composite manufacturing*. CRC Press.

Stefaniak, D., Kappel, E., Sprowitz, T. & Huhne, C. (2012). Experimental identification of process parameters inducing warpage of autoclave-processed CFRP parts, *Composites: Part A, 43*.

Twigg, G., Poursartip, A. & Fernlund, G. (2004). Tool–part interaction in composites processing. Part I: Experimental investigation and analytical model. *Composites: Part A, 35*, 121–133.

Author index